The Arid Frontier

The GeoJournal Library

Volume 41

The titles published in this series are listed at the end of this volume.

The Arid Frontier

Interactive Management of Environment and Development

edited by

HENDRIK J. BRUINS

Negev Center for Regional Development and Jacob Blaustein Institute for Desert Research,
and Department of Geography and Environmental Development,
Ben-Gurion University of the Negev,
Beer-Sheva and Sede Boker Campus, Israel

HARVEY LITHWICK

Negev Center for Regional Development,
Ben-Gurion University of the Negev,
Beer-Sheva, Israel
and
School of Public Administration,
Carleton University,
Ottawa, Canada

KLUWER ACADEMIC PUBLISHERS
DORDRECHT / BOSTON / LONDON

Negev Center for Regional Development
Ben-Gurion University of the Negev
Beer-Sheva, Israel

A C.I.P. Catalogue record for this book is available from the Library of Congress.

ISBN 0-7923-4227-5

Published by Kluwer Academic Publishers,
P.O. Box 17, 3300 AA Dordrecht, The Netherlands.

Sold and distributed in the U.S.A. and Canada
by Kluwer Academic Publishers,
101 Philip Drive, Norwell, MA 02061, U.S.A.

In all other countries, sold and distributed
by Kluwer Academic Publishers,
P.O. Box 322, 3300 AH Dordrecht, The Netherlands.

Printed on acid-free paper

Printed in the Netherlands

This book is dedicated to the memory of

Harry Levy,

an extraordinary friend

Contents

Part Four: Drought, Food and Land Management in Africa

Part Five: Management of Water Resources

Part Six: Renewable Energy for Development of Arid Lands

Preface and Acknowledgements

The arid frontier has been a challenge for humankind from time immemorial. Development of drylands involves the formidable task of finding sustainable interactions between human activities and arid environments, which cover more than one-third of the global land surface, distributed over Africa, Asia, Australia, North & South America, and Southern Europe. Capricious rainfall variations and drought are the main environmental causes of human vulnerability in drylands. Disasters may develop from the complex interaction between drought, desertification and society, as occurred in the Sahel during the 1970s and 1980s. Millions of grazing animals died and large numbers of peasants and pastoralists suffered economic hardship, hunger and starvation, while 10 million people migrated from their traditional habitats and became refugees.

The international community reacted by organizing the United Nations Conference on Desertification (UNCOD), held in 1977 in Nairobi, where a plan of action was adopted to combat desertification. This eventually led to the United Nations Convention to Combat Desertification in those Countries experiencing Serious Drought and/or Desertification, particularly in Africa (UNCCD). The convention was adopted in Paris in 1994 and was ratified by more than 50 nations in 1996, gaining legal force on December 26, 1996.

In this volume we propose a standardized system to define and compare bioclimatic aridity on a world-wide basis. Moreover, we propose to include the hyper-arid zone in the concept and definition of desertification, because land degradation in drylands due to climatic variations and human activities does not stop at the frontier between the arid and hyper-arid zone.

This book addresses urgent issues in arid frontier development. Part One opens with the central theme and conceptual paradigm of proactive planning and interactive management (1). The conceptual framework aims to guide development in drylands in a way that is beneficial both for society and the environment. We emphasize the option of a rational and ethical discourse for development, avoiding extreme environmentalism *and* human destructionism. Such development has to be based on proper ethics, legislation, policy, proactive planning and interactive management. The latter includes monitoring, evaluation, readjustment and maintenance. As even proactive planning cannot foresee every problem that may arise in arid-frontier development, the role of interactive management is crucial to maintain societal and environmental quality.

The various chapters provide building blocks and regional experiences related to the conceptual framework. Part Two focusses on national, regional and multinational dimensions of environmental policy, planning and management: (2) Drought planning in the U.S.A; (3) Environmental policy in Australia's Murray-Darling basin; (4) Environmental law and management in the Gulf of Aqaba, bordered by Jordan, Israel, Egypt and Saudi Arabia.

Part Three adresses issues of desertification and climatic variability in arid environments: (5) Definitions and management options for bioclimatic aridity, climatic variability, drought and desertification; (6) Relations between policy, perception and environmental science to tackle desertification problems in arid frontiers; (7) Human-made desertification in the Aral Sea basin; (8) Drought risk assessment for dryland agriculture in Andhra Pradesh, India.

Part Four offers regional studies concerning drought, food and land management in Africa, a continent that has suffered most from drought in recent years: (9) Traditional management of drought and famine in Kenya; (10) Land tenure frontiers and related food security among Maasai pastoralists in Kenya; (11) Village land management of peasants and pastoralists in Benin; (12) The use of wild plants as food security in Namibia and Senegal.

Part Five focusses on management of water resources, which is a crucial issue in arid frontier development: (13) Proactive planning and interactive management in irrigation development, with detailed examples from Sudan, India and Israel. (14) Drought and water management in California, economically one of the most developed regions in the world; (15) Cost-benefit analysis of the world's largest water management projects, in India and China.

Part Six adresses important aspects of renewable energy potential for development of arid lands: (16) The use of photovoltaic power to supply solar energy to a rural village in Israel; (17) The potential of geothermal power as clean energy in rural electrification; (18) Production of fuelwood in arid lands with runoff agriculture; (19) Energy use and thermal comfort of housing in arid zones, in relation to building materials and design.

This is the second volume to have been influenced by the international conference entitled *Regional Development: The Challenge of the Frontier*, which was held at the Dead Sea, Israel, in December 1993, under the auspices of the Negev Center for Regional Development, headed by Prof. Yehuda Gradus. The first volume, *Frontiers in Regional Development*, edited by Yehuda Gradus and Harvey Lithwick was published in 1996 by Rowman & Littlefield. A third volume, being edited by Avinoam Meir and Oren Yiftachel, entitled *Ethnic Frontiers and Peripheries*, to be published by Westview Press, is expected towards the end of 1997.

The above Conference has provided both the inspiration and initiative for this volume *The Arid Frontier: Interactive Management of Environment and Development*. As we developed the theme for this book, we invited some of the conference participants to contribute from their field of expertise. In addition, we have attempted to bring together experts from a wide variety of disciplines and regions of the globe, to share their concerns and research findings on arid frontiers with a wider audience. The task has been at once daunting and satisfying, as the attempt to link environment with development in a variety of arid frontiers, has opened new perspectives on policy, planning and management in this vital area.

An ongoing project sponsored by The Netherlands-Israel Development Research Programme (NIRP), entitled *Drought Planning and Rainwater Harvesting for Arid-Zone Pastoralists: the Turkana and Maasai (Kenya) and the Negev Bedouin (Israel)—Social Constraints and Opportunities*, has also been a source of inspiration for certain parts of the book.

This publication has been made possible first and foremost by the generosity of the major supporters of the Negev Center, Els and Martin Wyler, as well as the recently deceased Harry Levy, to whom we dedicate this volume. A publication grant to help cover typesetting and related costs was generously provided by Prof. Jimmy Weinblatt, Dean of the Faculty of Social Sciences and Humanities, Ben-Gurion University of the Negev. In addition, the Social Studies Center, headed by Prof. Gideon M. Kressel, of the Jacob Blaustein Institute for Desert Research at the Sede Boker Campus, covered the remainder of these costs.

Our projects coordinator, Debbie Goldman Golan helped organize the assembly of material and proof-read some of the chapters. Initial typesetting was done expertly by Margo Tepper-Shotz, with valuable contributions in the final phase by Miriam Brodt. A number of maps and figures were designed by Pieter Louppen of the Department of Geography and Environmental Development, Ben Gurion University of the Negev.

Most important has been the guidance and support offered by Petra van Steenbergen of Kluwer Academic Publishers. She saw the potentials of such a volume very early, and has been an enthusiastic supporter throughout. If we are the initiators of this book, together with Prof. Yehuda Gradus, who is also a member of the Editorial Board of the GeoJournal Library, then in every way, Petra has been instrumental in its realization.

We conclude by thanking the many authors who have joined us in producing this original work. They have put up with deadlines and editing demands, and have helped us produce a work of which we are very proud. We look forward to many fruitful collaborations in the future.

Hendrik J. Bruins,
Research Fellow

Harvey Lithwick,
Senior Fellow

Negev Center for Regional Development
Ben-Gurion University of the Negev
Beer-Sheva
Israel

Senior Researcher/Lecturer
Jacob Blaustein Institute for Desert Research,
Social Studies Center
& Department of Geography and
Environmental Development,
Ben-Gurion University of the Negev

Professor of Regional Economics
School of Public Administration
Carleton University
Ottawa
Canada

Part One

A Management Approach to Environment and Development

Chapter 1

Proactive Planning and Interactive Management in Arid Frontier Development

Hendrik J. Bruins and Harvey Lithwick

Introduction

Deserts and drylands constitute about one-third of the terrestrial surface of the Earth. Nearly half the nations of the world have to cope, at least in part of their territory, with the constraining factors of aridity. Dryness negatively affects local availability of the most basic requirements for sustained human existence: water and food. A deeper understanding of these difficulties, as well as ways to alleviate them, is therefore of great importance to humanity, concerned as it is with environmental, social, political and economic development. It is possible to gain important insights into the problems of development in drylands by examining areas of transition, stress, and conflict. Such *arid frontiers* in the wide sense are the central focus of this volume.

The spatial characteristics of the frontier (Friedman, 1996) can be sharp and narrow, fuzzy and wide, static and dynamic. *Drought* is the most dynamic aspect of aridity, which may even sweep into humid areas. It affects both developed nations, like the U.S.A. (Wilhite & Hayes, Chapter 2), and developing countries, like Kenya (Akong'a & Kareithi, Chapter 9). 'Drought is the most complex and least understood of all natural hazards, affecting more people than any other hazard' (Wilhite, 1993a:3; Hagman, 1984). Drought in the mid-1980s in the Sahel was the main *environmental* factor that contributed to *desertification* and the displacement of some 10 million people, who became refugees, as stated by Warren (Chapter 6): 'no other environmental problem has caused so much migration'. The traditional drought preparedness of the people in the Sahel had declined markedly in the years preceding the drought as a result of social and political changes (Garcia, 1981) which made them more vulnerable. Environmental and human factors may be intertwined in complex manners, so that simplistic determinism, environmental *or* human, are both misleading. Humankind living in dry regions usually developed numerous

H. J. Bruins and H. Lithwick (eds.), The Arid Frontier, 3–29.

environmental and social mechanisms to cope with drought and hunger, including the use of wild plants (Van Damme, Chapter 12). Several factors determining the pattern of responsiveness are explored in this volume.

Any assessment of environmental resource potential and constraints, which is the essential basis for policy, planning and management of arid frontier development, requires a rigorous definition of bioclimatic aridity (Bruins & Berliner, Chapter 5). We strongly favor the approach for defining arid regions adopted by UNESCO (1979), differentiating four climatic zones: *hyper-arid, arid, semi-arid and sub-humid*. The main climatic characteristics of each zone and possible options for rainfed land-use are presented in Table 1. The drought and desertification hazards are the main reason for the addition of the *sub-humid* climatic zone on the map of the world distribution of arid regions (UNESCO, 1979). Rainfed farming is even more vulnerable to drought in the *semi-arid* zone (Dietz, Put & Subbiah, Chapter 8). The *arid* and *hyper-arid* zones, unsuited for agriculture in the absence of irrigation, are widely used for pastoralism (Bruins & Berliner, Chapter 5). Rainwater-harvesting or runoff agriculture is based on local rainfall to enable or enhance plant production in drylands (Evenari, Shanan & Tadmor, 1982; Bruins, Evenari & Nessler, 1986; Reij, Mulder & Begemann, 1988).

TABLE 1. The dryland zones and their characteristics

Climatic zone	*P/ETP ratio*	*Interannual rainfall variability*	*Possible rainfed land-use*
Hyper-arid	< 0.03	~ 100%	Limited livestock grazing, mainly by camels and goats.
Arid	0.03–0.20	50–100%	Livestock grazing (less suited for cattle); runoff agriculture.
Semi-arid	0.20–0.50	25–50%	Rainfed agriculture (more viable with additional runoff or irrigation water); livestock grazing.
Sub-humid	0.50–0.75	< 25%	Rainfed agriculture (more viable with additional runoff or irrigation water); livestock grazing.

Source: Based on UNESCO (1979); relationships in the last column have been added by the authors.

The classical arid frontier since time immemorial is the meeting place between pastoralists and farmers, between the desert and the sown (Reifenberg, 1955). Conflicting development interests, strife and conflict at this arid frontier occurred repeatedly throughout history, in the biblical lands of the Near East (Kressel, 1993), in the African Sahel, as well as in many other areas, such as Benin (De Haan, Chapter 11) and Kenya (Rutten, Chapter 10). One of the most famous landmarks in the world, the Great Wall in China, was built by settled farmers at the arid frontier as a line of defense against the penetration of nomadic pastoralists. Extensive livestock raising (pastoralism) is a rational system of land-use in regions too dry for rainfed agriculture, such as in large parts of Australia (Powell, Chapter 3). Domesticated grazing animals transform part of the natural vegetation, mostly inedible for human beings, into protein-rich food, *i.e.* milk and meat, as well as products such as skins (leather) and wool. Pastoralism has been sustainable for thousands of years in many arid parts of Africa and Asia.

Irrigation is an important form of development in the arid frontier. It is intricately linked with the parallel emergence of ancient and complex civilizations in Egypt, Mesopotamia, Pakistan, India and China. It has provided an important basis for development in modern times in many arid regions, such as in California (Roos, Chapter 4, this volume) and Israel (Shanan, Chapter 13). Huge irrigation projects in India and China are under consideration for possible implementation, subject to vigorous debate about the costs and benefits (El-Bihbety & Lithwick, Chapter 15). Irrigation development has many potential hazards (Shanan, Chapter 13) and may lead to severe environmental degradation and desertification (Mainguet & Létolle, Chapter 7). Combating both desertification *and* human livelihood insecurity are essential elements in arid-frontier development.

Limiting factors in arid lands often predominate in the perception of development planners. However, assets also exist: bountiful solar radiation (Faiman, Chapter 16), space, beautiful scenery, and, quite often, surprisingly pleasant climates for habitation, if proper housing materials are used (Pearlmutter & Meir, Chapter 19). Other renewable forms of energy such as geothermal power (Meidav, Chapter 17) may be available in arid zones for development of rural areas. Fuelwood can even be produced in very dry regions through runoff agriculture, *i.e.* intensified resource management of local rainfall (Sauerhaft, Berliner & Thurow, Chapter 18).

Flexible planning is required for arid frontier development to deal with a wide diversity of potential drought situations and environmental hazards related to irrigation. The very nature of the problems described above make it necessary to go beyond *conventional planning*. What is necessary is a framework that allows for *proactive planning*, which in advance develops a system of increased preparedness through appropriate coping mechanisms to deal with various issues that may arise. Such plans are activated by the occurrence of drought and other uncertainties inherent in drylands, as well as signs of environmental degradation, for example in irrigation development (Shanan, Chapter 13). Wilhite (1993b) developed an important methodology for proactive drought planning. Its current status and future directions in the U.S.A. are discussed in this volume (Wilhite & Hayes, Chapter 2). The unique tools of *disaster planning* are required to mitigate the impact of severe droughts in arid frontier regions, such as occurred in the Sahel in the 1970s and 1980s.

Proactive planning, including disaster planning, is a necessary but insufficient condition for ensuring effective action, as no plan can comprehend the future completely and accurately. The overseeing of the proper implementation of plans, which we would define as management, is no less important than the planning process itself. Management entails ensuring that the agents responsible for implementing the plan are induced and/or constrained to act in ways consistent with that plan. There is a potential conflict between environment and development in arid frontier regions, as in the Gulf of Aqaba (Sandler, Chapter 4), requiring environmental law in relation to management. The ramifications of development in terms of environmental and social impact are often unforeseen in the planning stage and not corrected in time in traditional project management, as exemplified by irrigation projects leading to human-made desertification in the Aral Sea basin (Mainguet & Létolle, Chapter 7) and other areas, including Sudan (Shanan, Chapter 13). An approach is required that goes beyond traditional management. Potential problems need to be anticipated, monitored and resolved in time. The plans developed in the planning stage require continual evaluation and flexible adaptation to ensure the achievement of development without harming the environment and society. It is this approach that we call *interactive management.*

Analytical Framework

Our conceptualization of *Arid Frontier Development* and the critical roles of *Proactive Planning* and *Interactive Management* are presented in Figure 1.The elements identified in Figure 1 are discussed in the following sections, elucidated by examples where suitable, as well as references to chapters in the volume where these elaborate upon essential points.

The complexity of the interaction between environmental and human resources in relation to both development and environmental quality preservation is lucidly expressed by Bennett (1984:289–290): 'While anthropologists perform archaistic studies of odds and ends of humanity, municipal authorities struggle with the chemical, geological, economic and political problems of toxic wastes, with little help from social scientists. Agricultural research provides clever answers to the problem of increasing yields without ruining the soil and water, but rarely concerns itself with the problem of decision-making among farmers who must make a living and function as members of communities. We talk about 'sustained yield' of resources: a fine conservationist principle that has attracted little work by the social sciences on how to put it into practice in a competitive and individualistic economy, with the partial exception of economics. Entrepreneurial production - rapidly becoming the dominant form of production in the world - is rarely compatible with ecosystemic principles, and hence one wonders, again, if ecosystem and other concepts like it have any relevance to the real world. This world includes humans who struggle to survive and to realize their goals in a human society and culture. If these factors are not somehow built into our ecological science, we are all monuments to futility and elitist pomposity.'

ENVIRONMENTAL, SOCIO-ECONOMIC AND TECHNOLOGICAL RESOURCE POTENTIAL AND CONSTRAINTS IN ARID FRONTIERS

The resource potentials and constraints in environmental, socio-economic and technological terms (Figure 1) determine the basic opportunities for development at the arid frontier. Grandiose government schemes and mistaken scientific concepts about conquering the desert, based on misreadings of the environmental and technological resource potential, have been repeated this century in different dry regions, for example in the U.S.A., the former Soviet Union and in Africa. Some of these fallacies have been analyzed and exposed by Warren (Chapter 6), such as the 'expert' opinion in the recent UNEP (1992) atlas of desertification, in which grazing lands are *a priori* associated with desertification. Expert opinions are no substitute for actual measurements of land degradation and desertification. If pastoral land-use were synonymous with desertification, how could pastoralism have continued as a sustainable system for thousands of years in many arid parts of the world, in the tropical drylands of the Sahel, in East and Southern Africa, in the subtropical Near East and in the temperate drylands of Mongolia and China?

Important aspects of the ecological role in desert development are presented by Safriel (1995). The tension between nature conservation and livestock grazing is discussed by Perevolotsky (1995). He points out that grazing has been going on for thousands of years in the drylands of the Mediterranean 'Old World', conferring on it a 'natural' quality. Conservation and grazing, therefore, are not necessarily contradictory but can be complementary concepts in ecosystem management. He reaches the conclusion that claims of overgrazing and resource degradation *a priori* can be suspect 'and should be substantiated professionally, rather than manipulated politically' (Perevolotsky, 1995:15).

Failure to appreciate the dynamics of the development process in arid frontiers, whether conducted by private or public agencies, can have devastating consequences. Most striking has been human-made desertification, such as occurred in the Aral Sea basin (Mainguet & Létolle, Chapter 7). These situations reflect two complex phenomena: the interaction between competing goals of environmental sustainability and development, and a failure to account for social costs created by

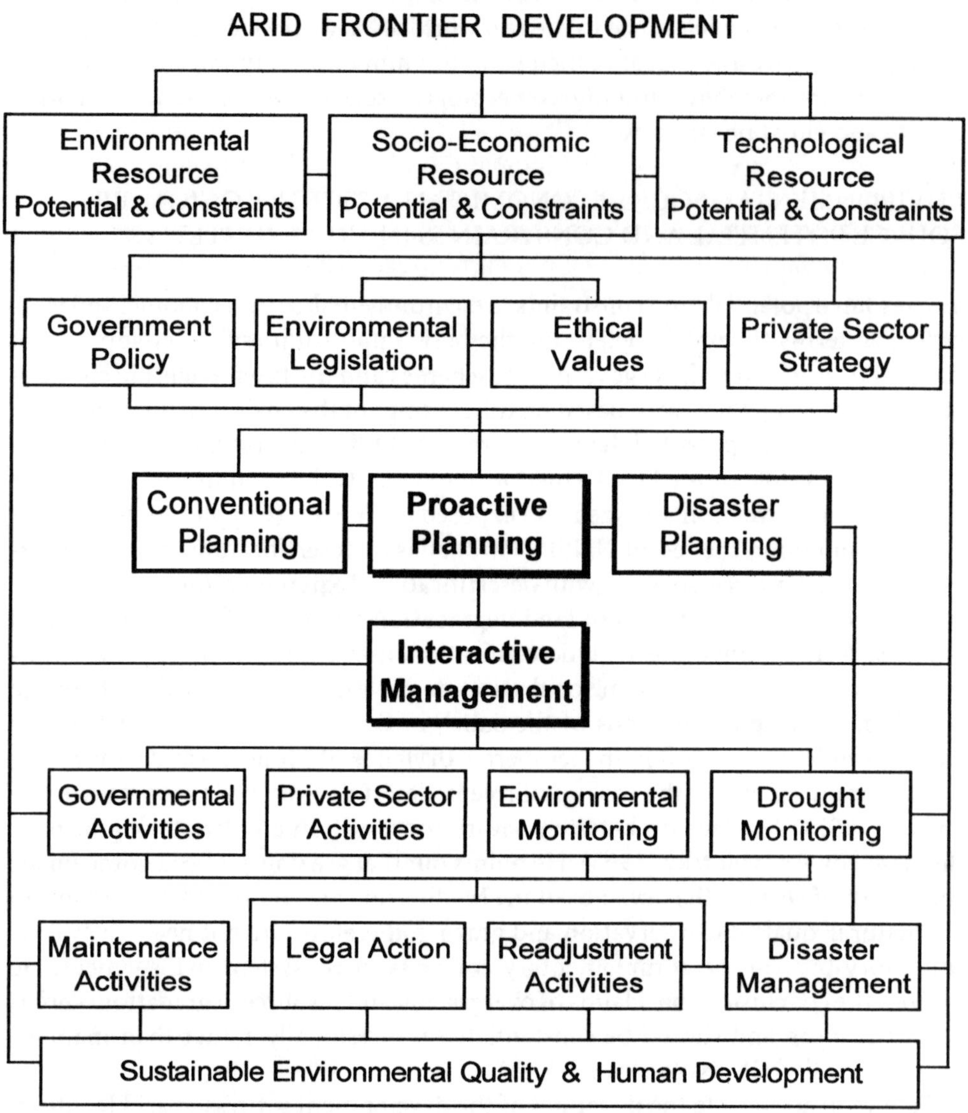

Figure 1. Conceptualization of Arid Frontier Development and the critical roles of Proactive Planning and Interactive Management.

negative externalities or public disamenities, an issue that is raised explicitly in the field of project analysis by El-Bihbety & Lithwick (Chapter 15).

To deal with complex problems such as these, planners require a problem-oriented systems analysis approach. The so-called 'soft' and 'hard' sciences are equally important in such approaches because both represent part of the 'reality'. An irrigation system well designed in socio-economic, anthropological and environmental terms but weak in technical terms is as bad as a technically well designed livestock raising system which does not suit the population for which it is designed in relation to their socio-economic structure and land rights.

In addition to the usual factors that make up the tool kit of systems analysis, two should be given special attention, although in practice, they rarely are. They are: ethical and distributional concerns. Proper ethical conduct, including scientific and financial honesty, are vital for success at all levels of development project planning as well as implementation. Many if not most projects have been negatively affected or even ruined through corruption. Nor is it appropriate to assume that such behavior is characteristic of private agents alone. Both the government and private firms have been found wanting in this field. The infamous Bhopal disaster in India, causing environmental pollution with terrible consequences, shows that the dividing line between the culpability of transnational corporations and local governments cannot always be clearly demarcated (Dias, 1993).

The subject of ethical conduct seems to be absent in most studies dealing with the environment and development planning. A search in the index or contents of books or articles, even in apparently extensive works, rarely if ever reveals the word ethics or related terms. But sometimes the subject appears, as in the proceedings of the First Regional Conference on Afro-Asian Studies on Social Systems and Food Crisis, held in Delhi in 1988, and published by the Indian Law Institute under the title *Famine and Society* (Floud & Rangasami, 1993). In its first chapter, entitled State, Law and Famine, Dias (1993) deals with the manifold problem of unethical behavior and its ramifications for famine, environment and development. 'In my view, all too often, famine is a euphemism for the planned, foreseeable, often deliberate genocide of certain powerless and disadvantaged populations within society. Astonishingly, this planned genocide is not usually conducted as a covert operation involving clandestine activity' (Dias, 1993:23–24).

While mentioning demographic, technological, food production, food distribution and environmental explanations of famine, Dias (1993:24) focuses 'attention to the fact that very often the kind of combination of poverty, powerlessness, exclusion, deprivation, hunger and starvation that leads to famine-type conditions is really a product of development policies pursued by state actors, often at the instigation of international and transnational actors ... who are often rapacious and ruthlessly exploitive of ecology in order to support the high-consumption lifestyles of an elite few and to support profit-making at utterly obscene levels. These kinds of development policies often result in prodigal and profligate mismanagement of renewable and non-renewable resources in order to achieve a model of parasitic development for a few. But the development of a few is often achieved at the subjugation, impoverishment and exploitation of the many.'

Dias (1993) gives various examples of what he terms *perverse development* which has caused:

1. Changes in land rights and land-use.
2. The dismantling of traditional food production and food security systems.
3. The destruction of traditional mechanisms and strategies for the survival of subsistence communities.
4. The destruction of traditional food cultures.

The studies by Akong'a & Kareithi (Chapter 9) and by Rutten (Chapter 10) also show how development changes in traditional social relationships and land rights have made local communities more vulnerable to the impact of drought. This theme is also explored in depth by Garcia (1981) and by Garcia & Spitz (1986). The problems of secure livelihoods in African drylands, in terms of research, development and policy, are investigated by Hjort af Ornäs (1992) and other contributing authors.

The consequence of ignoring distributional matters which are so often the result of ethical lapses can be illustrated anecdotally. During an invited lecture presented by the first author at a meeting of a nature protection society, a question was raised about the wisdom of providing development aid to pastoralists in the Sahel. It was suggested that it would be better to let the people die during times of drought and famine, in order to preserve the ecological balance. This represents a dangerous kind of environmental thinking. Dietz (1996), in his inaugural lecture about entitlements to natural resources, discussed the ecototalitarian approach, as he developed the academic theme of political environmental geography. He remarked that authoritarian elements in ecopolicy were already noted by the geographer Harvey (1974) and the philosopher Enzensberger (1974), who relate these elements to neo-Malthusianism. The dangerous and false logic of extreme environmentalism is exposed by Adams (1990), quoted by Dietz, 'environmentalism offers an essentially determinist analysis based on the principle of unchanging limits on human action, and a pessimistic view of the potential impact of social reform.... this generates the deeply conservative ideology and reactionary and repressive politics of 'eco-fascism''. In a historical and ideological study of modern environmentalism, Pepper (1996) shows links between New Age philosophies and radical environmentalism. There has been a tendency among environmentalists, according to Dias (1993), to ignore the link between environmental devastation and human degradation in developing countries. The extent to which the Indian authorities chose to ignore that link with respect to the Narmada Dam Project is explored in El-Bihbety & Lithwick (Chapter 15).

As a result of the above pessimism, all too often the positive impact of environmental potential on societal development is ignored, such that the two are seen as mutually exclusive. Yet a positive, sustainable environment can be one of the most important sources of human well-being, contributing to health, recreation, and long term economic improvement. And balanced development can provide the resources and incentive to invest in environmental sustainability. Unbalanced viewpoints occur at both extremes: in extreme environmentalism as well as in extreme human-centered developmentalism, in which humankind is regarded as the sole determiner of its development, irrespective of the environment. The origins of anthropology are associated, according to Moran (1984), with an intellectual thrust to reject various forms of environmental determinism and the over-generalizations of anthropo-geography in the late 19th century (Moran, 1982). In the view of 'historical possibilism, environment was seen as a passive force in the emergence of observable human traits or institutions' (Moran, 1984:7). However, segments of anthropology gradually changed their perception and assigned a more realistic role to environmental forces via 'cultural ecology' (Steward, 1955) and the ecosystem approach (Moran, 1984), which is based upon systems theory. The significant role for anthropology in development has been underlined by the Institute for Development Anthropology at the State University of New York at Binghamton (Horowitz, 1986; Little & Horowitz, 1987).

The complex relationship between environmental and human stress has been investigated by Garcia (1981), who warns against a simplistic causal relationship between drought and famine. He conducted a detailed and comprehensive study about widespread food shortages and famines in the year 1972. Society should not be conceived as a passive receptor of the 'impact' of climate with only limited possibilities of 'adapting' itself to a climatic variability. The worst famine situation in 1972 was located in the Sahel, following a drought which began in 1968. Garcia (1981:3) notes: 'But there have been other similar droughts in the past and their consequences have not been so dramatic. Why was this one so serious?'

The explanation usually consists of three main points: (1) the destruction of the natural environment; (2) soil depletion by an excessive number of animals, causing overgrazing and desertification; and (3) relentless demographic pressure. Garcia attacks these 'expert' explanations as superficial, incomplete and misleading. What is lacking is an appreciation of the social dimensions of drought: 'The pre-colonial pastoral and agro-pastoral ethnics had response mechanisms ingrained in the structure of social relations which were at the basis of the production systems. An analysis of the historical evolution of the system of social relations and the disruption of the response mechanism provides a solid foundation for the diagnosis of the Sahelian catastrophe in the early 1970s' (Garcia, 1981:226). Similar disruption of traditional drought coping mechanisms in Kenya, as a result of social changes, are noted in Kenya (Akong'a & Kareithi, Chapter 9). Garcia may have overstated his excellent analysis in the title of his book 'Nature pleads not guilty', which may wrongly suggest that the impact of drought was negligible in comparison to the inability of society to absorb the shock.

Unlike the dangerous view of radical environmentalism, we want to emphasize that human carrying capacity is not a fixed quantum solely determined by supreme and unchanging environmental factors. The interaction between socio-economic, technological and environmental resource potentials have changed human carrying capacity continuously throughout history. Hunting and gathering, situated at the bottom end of human carrying capacity systems, would only allow for a very small human population in arid regions, as well as in the world at large. A change in socio-economic and technological resource potential can change the human carrying capacity dramatically without any change in the climate and the degree of aridity. Irrigation development in Egypt and Mesopotamia in antiquity, as well as modern introduction of imported water through large national conveyance systems in California (Roos, Chapter 14) and Israel (Shanan, Chapter 13), greatly increased the carrying capacity and development in these arid regions.

In addition, the potential of solar and other renewable forms of clean energy, made possible through technological progress on the part of humankind (Faiman, Chapter 16; Meidav, Chapter 17) will in time produce relatively cheap desalination of sea and brackish water to enable a much higher level of sustainable food production in many arid regions. But we must avoid the technology trap that asserts that all problems can and will be solved through future technological improvements. Man-made desertification resulting from bad irrigation development, such as in the Aral Sea basin (Mainguet & Létolle, Chapter 7) and other regions (Shanan, Chapter 13), is a continuous warning that grandiose development planning in arid regions may go grandiosely wrong.

GOVERNMENT POLICY AND ENVIRONMENTAL LEGISLATION

Development planning does not emerge directly from the environmental, socio-economic and technological resource potential, as illustrated in Figure 1. Planning is usually guided by the goals of government policy and private-sector strategies, which influence one another. However, the planning process must also be influenced and controlled by legislation to protect the environment as well as the people against harmful exploitation, while encouraging social and economic development.

Government policy, environmental legislation and enterprise strategy influence one another and together they determine the content, shape, direction and goal of development plans, while environmental legislation sets limits and constraints to preserve environmental quality. Sandler (Chapter 4) demonstrates the significance of environmental law to protect and manage the beautiful but fragile ecosystem in the Gulf of Aqaba, shared by Jordan, Israel, Egypt and Saudi Arabia. She discusses the difficulties of arriving at mutually acceptable laws and management practices at an international level in relation to industrial and tourism development. The role of both government and private enterprise are important. The former has to create a good infrastructure, proper legislation, and security, including environmental security through proactive drought and disaster planning, in order to enable the latter to develop in a constructive, accountable, non-exploitive and non-destructive manner.

While it is not necessarily the case that the pursuit of human development activities and the preservation of the environment are inherently in conflict, the reality is sometimes even worse. Both environmental *and* human resources may be exploited, causing misery to many people. Dias (1993) reached the conclusion that the (unstated) policy and strategy of both state governments and large multinational enterprises sometimes involves the exploitation of the environment and masses of poor, powerless, cheap laborers and primary producers. Agricultural development policies and projects may have admirable stated objectives, but are often designed for selfish national or corporate reasons. This leads, in the words of Dias (1993), to:

(1) Exploitation of natural resources in developing countries, including lands for transnational agri-business plantations, hazardous industries, and dump sites for toxic waste. Ruling elites in developing countries are willing accomplices preferring short-term profits and wealth for long-term pauperization of the human and natural environment.

(2) Exploitation of labor in developing countries, both cheap unskilled and skilled (brain drain) labor.

(3) Exploitation of markets in developing countries, with the introduction of junk foods, to sustain levels of economic growth in industrialized countries.

Development policy within the developing countries is often characterized, unfortunately, by a perpetuation of colonial-type exploitation of the primary producers (the vast majority of the population) by a small urban-industrial elite and its client class of dependent rural elite - a dynamic that has been analyzed under the general concept of 'Dependency Theory' (Matthews, 1983).

One of the great observed dangers is that if the value of the environment does not get built into the overall framework of public policy and hence planning for development, it may be ignored. More recently, environmentalists have become more vociferous in pursuing their agenda, but they too have neglected to take into account the potential consequences for humans, in terms of loss of jobs, income and living space as a result of pursuing their agenda through public policy. What is required to shift the balance of outcomes to a positive sum situation, is to recognize that there are a wide range of possibilities that permit the attainment of both sets of values. Since these possibilities will not tend to arise when the various interest groups pursue their respective goals autonomously, mechanisms are required for ensuring that they are jointly pursued. Legislation to protect both the environment and the people against exploitation is of utmost importance to help attain such goals.

An important role for government policy is to create an infrastructure, in spatial, transport and legal terms, which gives incentives, confidence and security for arid frontier development by rural enterprises, whether individual farmers and livestock producers or larger companies. The remarkable agricultural development at the arid frontier in Israel by the private sector is partly due to the policy of intervention by the government. There are quite stringent state water laws in Israel about water use, its compulsory metering by users, specific allocations, as well as continuous monitoring of water quality and quantity (Bruins, 1993; Shanan, Chapter 13). Such environmental issues cannot be left, *laisser faire laisser allez,* to free market forces. Nor should they be left to the soaring ambitions of individual politicians or their bureaucrats. It cannot be expected that the cumulative effect of private production by individual firms, or public decision making by individual officials each one mainly interested in profitability, income and/or power, can direct long-term environmental policy and preserve environmental quality. These are social matters to be directed by the democratically responsible state, which alone has the legislative, organizational and financial power to manage such matters on a large scale.

Applied policy research has been explored very creatively in the work of Stockey and Zeckhauser (1978). Perhaps the most widely-used tool that represents this approach is Social Cost Benefit analysis (CBA) (Little & Mirlees, 1974). Although widely criticized for its attempt to quantify things which are difficult to quantify, the criticisms generally are inappropriate, reflecting either a misunderstanding of the vocation of this tool, or unhappiness with the sorts of decisions it points to. The fact is that any managerial judgment contains implicit evaluations of all costs and benefits, and CBA simply forces the analyst to be explicit about those judgments. Needless to say, many practitioners of CBA tend to be much too cavalier in their assignments of costs and benefits, but this reflects poor practice, rather than a problem intrinsic to the tool itself.

This criticism about quantification has been especially prevalent among environmentalists, many of whom are unwilling to concede that there are opportunity costs to the policies they are advocating, even when those costs are obvious and serious. Preserving a forest may have major implications for the livelihood of loggers, for example. While not denying the importance of that preservation, the environmentalist must, if honest, indicate what the human costs will be, who will bear them, and whether from society's point of view, those costs are tolerable. Moreover, the benefits derived from such a policy are often asserted, rather than proven. Indeed, on many issues, there are serious differences of opinions among knowledgeable scientists about those benefits. Again, being forced to identify those benefits, with some degree of intellectual support, is very difficult for the advocacy lobby in particular.

But it is unfair to single out the environmental lobby. Advocates of economic development are no less prone to oppose the rigor of CBA. They will assert that the benefits are obvious, when they are indeed questionable, and they will downplay the human and social costs, which are often intolerable, asserting that, in the long run, society will be better off. Most problematical is the distributional issue, the weighing of the benefits and costs of any policy initiative among different groups in society. There are techniques for so doing, but they remain problematical, and require explicit welfare judgments about the distribution of income. Again, this is not a problem intrinsic to CBA, but rather to the issue itself, about which philosophers no less than economists remain at odds.

Although this volume does not stress CBA techniques, with only one chapter (El-Bihbety & Lithwick, Chapter 15) devoted to it, the management planning oriented logic of CBA enters into a number of chapters, which attempt to introduce analytical techniques that consider and where possible quantify benefits and costs and discuss lost opportunities as well as distributional issues. It is recognized by the authors that such information and such an orientation is essential for managing the process of development in various areas of the environment, especially the fragile arid frontier environment with its unique sets of benefits and costs. It leads in virtually all cases to an understanding that the simplistic juxtaposition of alternatives, of environmental preservation *vs.* development, is neither realistic nor necessary.

Rather, we demonstrate that it is not only feasible, but optimal, to pursue a harmonization of these two domains, to enable humankind to live in and with this unique environment. The tools that enable this harmonization, or synergy, are proactive planning and interactive management.

PRIVATE SECTOR STRATEGY

Economists work with the strong assumption that all private actors behave in their self interest. While not an attempt to describe the behavior of each and every actor, this assumption leads to robust predictions about market performance and the ability to attain efficient production and rapid development. The qualifications to the theory are well-known:

1. The weakening of market performance and hence the social benefits as a result of monopolistic structures.
2. Externalities or spill-over effects (especially environmental).
3. The existence of collective or public goods (again, many of which are environmental in nature).

But these qualifications tend to be lost sight of when policies are being designed for the real world.

In arid zones, virtually all these qualifications have unique saliency, so that the easy assumptions about 'greed is good' or keeping governments out of marketplaces must be severely qualified if planning is to address real world concerns. Appropriate policy responses to such situations are government regulation or assumption of control over enterprises—but the very circumstances that create these distortions are likely to ensure that government involvement is likely, if anything, to make matters worse.

One solution, favored by Gunnar Myrdal, may have unique relevance to these circumstances. Concerned precisely with government inefficiency, Myrdal advocated a role for government that was largely restricted to 'framework planning', that is setting appropriate rules of the game (Myrdal, 1970). The government should be the referee, not a direct player. In circumstances where government is limited, in terms of talent, finance and control, such an approach, which would be defined by a proactive planning process, and supported by interactive management, could yield better if not superlative results.

PROACTIVE AND DISASTER PLANNING

Planning may be defined as a bridge between knowledge and action (Faludi, 1987:7). Classic planning thought, relying largely on the intuitive understanding of experts, is considered inferior to rational planning as developed by the 'Chicago School' (Meyerson & Banfield, 1955; Faludi, 1987). Such rational planning is generic, relating to public policy, and involving decision-making and implementation. Rational planning in theory considers all the alternative courses of action within existing conditions and in view of the goals to attain, while identifying and evaluating the consequences of each alternative. It finally selects the most preferable alternative. Even 'rational' planning in practice cannot be perfectly rational as no human being 'can ever know all of the alternatives open to him at any moment or all the consequences which would follow from any action' (Faludi, 1987:315). Indeed, there are scholars such as Wildavsky (1979) who argue that in the real world, the rational planning model is inherently flawed. While recognizing the limits of obsessive rationalism, we do believe that there is no alternative to structured thinking in the face of complex problems. It is here that the significance of interactive management becomes clear, as such management can adapt to unforeseen events by changing and readjusting planning elements which appear inferior during planning implementation.

Planning for development in arid lands therefore requires an appreciation of both areas of planning, environmental and developmental. Given their distinct characteristics, it is useful to examine them in turn, and then see how they might be integrated. Environmental planning is concerned with designing ways to improve areas of environmental degradation based on an understanding of the consequences of failure to act and to anticipate future problems often random in nature and long before the public is aware of their potential consequences. In contrast, human development planning aims to correct distortions that prevent societies from progressing, and designing programs that can accelerate the pace and scope of development. For our overriding concern with the potential conflict between development and the environment in arid frontiers, a more complex kind of planning is therefore required, one that can articulate policy options that balance these two planning concerns of the environment and development in a way that reflects society's valuation of both these goals.

Both arid environments and human societies living in dry regions tend to be vulnerable (Downing, 1991; Cannon, 1991; Bohle *et al.*, 1993) to the impact of increased marginality resulting from climatic or man-made causes. Landscapes, livelihood systems and survival strategies may all be adversely affected as a result of drought, desertification or social upheavals. Development planning at arid frontiers has to include special elements to cope with high-risk environments (Maxwell, 1991; Blaikie *et al.*, 1994), in which drought, being rather unpredictable in temporal and spatial terms, is the most hazardous part of climatic variability (Wilhite & Easterling, 1987; Glantz, 1987; Wilhite, 1993a). Drought causes recurring insecurity in economic activities like rainfed agriculture, livestock herding, and related enterprises, as well as in the society at large, for example in relation to rural and urban water supply.

There is a need for planning to be proactive in order to be prepared for the impact of light and moderate drought. Proactive planning takes such environmental uncertainties into consideration by preparing plans for ephemeral stochastic events, which become activated when such events (*i.e.* droughts) occur. This requires regional development planning with interlocking bottom-up and top-down mechanisms in arid frontier conditions, in which environmental resource management has to accumulate reserves in food, water or money (Bruins *et al.*, 1987) during the good years for use and survival in drought years. Severe drought, though not occurring usually at frequent or predictable intervals, requires its own set of strategies in the sphere of disaster planning, which can be regarded as a component of proactive planning (Figure 1).

Proactive planning for drought as well as rainwater harvesting have a great potential to increase system viability and livelihood security for arid-zone pastoralists under more limited grazing conditions. The free movement of nomadic and semi-nomadic pastoralists has been considerably curtailed in many arid regions as a result of new state boundaries, population pressure and the expansion of agricultural lands (Kressel, 1993). An increasingly sedentary lifestyle by nomadic pastoralists is accompanying current changes in parts of Africa as well as among the Bedouin in south-west Asia (Kressel & Ben-David, 1991; Meir, 1997). Consequently, new approaches in land-use and resource management need to be developed in an acceptable social context. The human factor and the aspirations of the people are, however, critical in development.

INTERACTIVE MANAGEMENT

Both management and planning are concerned with providing frameworks that direct activities towards specific goals. In theory and practice, there is a fuzzy dividing line between these two phenomena. Planners tend to extend the scope of their activities to include supervising implementation, which comes very close to managerial activity. And most managers would include within their brief, planning at the front end and evaluation at the back end. While such 'imperialism' is normal, it is useful to try to find a working definition that identifies these tasks and the boundary between them more sharply. The working definition we have provided of planning, as concerned with framing the rules of the game, leaves to management responsibility for implementation, and, through monitoring, providing feedback to adapt the planning process as required.

From this perspective, it is the task of the manager to ensure that there is constant reinforcement of the planned strategy. The executive character of management places it in a position of crucial importance that may determine success or failure. We would go even further than that: there has to be constant interaction through monitoring of planning implementation and the results, in the two domains of the environment and development. If the results are unwanted or negative, the interactive managerial process should have the ability to revise the original plans accordingly in order to attempt to arrive at ongoing environmental quality as well as development activity sustainability (Figure 1). That means that the manager must be in constant touch with the actual implementers, monitoring their actions and receiving feedback on potential trouble spots. This information can then be fed back into the overall program management system, to revise targets and tasks consistent with the overall objectives. It is this type of management activity that we designate as interactive management. Only such an approach can ensure that the joint goals of environmental enhancement and human development can be kept on the same trajectory, particularly in sensitive arid frontier situations (Rondinelli, 1983).

Good proactive planning may fail as a result of bad management, but it is also the case that potentially harmful results from bad planning may be averted through interactive management, by which defective planning elements are changed and adjusted in the course of plan implementation and subsequent project continuation.

But that advantage also poses a problem, for the interactive approach requires expertise generally far broader than that available to the typical manager. To build on a wider frame of reference, it thus becomes necessary to work in teams, or at least to draw on works from multiple disciplines. It is nearly impossible to do so exhaustively. As a result, the practical managerial solution is to proceed iteratively, to start with a clear understanding of management needs, to draw on as many available sources as possible, identifying the major gaps, and then actively solicit additional input to broaden and deepen the perspective. Over time, this approach will evolve into increasingly relevant diagnoses of problem areas, leading to more appropriate solutions.

Interactive management includes environmental and enterprise monitoring, while adhering to ethical norms and conduct are of crucial significance at all levels of management. Interactive management also includes maintenance and readjustment activities in areas for which the government or state is responsible no less than in areas for which the enterprise is responsible, be it a small farmer or a large multinational corporation. The effect of climatic variability and drought at the arid frontier requires interactive management of environment and development, to ensure that plans are implemented, and adjusted in line with the great variability and uncertainty inherent in such phenomena.

Proper management, therefore, is the critical link between environment and development in arid frontiers. The environment provides opportunities for human existence, be it in the realm of water supplies, food production, energy or shelter. This book has been conceived to emphasize how proactive planning and interactive management enables man to be a constructive partner and not a destructive agent. A number of excellent scholars have come together to stress this management necessity with respect to ensuring food security, dealing with desertification, preserving the environment while using its qualities for development, and harnessing energy potential, all in arid regions.

Limiting factors in arid land development, such as the lack of potable water resources and excessive heat, often predominate in the perception of planners. One has to learn to see the other side of the coin: bountiful solar radiation, surprisingly pleasant climates for habitation, available space and beautiful scenery constitute real assets for development. The potential use of solar power for energy and water resources development is probably the most promising factor in arid-frontier development.

It is therefore the joint application of proactive planning and interactive management that we believe is capable of ensuring patterns of action that yield positive synergy between development and the environment, as this volume sets out to emphasize. The value of this approach is strongly underlined by evidence from the field of irrigation development, as presented by Shanan (Chapter 13, this volume), and illustrated with examples. The managerial or management approach to problems is based on a rationalistic paradigm. That paradigm states that full information and knowledge about processes underlying public and private activities as well as their impacts can optimize the efficiency with which resources are used, and help resolve conflicts that are inherent in the distribution of costs and benefits associated with those activities.

As such, the management approach is very much a hands-on orientation towards problem solving. The goal is not to provide new information per se, although identifying what is needed and setting priorities for obtaining it is a direct offshoot from this approach. Rather, the goal is to use existing information with a view to moving activities ahead in a direction that is more optimal and more efficient.

Perhaps the main advantage of this approach is that it requires the inclusion of information from multiple sources. It is insufficient to work within narrow academic boundaries, for example. Water resource management requires an appreciation not only of the science of hydrology, but the economics of water pricing, the politics of assigning water rights, and the art of conflict resolution. Similarly, environmental management draws not only on ecology, but anthropology, law and the social sciences.

GOVERNMENT AND PRIVATE SECTOR ACTIVITIES, ENVIRONMENTAL AND DROUGHT MONITORING

The core activity of interactive management is 'monitoring', for only accurate and timely information can provide the manager with evidence as to the unfolding of the planning process. A good plan has built-in from the outset a family of performance indicators, which provide substantive, quantitative guidelines for the key plan variables. The chief task of the manager is to ensure that the implementers of the plan produce the information upon which the relationship of results to plans can be compared. Naturally, there will be strong resistance to providing such information, especially if the information indicates under-achievements. As a result, the manager may have to take certain areas of monitoring out of the hands of the interested parties, and delegate them to more independent, objective evaluators. This problem is no less prevalent in public than in private activities, as has been revealed by mountains of phony data that both sectors managed to produce for Soviet and Chinese planners.

One way to facilitate cooperation, and hence more accurate monitoring is to have implementers involved in designing the plan targets from the outset. They will then have some sense of their relationship to the overall process, and some interest in contributing to collective goals. Of course this approach entails an element of delegating power from the top down, and only an enlightened planning process, that also values bottom-up efforts, will be receptive to it (Stöhr & Taylor, 1981).

Environmental monitoring in general and drought monitoring in particular form core activities of interactive management in arid frontiers. The type of environmental parameters to be monitored obviously depends on the characteristics of both development and the environment in which it takes place. For example interactive management of irrigation development in lowland basins requires the monitoring of groundwater depth, soil salinity, sedimentation in conduit channels, and other aspects as presented by Shanan (Chapter 13). Desertification indicators include actual measurements of landscape degradation, soil erosion, changes in plant cover and biodiversity. The monitoring of various chemical and hazardous materials in the air, water and soil, may be required in industrial forms of arid frontier development.

Drought monitoring follows the amount of water available through natural and managed systems (Hubbard, 1993), which include precipitation, streamflow, reservoir storage, groundwater, soil water, evaporation and evapotranspiration. Such data help to assess the development of the drought, which can be expressed in drought severity indices. Concrete examples of drought monitoring and water management in California are presented by Roos (Chapter, 14). Satellite data are important in drought monitoring (Johnson *et al.*, 1993) in addition to meteorological and hydrological data, for example to assess the condition of agricultural crops, pasture and natural vegetation through time series of Normalized Difference Vegetation Index (NDVI) data. Drought monitoring and impact assessment data are to be fed, interactively, in the proactive planning system to guide responsive management decisions to mitigate the effect of the drought.

DISASTER MANAGEMENT

A disaster is characterized by a drastic change of the physical and technological foundations of social regularity (Kreps, 1983). 'A *hazard* may be regarded as the predisaster situation, in which some *risk* of disaster exists, principally because the human population has placed itself in a situation of *vulnerability*' (Alexander, 1993:7). Humankind living in arid regions is vulnerable to the risk of drought, which is a natural hazard that may *indirectly* develop into a disaster through the inability of society to cope with its impact on food and water resources. Disasters are not always neatly related to a single hazard. They may be a complex mix of natural and man-made causes. War may lead to various disasters including famine. If war coincides with drought, it will be more difficult for the people to cope with the impact of the latter, which might have been managed satisfactorily without war (Blaikie *et al.*, 1994).

The first requirement of disaster management is correct and detailed information about the disaster (Alexander, 1993), so that emergency measures can be organized and directed accordingly. However, without the availability of proactive disaster plans, disaster management can only be *ad-hoc* crisis management, which is always inferior in economic and efficiency terms. In our model the disaster information, *i.e.* related to drought, is inter-actively fed into existing proactive disaster plans, which will activate the proper response implementation through prescribed disaster management.

Removing oneself physically from a disaster situation is sometimes a rational action. The drought related disaster problems in the Sahel in the 1980s led to decisions by some 10 million people to move away and flee from their original habitat (Warren, Chapter 6). Emergency shelter, therefore, can be a very important aspect of successful disaster management. A refugee camp has to be composed of elements for housing, sanitation, water supply points, clinics and feeding centers (Holt, 1981). The proper design of refugee camps is very important, as outlined by Alexander (1993:520): 'One aspect of the relief and management of refugee situations that is often lamentably misunderstood is the design of camps and the provision of sanitation, shelter, clinics and feeding arrangements. According to one expert in the field (Cuny, 1977), the total costs of designing and constructing a refugee camp properly and efficiently are less than the extra operating costs of a substandard one.'

Psychological aspects and social organization of disaster victims in refugee camps need to be addressed as much as the physical needs. Rehabilitation is also a crucial aspect of disaster management. Severe droughts in the Turkana District of Kenya in 1979/1980 and 1984/1985, coupled with a decline in traditional drought preparedness, resulted in severe loss for the local pastoralists. A large portion of the population lost their livelihood, *i.e.* their entire herd, and survived in refugee camps, where they received food aid from the Kenyan government and international aid agencies. However, if they stay too long in such camps, the younger generation may not learn the knowledge and expertise of grazing management, making rehabilitation more difficult. Landscape degradation and desertification may also occur in and around refugee camps. Hence educational, environmental, training and rehabilitation aspects of refugee camps are equally relevant in proactive design for disaster management.

Concluding Remarks and Book Overview

The ultimate objective of proactive planning and interactive management is the realization of the joint goals of sustainable environmental quality and human development in arid frontier regions. It is abundantly clear that the mindless pursuit of human development will invariably cause environmental devastation, like in the Aral Sea basin (Mainguet & Létolle, Chapter 7), while extreme environmentalism will erode the possibility for human improvement. These extremes need not be the only set of policy options.

We favor even-handed policies to be realized through proactive planning and interactive management. Areas of conflict or tension, like drought and its ramifications, which can lead to environmental stress and desertification, as well as to human stress and disaster, need to be identified in advance to be included in the planning *proactively:* development of proactive coping and mitigation mechanisms. The risk of severe drought in arid zones requires the extension of proactive planning into *disaster planning*, based on vulnerability and impact studies, in order to cope with major system collapse. Since rational planning can never be perfectly rational, due to intrinsic lack of knowledge about the future, it is inevitable that unimagined areas of conflict or stress will emerge as the planning is implemented, be it in the environmental sphere, the human sphere, or both. This is to be discovered and corrected by *interactive management* through monitoring and the capacity to readjust plans accordingly, in order to achieve more compatible and meritorious ends. In the event of severe stress or catastrophe, it is clear that disaster planning needs to be implemented through *disaster management*, which is another building block in our paradigm (Figure 1).

It is our view that the dual combination of proactive planning and interactive management is theoretically capable of enhancing both the environment and living standards. How to do this for each circumstance is a major project, one that will have to be calibrated independently. Nevertheless, the evidence we have accumulated to date, some of which is incorporated in this volume, has led us to believe that the approach advocated has great potential value, and we hope it will be developed and extended with as much haste as the urgency of the situation of peoples and environments in arid frontier regions demand.

In approaching the subject of Arid Frontier Development in this volume, we have attempted to impose a policy-oriented framework on most of the chapters, in the belief that such a framework is essential to move from theory to solutions. This book introduces a number of studies of key issues pertaining to arid frontier development from this perspective. The varied nature of the problems and solutions addressed by the authors preclude a rigid, common approach, but the elements of rational policy analysis are teased out in virtually all the chapters, while others provide building blocks for development policy on certain issues.

This, our first chapter, provides the overall rationale and focus of the volume, emphasizing the importance of proactive planning and interactive management in arid frontier development.

Part 2 of the book builds upon the basic paradigm of rational policy making, by introducing additional spatial dimensions of environmental and development policy: national, regional and multinational. They demonstrate the sort of modifications that are required when we shift our focus from the national unit, both inwards and outwards. Wilhite & Hayes (Chapter 2) focus on complex national (state and federal) dimensions of proactive drought planning in the United States. Powell (Chapter 3) analyzes the Murray-Darling Basin in Australia, covering several states, and gives us a view of the actors and environmental policy making in a regional framework. Deborah Sandler (Chapter 4) explores the interaction between national and international requirements of environmental law and management in the Gulf of Aqaba, situated in a hyper-arid region, shared by Jordan, Israel, Egypt and Saudi-Arabia.

In part 3, we turn to specific environmental problems for planning and management in arid frontier regions: desertification and climatic variability. Bruins & Berliner (Chapter 5) discuss and define these environmental concepts, suggesting a standardized way to express bioclimatic aridity on a world-wide comparable basis. They also advocate the inclusion of the hyper-arid zone in the desertification concept. Warren (Chapter 6) provides a vigorous but much needed weeding of the rather jumbled garden of concepts surrounding the role of environmental science in arid-frontier development and combating desertification. Mainguet & Létolle (Chapter 7) move on from concepts to praxis, and analyze one of the most alarming cases of human-made desertification in the Aral Sea Basin. A planning dimension for rainfed agriculture in semi-arid regions is provided by Dietz, Put & Subbiah (Chapter 8) who analyze in detail techniques of drought risk assessment, focusing on Andhra Pradesh (India).

The complex environmental and human relationships in drought, food and land management is the theme of Part 4, with particular reference to the most challenged of all regions, Africa. A review of indigenous social-environmental systems for managing drought and famine in Kenya is provided by Akong'a and Kareithi (Chapter 9), while Rutten (Chapter 10) focuses in a detailed case study on the process of changes in land rights among Maasai pastoralists in Kajiado District (Kenya) and its bearing on food security. Land management in Benin is explored by De Haan (Chapter 11) in a classical dryland frontier, shared by peasants and pastoralists. The role of wild plants as hunger food, providing some measure of food security in times of stress, is explored by Van Damme (Chapter 12) in ethnobotanical studies focusing on Namibia and Senegal.

Part 4 addresses the issue of development planning and management of the most limited or critical environmental resource in arid zones: water. Shanan (Chapter 13) emphasizes the immense significance of proactive planning and interactive management in irrigation development. The potential environmental hazards of irrigation are presented in several case studies, showing examples of failure and success regarding the sustainability of irrigation. Shanan gives examples of readjustment of failing projects through new forms of management, in which the interactive approach, with built-in equity and anti-corruption mechanisms, is able to improve the situation drastically. The particular problems of drought and water management in California, which is one of the most developed arid-frontier regions in the world, are presented by Roos (Chapter 14). The mother of rational planning techniques, Cost-Benefit Analysis (CBA), is applied by El-Bihbety & Lithwick (Chapter 15) to an appraisal of the world's two largest water projects, in China and India.

Part 5 deals with the renewable energy potential for development of arid lands, including innovative architectural aspects. Solar energy is perhaps the greatest environmental resource in arid lands; Faiman (Chapter 16) presents a case study for the design of a photovoltaic power plant in a hyper-arid frontier region of Israel.

Meidav (Chapter 17) compares some options for the development of rural electrification with renewable energy, focusing on geothermal power. Sauerhaft, Berliner & Thurow (Chapter 18) present innovative research for the production of fuelwood, which is the most important source of household energy in many countries, but shortages and resulting environmental degradation are a problem in various places. The book ends with habitation in arid zones, as Pearlmutter & Meir (Chapter 19) present a study about thermal comfort and energy use in two different climatic regions of Israel, pointing to passive cooling and heating potential in architecture by means of appropriate technology for housing construction.

Consistent with our approach to policy, this book is not the last word on the subject. Subsequent rounds are called for to develop, apply and reevaluate the paradigm together with investigations of the core problems in further depth. By means of such an iterative process, we sincerely believe that there is the potential for great improvement in dealing with Arid Frontier Development, one of the Earth's most compelling challenges.

References

Adams, W.M. (1992) *Wasting the Rain: Rivers, People and Planning in Africa.* Earthscan, London.

Akong'a, J.J. and J. N. Kareithi (1997) Traditional management of drought and famine in Kenya. *In* H.J. Bruins and H. Lithwick (eds.) *The Arid Frontier: Interactive Management of Environment and Development.* Kluwer Academic Publishers, Dordrecht, pp. 165-184.

Alexander, D. (1993) *Natural Disasters.* UCL Press, London.

Bennett, J.W. (1984) Ecosystems, environmentalism, resource conservation, and anthropological research. *In* E.F. Moran (ed.) *The Ecosystem Concept in Anthropology, American Association for the Advancement of Science, AAAS Selected Symposium 92.* Westview Press, Boulder, pp. 289–310.

Blaikie, P., T. Cannon, I. Davis, and B. Wisner (1994) *At Risk: Natural Hazards, People's Vulnerability and Disasters.* Routledge, London.

Bohle, H.G., T.E. Downing, J.O. Field, and F.N. Ibrahim (eds.) (1993) *Coping with Vulnerability and Criticality.* Breitenbach Publishers, Saarbrücken.

Bruins, H.J (1993) Drought risk and water management in Israel: planning for the future. *In* D.A. Wilhite (ed.) *Drought Assessment, Management and Planning: Theory and Case Studies.* Kluwer Academic Publishers, Dordrecht, Chapter 8, pp. 133–155.

Bruins, H.J. (1996) Drought and Fallout: Problem-Oriented Food Crisis Research. 28th International Geographic Conference (IGU), Abstract Book, Stichting International Geographic Congress. Utrecht, pp. 71

Bruins, H.J., M. Evenari, and U. Nessler (1986) Rainwater-harvesting agriculture for food production in arid zones: the challenge of the African famine. *Applied Geography* 6:13–32.

Bruins, H.J., M. Evenari, and A. Rogel (1987) Run-off farming management and climate. *In* L. Berkofsky and M.G. Wurtele (eds.) *Progress in Desert Research.* Rowman & Littlefield, Totowa, N.J., pp. 3–14.

Bruins, H.J. and P. Berliner (1997) Bioclimatic aridity, climatic variability, drought and desertification: Definitions and management options. *In* H.J. Bruins and H. Lithwick (eds.) *The Arid*

Frontier: Interactive Management of Environment and Development. Kluwer Academic Publishers, Dordrecht, pp. 97-116.

Cannon, T. (1991) Hunger and famine: Using a food systems model to analyse vulnerability. *In* H.G. Bohle, T. Cannon, G. Hugo and F.N. Ibrahim (eds.) *Famine and Food Scarcity in Africa and Asia: Indigenous Responses and External Intervention to Avoid Hunger*. Bayreuther Geowissenschaftliche Arbeiten, Naturwissenschaftliche Gesellschaft Bayreuth 15:291–312.

Cuny, F.C. (1977) Refugee camps and camp planning: the state of the art. *Disasters* 1:125–144.

De Haan, L. (1997) Gestion de terroir at the frontier: Village land management of peasants and pastoralists in Benin. *In* H.J. Bruins and H. Lithwick (eds.) *The Arid Frontier: Interactive Management of Environment and Development.* Kluwer Academic Publishers, Dordrecht, pp. 209-227.

Dias, C.J. (1993) State, law and famine. *In* J. Floud and A. Rangasami (eds.) *Famine and Society, The Indian Law Institute.* New Delhi, pp. 23–31.

Dietz, T.J. (1997) *Entitlements to Natural Resources: Contours of Political Environmental Geography.* International Books, Utrecht.

Dietz, T.J., M. Put, and S. Subbiah (1997) Drought risk assessment for dryland agriculture in semi-arid Telengana, Andhra Pradesh, India. *In* H.J. Bruins and H. Lithwick (eds.) *The Arid Frontier: Interactive Management of Environment and Development.* Kluwer Academic Publishers, Dordrecht, pp. 143-161.

Downing, T.E. (1991) *Assessing Socioeconomic Vulnerability to Famine: Frameworks, Concepts and Applications.* Alan Shawn Feinstein World Hunger Programme, Brown University, Providence RI.

Enzensberger, H.M. (1974) A critique of political ecology. *New Left Review* 8:3–32.

El-Bihbety, H. and H. Lithwick (1997) Cost-benefit analysis of water management mega-projects in India and China. *In* H.J. Bruins and H. Lithwick (eds.) *The Arid Frontier: Interactive Management of Environment and Development.* Kluwer Academic Publishers, Dordrecht, pp. 295-317.

Evenari, M., L. Shanan, and N. Tadmor (1982) *The Negev: The Challenge of a Desert.* Harvard University Press, Cambridge, Massachusetts.

Faiman, D. (1997) Solar energy in arid frontiers: Designing a photovoltaic power plant for Kibbutz Samar, Israel. *In* H.J. Bruins and H. Lithwick (eds.) *The Arid Frontier: Interactive Management of Environment and Development.* Kluwer Academic Publishers, Dordrecht, pp. 321-336.

Faludi, A. (1987) *A Decision-Centred View of Environmental Planning.* Pergamon Press, Oxford.

Floud, J. and A. Rangasami (eds.) (1993) *Famine and Society.* The Indian Law Institute, New Delhi.

Friedman, J. (1996) Borders, margins, and frontiers: Myth and metaphor. *In* Y. Gradus and H. Lithwick (eds.) *Frontiers in Regional Development.* Rowman & Littlefield Publishers, Lanham, MA, pp.1–20.

Garcia, R.V. (1981) *Drought & man: The 1972 case history. Volume 1. Nature pleads not guilty.* Pergamon Press, Oxford.

Garcia, R.V. and P. Spitz (1986) *Drought & man: The 1972 case history. Volume 3. The roots of catastrophe.* Pergamon Press, Oxford.

Glantz, M.H. (ed.) (1987) *Drought and Hunger in Africa: Denying Famine a Future.* Cambridge University Press, Cambridge.

Gradus, Y., and E. Stern (1980) From preconceived to responsive planning: Cases of settlement design in arid environments. *In* Y. Gradus (ed.) *Desert Development: Man and Technology in Sparselands.* D. Reidel, Dordecht, pp. 41-59.

Harvey, D. (1974) Population, resources and the ideology of science. *Economic Geography* 50:256–277.

Holt, J. (1981) Camps as communities. *Disasters* 5:176–179.

Hjort af Ornäs, A. (ed.) (1992) *Security in African Drylands: Research, Development and Policy.* Research Programme on Environment and International Security, Department of Human and Physical Geography, Uppsala University, Uppsala.

Horowitz, M.M. (1986) Ideology, policy, and praxis in pastoral livestock development. *In* M.M. Horowitz and T. Painter (eds.) *The Anthropology of Rural Development in West Africa.* Westview Press, Boulder, CO, pp. 251–272.

Hubbard, K.G. (1993) Monitoring regional drought conditions. *In* D.A. Wilhite (ed.) *Drought Assessment, Management and Planning: Theory and Case Studies.* Kluwer Academic Publishers, Dordrecht, Chapter 2, pp. 17–30.

Johnson, G.E., V.R. Achutuni, S. Thiruvengadachari, and F. Kogan (1993) The role of NOAA satellite data in drought early warning and monitoring: Selected case studies. *In* D.A. Wilhite (ed.) *Drought Assessment, Management and Planning: Theory and Case Studies.* Kluwer Academic Publishers, Dordrecht, Chapter 3, pp. 31–47.

Kreps, G.A. (1983) The organization of disaster response: core, concepts and processes. *International Journal of Mass Emergencies and Disasters* 1:439–466.

Kressel, G.M. (1993) Nomadic pastoralists, agriculturists and the state: self-sufficiency and dependence in the Middle-East. *Journal of Rural Cooperation* 21(1):33–49.

Kressel, G.M. and J. Ben-David (1991) Changes in land usage by the Negev Bedouin since the mid-19th century. *Nomadic Peoples* 28:28–55.

Little, I. and J. Mirlees (1974) *Project Appraisal and Planning for Developing Countries.* Basic Books, New York.

Little, P.D. and M.M. Horowitz, with A.E. Nyerges (eds.) (1987) *Lands at Risk in the Third World: Local-Level Perspectives, IDA Monographs in Development Anthropology.* Westview Press, Boulder.

Mainguet, M. and R. Létolle (1997) Human-made desertification in the Aral Sea basin: Planning and management failures. *In* H.J. Bruins and H. Lithwick (eds.) *The Arid Frontier: Interactive Management of Environment and Development.* Kluwer Academic Publishers, Dordrecht, pp. 129-142.

Matthews, R. (1983) *The Creation of Regional Dependency.* University of Toronto Press, Toronto.

Maxwell, S. (1991) *To Cure all Hunger: Food Policy and Food Security in Sudan.* Intermediate Technology Publications, London.

Meidav, T. (1997) Rural electrification with renewable energy: Geothermal power in arid regions. *In* H.J. Bruins and H. Lithwick (eds.) *The Arid Frontier: Interactive Management of Environment and Development.* Kluwer Academic Publishers, Dordrecht, pp. 337-350.

Meir, A. (1997) *As Nomadism Ends: The Israeli Bedouin of the Negev.* Westview Press, Boulder.

Meyerson, M.M. and E.C. Banfield (1955) *Politics, Planning and the Public Interests: The Case of Public Housing at Chicago.* Free Press, New York.

Moran, E.F. (1982) *Human Adaptability: An Introduction to Ecological Anthropology.* Westview Press, Boulder.

Moran, E.F. (1984) Limitations and advances in ecosystems research. *In* E.F. Moran (ed.) *The Ecosystem Concept in Anthropology.* American Association for the Advancement of Science, AAAS Selected Symposium 92, Westview Press, Boulder, pp. 3–32.

Myrdal, G. (1970) *Objectivity in Social Research.* Duckworth, London.

Pearlmutter, D. and I.A. Meir (1997) Lightweight housing in arid zones: Thermal comfort and energy use. *In* H.J. Bruins and H. Lithwick (eds.) *The Arid Frontier: Interactive Management of Environment and Development.* Kluwer Academic Publishers, Dordrecht, pp. 365-381.

Penman, H.L. (1948) Natural evaporation from open water, bare soil and grass. *Proceedings of the Royal Society.* Section A, 193:120–145.

Pepper, D. (1996) *Modern Environmentalism: An Introduction.* Routledge, London.

Perevolotsky, A. (1995) Conservation, Reclamation and Grazing in the Northern Negev: Contradictory or Complementary Concepts? Overseas Development Institute (ODI) Pastoral Development Network Paper 38a.

Powell, J.M. (1997) Interregional environmental policy in Australia's Murray-Darling basin. *In* H.J. Bruins and H. Lithwick (eds.) *The Arid Frontier: Interactive Management of Environment and Development.* Kluwer Academic Publishers, Dordrecht, pp. 55-73.

Reifenberg, A. (1955) *The Struggle between the Desert and the Sown: Rise and Fall of Agriculture in the Levant.* Publishing Department of the Jewish Agency, Jerusalem.

Reij, C., P. Mulder, and L. Begemann (1988) Water harvesting for plant production. World Bank Technical Paper, No. 91, The World Bank, Washington, D.C.

Rondinelli, D.A. (1983) *Development Prospects as Policy Experiments.* Methuen, Toronto.

Roos, M. (1997) Drought and water management in California. *In* H.J. Bruins and H. Lithwick (eds.) *The Arid Frontier: Interactive Management of Environment and Development.* Kluwer Academic Publishers, Dordrecht, pp. 277-293.

Rutten, M.M.E.M. (1997) Land tenure frontiers and food security among Maasai pastoralists in Kenya. *In* H.J. Bruins and H. Lithwick (eds.) *The Arid Frontier: Interactive Management of Environment and Development.* Kluwer Academic Publishers, Dordrecht, pp. 185-208.

Safriel, U.N. (1995) The role of ecology in desert development. *Journal of Arid Land Studies* 5(Suppl.):351–354.

Sandler, D. (1997) Environmental law and management in the Gulf of Aqaba. *In* H.J. Bruins and H. Lithwick (eds.) *The Arid Frontier: Interactive Management of Environment and Development.* Kluwer Academic Publishers, Dordrecht, pp. 75-93.

Sauerhaft, B., P.R. Berliner, and T.L. Thurow (1997) The fuelwood crisis in arid zones: Runoff agriculture for renewable energy production. *In* H.J. Bruins and H. Lithwick (eds.) *The Arid Frontier: Interactive Management of Environment and Development.* Kluwer Academic Publishers, Dordrecht, pp. 351-363.

Shanan, L. (1997) Environmental hazards of irrigation development: The need for proactive planning and interactive management. *In* H.J. Bruins and H. Lithwick (eds.) *The Arid Frontier: Interactive Management of Environment and Development.* Kluwer Academic Publishers, Dordrecht, pp. 251-276.

Stern, P.H. (1979) *Small Scale Irrigation: A Manual of Low-cost Water Technology.* Intermediate Technology Publications Ltd, London & International Irrigation Information Center, Volcani Center, Bet Dagan, Israel.

Steward, J.H. (1955) The concept and method of cultural ecology. *In Theory of Culture Change.* University of Illinois Press, Urbana (4th paperback printing, 1979).

Stöhr, W.B. and D.R.F. Taylor(eds.) (1981) *Development from Above or Below? The Dialectics of Regional Planning in Developing Countries.* John Wiley & Sons, New York.

Stockey, E. and R. Zeckhauser (1978) *A Primer for Policy Analysis.* Norton, New York.

Thorthwaite, C.W. (1948) An approach towards a rational classification of climate. *Geographical Review* 38:55–94.

UNEP (1992) World Atlas of Desertification, Edward Arnold, London.

UNESCO (1979) Map of the World Distribution of Arid Regions. UNESCO, Man and the Biosphere (MAB) Technical Notes 7, Paris.

Van Damme, P. (1997) Wild plants as food security in Namibia and Senegal. *In* H.J. Bruins and H. Lithwick (eds.) *The Arid Frontier: Interactive Management of Environment and Development.* Kluwer Academic Publishers, Dordrecht, pp. 229-247.

Warren, A. (1997) Environmental science and desertification at the arid frontier. *In* H.J. Bruins and H. Lithwick (eds.) *The Arid Frontier: Interactive Management of Environment and Development.* Kluwer Academic Publishers, Dordrecht, pp. 117-127.

Wildavsky, A. (1979) *Speaking Truth to Power*. Little-Brown, Boston.

Wilhite, D.A. (1993a) The enigma of drought. *In* D.A. Wilhite (ed.) *Drought Assessment, Management and Planning: Theory and Case Studies*. Kluwer Academic Publishers, Dordrecht, Chapter 1, pp. 3–15.

Wilhite, D.A. (1993b) Planning for drought: A methodology. *In* D.A. Wilhite (ed.) *Drought Assessment, Management and Planning: Theory and Case Studies*. Kluwer Academic Publishers, Dordrecht, Chapter 6, pp. 87–108.

Wilhite, D.A. and W.E. Easterling (eds.) (1987) *Planning for Drought: Toward a Reduction of Societal Vulnerability*. Westview Press, Boulder.

Wilhite, D.A. and M.J. Hayes (1997) Drought planning in the United States: Status and future directions. *In* H.J. Bruins and H. Lithwick (eds.) *The Arid Frontier: Interactive Management of Environment and Development*. Kluwer Academic Publishers, Dordrecht, pp. 33-54.

ABOUT THE AUTHORS

Hendrik J. Bruins is Senior Researcher at the Jacob Blaustein Institute for Desert Research, in the Social Studies Center; he lectures at the Department of Geography and Environmental Development, and is Research Fellow at the Negev Center for Regional Development (Ben-Gurion University of the Negev). He obtained his M.Sc. *cum laude* at The Hebrew University of Jerusalem at the Institute of Earth Sciences, receiving the Dicker-Shraga Award. He obtained his Ph.D. at the Agricultural University of Wageningen (The Netherlands). His multi-disciplinary experience in the sphere of *Man and Desert* is directed at both historical and modern issues. New insights concerning the unresolved Exodus enigma were recently published in *Nature* (1996, Vol. 382, No. 6588). He is active in multilateral cooperation to combat desertification. Dr. Bruins heads a tripartite project on Drought Planning for Arid-Zone Pastoralists, funded by the Netherlands-Israel Development Research Programme (NIRP). The relations between environmental and social aspects in disaster planning, policy and management have his keen interest.

Dr. Hendrik J. Bruins
Ben-Gurion University of the Negev
Jacob Blaustein Institute for Desert Research, Social Studies Center
Sede Boker Campus, 84990
Israel (fax: 972–7–6596863; e-mail: hjbruins@bgumail.bgu.ac.il)

Harvey Lithwick is Professor of Economics at the School of Public Administration, Carleton University, Ottawa, Canada, and Senior Fellow at the Negev Center for Regional Development at Ben-Gurion University of the Negev, Beer-Sheva, Israel. He has published widely on the subjects of economic growth, project evaluations, urban economic development, and regional economic policy. He has served as an Assistant Secretary to the Ministry of Urban Affairs of the Government of Canada, and economic adviser to the Prime Minister of Canada. Among his most recent publications is *Frontiers in Regional Development*, co-edited with Prof. Yehuda Gradus, and published in 1996 by Rowman & Littlefield.

Prof. Harvey Lithwick
Negev Center for Regional Development
Ben-Gurion University of the Negev,
Beer-Sheva, 84105
Israel (fax: 972-7-6278991; e-mail: lithwick@bgumail.bgu.ac.il)

Part Two

Environmental Policy, Planning and Management: National, Regional, and Multinational Dimensions

Chapter 2

Drought Planning in the United States: Status and Future Directions

Donald A. Wilhite and Michael J. Hayes

Introduction

Drought has been a common feature in large portions of the United States during the past decade. To the casual observer, the frequency of recent droughts may signal a change in climate, a possible forewarning of global warming. Viewed in the context of the nation's climatic history, the pattern of the past decade is not unusual.

Recent droughts in the United States have resulted in significant economic, environmental, and social impacts. Many people are well acquainted with the serious impacts of the droughts of the 1930s and 1950s. The Dust Bowl of the 1930s is known worldwide as a classic example of the disharmony that can exist between human activities and nature, a pattern that has been repeated again and again in other areas of the world in the past several decades. The images of the Dust Bowl in the United States are now fading because a large percentage of the populace is too young to remember the hardships and suffering that resulted from this nearly decadal episode of drought. Droughts of the past decade have reminded us of our continuing vulnerability to extended periods of water shortage and the urgent need to prepare for their recurrence. Although technology has reduced our vulnerability to drought to some degree, it has also exposed new sensitivities.

Climate and, more specifically, drought were significant factors in the settlement of the western United States. As the wave of settlement moved westward, settlers were repelled repeatedly by drought. Some endured conditions at the arid frontier, but many returned to more humid eastern locations or continued westward to California, Oregon, and other seemingly more hospitable environments. The settlement of the more arid west represents an ongoing struggle with climatic elements and a never-ending attempt to adapt to the harsh realities of nature. This adaptation process has proceeded undaunted, assisted by the development of many new technologies and government programs over more than a century that have helped to make this region more resilient to the vicissitudes of climate. However, vulner-

H. J. Bruins and H. Lithwick (eds.), The Arid Frontier, 33–54.

ability to drought remains and may, in fact, be increasing because of increasing and shifting population, industrial growth, the ever-increasing demands of large urban areas, pollution of many water sources, and the value now placed on water to ensure the health of natural ecosystems.

The primary purpose of this chapter is to provide an overview of drought planning and mitigation activities that have occurred in recent years in the United States. This discussion will be preceded by a brief description of the climatology of drought during the period 1986–1995 and a discussion of the deficiencies that have characterized the traditional reactive or crisis management approach to drought management. The development and implementation of state drought plans in the United States represents an important first step toward a more proactive, anticipatory approach to drought management. The transition from crisis management to risk management can only be achieved through a radical philosophical change in how drought is viewed by policy makers at all levels of government.

The Climatology of Drought, 1986–1995

Drought is a normal, recurrent feature of the climate of virtually all portions of the United States. Because of the country's size and the wide range of climatic regimes present, it is rare for drought not to exist somewhere in the country each year. Figure 1 provides a historical perspective of the percent area of the country (48 contiguous states) in severe to extreme drought, according to the Palmer Drought Severity Index (PDSI) (Palmer, 1965) from 1895 to 1995.[1] Severe and extreme drought represent values of ≤ -3.0 on the PDSI scale. PDSI values commonly range from +4.0 (extreme wetness) to –4.0 (extreme drought), although values above and below these levels are often computed. For example, during August 1977, PDSI values reached –7.0 in parts of the upper midwest and –9.0 in eastern Oregon and Washington. The PDSI, computed for each climatic division, has been until recently the only index used nationally to monitor or assess climatic conditions.

Figure 1 reveals two features of drought in the United States: the variable but recurrent nature of drought and the magnitude and duration of the droughts of the 1930s and 1950s in comparison to other episodes during the remainder of the century. Drought frequently affects more than 10% of the nation, and it is not uncommon for more than 30% of the nation to be affected. The most benign periods occurred around the mid-1940s, between the late 1960s and mid-1970s, and from 1978 to 1985. In contrast, the 1930s drought continued for nearly a decade; PDSI values of ≤ -3.0 (severe and extreme drought) affected approximately 65% of the country and more than 95% of the Great Plains at the peak of the drought in 1934. The 1950s drought began in the southwest and southern Great Plains states in the late 1940s and persisted through 1957. At its peak, severe to extreme drought affected nearly 50% of the nation.

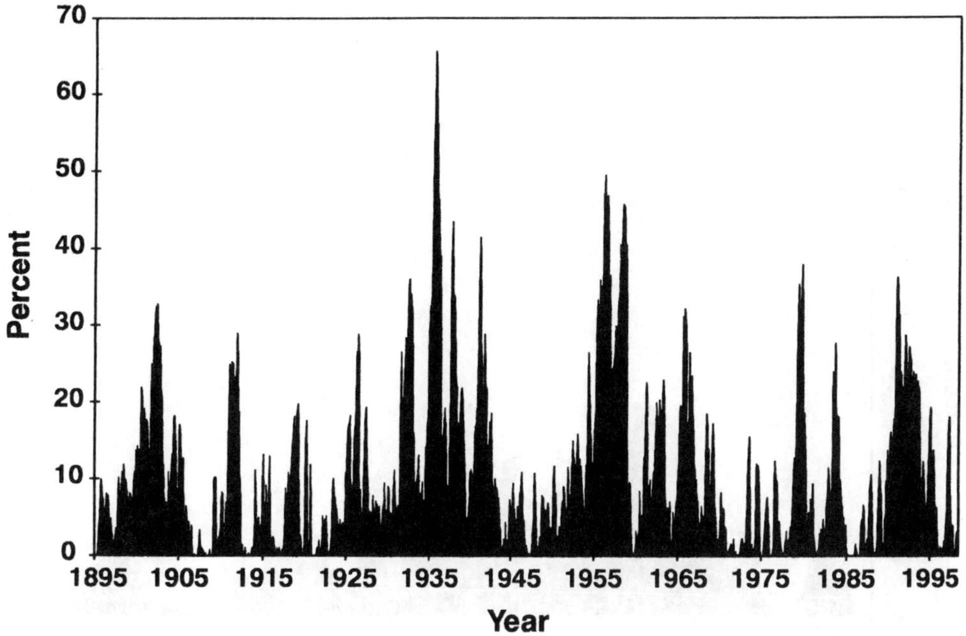

Figure 1. Percent area of the United States (48 contiguous states) in severe and extreme drought (*i.e.* ≤ –3.0), 1895–1995.

Significant periods of drought affected large portions of the northeast in the mid-1960s and extensive portions of the West Coast, Great Plains, and midwestern states in 1976–77. In fact, 1976–77 remains the most severe short-term drought for parts of the far western states of California, Oregon, and Washington and sections of the Midwest. For this reason, this drought is often used by planners and water managers in assessing the vulnerability of water supply systems.

The most recent series of drought years began in 1986 (Figure 2). Dry conditions in late 1985 persisted through the midsummer months. Drought conditions extended from the south central states to the Atlantic Coast. The epicenter of the drought extended from central Tennessee and Kentucky to central South Carolina and from Virginia to central Georgia. Precipitation in the core area was < 40% of normal during the period from December 1985 to July 1986. In early August 1986, moderate (PDSI ≤ –2.0) to extreme drought extended from southern Pennsylvania to central Florida and west to central Louisiana. Moderate to extreme conditions also existed in California and the Pacific Northwest and in parts of the northern Rocky Mountain states and central Great Plains states.

Drought conditions abated in early 1987 but reappeared swiftly in the spring. Moderate to extreme drought was widespread over the western United States by

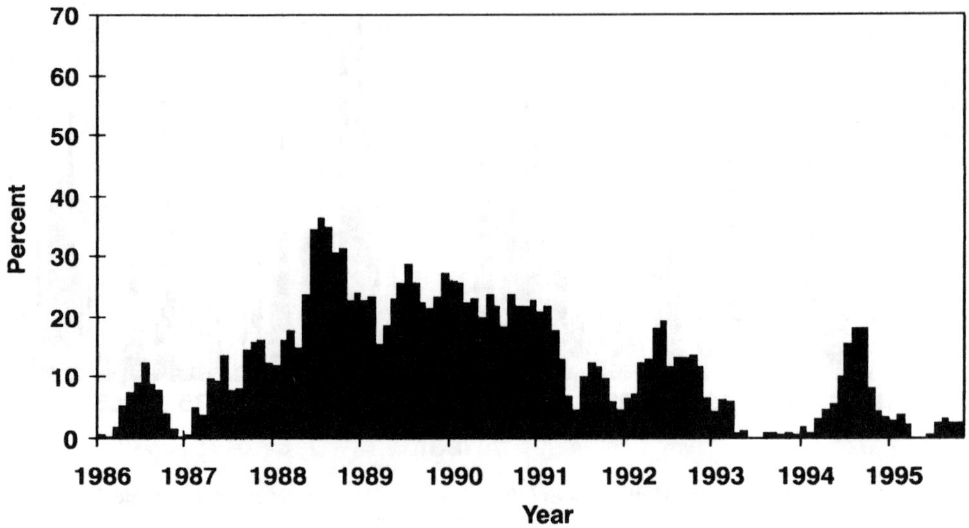

Figure 2. Percent area of the United States (48 contiguous states) in severe and extreme drought (i.e., ≤ –3.0), January 1986–October 1995.

August, and moderate drought reappeared in the southeast and spread west into the midwestern Corn Belt states. Pockets of severe to extreme drought developed in portions of the Tennessee, Ohio, and Mississippi River valleys, a forewarning of events to occur. By the end of 1987, approximately 17% of the nation was experiencing severe to extreme drought.

By April 1988, drought conditions in the west had deteriorated significantly. Severe to extreme drought affected all of California, Oregon, Washington, and Idaho; northern portions of Utah; and western portions of Wyoming and Montana. In addition, moderate drought had spread into eastern Wyoming and Montana. Significant pockets of moderate to severe drought had formed in Minnesota and eastern portions of North and South Dakota and in New England. Moderate to extreme drought persisted in the southeast. By late May, drought in the west, northern Great Plains, and upper midwest had intensified and spread into adjacent states. The drought area in the southeast also began to spread northward into Indiana, Ohio, Illinois, and Iowa. By late July, the intensity had worsened and the spatial coverage had increased. The drought areas in the west and midwest were joined, and moderate drought had spread throughout the south and into eastern Texas. Severe drought continued in New England, and parts of the mid-Atlantic coast were also affected. By August 1988 more than 35% of the nation was experiencing severe to extreme drought.

Dry conditions moderated during the fall and winter months as precipitation returned to normal for some areas and water demand subsided. By May 1989, the drought area was reduced to under 20%, but quickly increased to nearly 30% by late summer. The most severely affected areas were California, Nevada, Arizona, Utah, Wyoming, and parts of Colorado and New Mexico. Portions of the Pacific Northwest were also affected. The drought in the midwest and northern plains states shifted south and west to affect Kansas, eastern Nebraska, Iowa, and northern Missouri. For the most part, drought conditions in the southeast, mid-Atlantic, and New England states disappeared. The area in severe to extreme drought leveled off at about 25% in 1990 and continued at that level through 1991. The principal areas affected were the western states and portions of the northern Great Plains. A significant decline in the drought area occurred in early 1991 (to under 10% of the nation), peaking at 15% in July and rising slightly to about 17% in July 1992. Again, the drought area was confined mainly to the western states, including portions of the western Great Plains.

Drought conditions abated by mid-1993 for virtually all portions of the nation. Portions of the western states experienced what they believed to be an end to the drought that had been ongoing since 1986. However, much below normal winter precipitation over most of the region resulted in the return of severe to extreme drought conditions in 1994. By May, extreme drought extended from California and the Pacific Northwest to western Nebraska and northern Colorado. During 1995, drought conditions were widely scattered and principally in the moderate category for most of the nation. The primary areas of concern were the northeast, portions of the southeast, and portions of west Texas and eastern New Mexico. The wet winter of 1995–96 in the northeast effectively ended concerns about an inadequate water supply in this region during the 1996 summer.

The drought that affected western Texas and New Mexico in 1995 expanded into southern California, Arizona, and New Mexico and parts of Nevada, Utah, Colorado, Texas, Oklahoma, Kansas, and Nebraska in 1996. In April, most of the drought area of the Southwest and Great Plains regions was classified in the moderate category (PDSI values between –2.0 and –2.99), with only small areas classified in severe or extreme drought. By May, the PDSI showed an intensification of drought in this region, including most of the climatic divisions from southern California on the west to Arkansas and Louisiana on the east and portions of Colorado, Kansas, and Nebraska on the north. By mid-May, large areas were classified in the extreme drought category (≤ -4.0); these areas continued to expand in spatial extent through the early part of July.

Recent Droughts and Vulnerability

What lessons can be learned from this series of drought years in the United States? First, it reinforces the notion that drought is a normal, recurrent feature of the climate of the United States. Second, it suggests that no portion of the United States is immune to extended periods of precipitation deficiency. Some areas of the United States were affected by drought during seven consecutive years, and nearly all portions of the nation experienced drought of sufficient intensity and duration to result in serious impacts on more than one occasion during the past decade. Third, the character of drought differs considerably from one climatic region to another, resulting in profound impacts in many economic, social, and environmental sectors. Fourth, the series of drought years between 1986 and 1995 (Figure 2) is not unusual when examined in its historical context (Figure 1). Fifth, no apparent trend exists in the percent area of the United States affected by severe to extreme drought during the past century, although a local or regional scale analysis might produce different results.

What about future occurrences of drought in the United States? Is the frequency and intensity of drought likely to increase or decrease? How might projected changes of climate because of increasing concentrations of greenhouse gases alter climate variability and the incidence of extreme climate events? There is considerable uncertainty surrounding the debate on these issues (Houghton *et al.*, 1990). It is important to note, however, that the impacts of drought are the product of both the natural event and the vulnerability of society to extended periods of precipitation deficiency. In other words, should drought be defined by the natural event (*i.e.* the intensity, duration, and spatial extent of meteorological drought), the impacts of that natural event, or both? In our view, since physical and social factors determine risk, both should define drought. Therefore, the impacts that result from future drought occurrences will be determined not only by the frequency and severity of meteorological drought but also by the number of people at risk and their degree of vulnerability.

Available statistics on natural disasters indicate that societies are becoming more vulnerable to drought and other natural hazards. The number of drought-related disasters reported globally increased from 62 during the 1960s to 237 during the 1980s (Blaikie *et al.*, 1994). It is worth mentioning that this total only includes droughts associated with appeals for assistance to donor governments, international organizations, or nongovernmental organizations. Many droughts (*e.g.*, the recent droughts in the United States, Canada, Australia, United Kingdom, and Spain) are not included in this total.

Worldwide, economic damages attributed to natural disasters have tripled in the last three decades, from an estimated $40 billion in the 1960s to $120 billion by the end of the 1980s (Domeisen, 1995). It is estimated that the number of people vulnerable to natural disasters is increasing at a rate of about 10 million per year (International Federation of the Red Cross and Red Crescent Societies, 1995). Until 1987, only one natural disaster had resulted in losses exceeding $1 billion. Since 1987, there have been thirteen such disasters (Domeisen, 1995). This figure does not include drought-induced disasters; these are often excluded from the assessments because of their slow-onset character and nonstructural impacts. However, drought frequently results in costs and losses in excess of $1 billion in the United States (Riebsame *et al.*, 1990) and elsewhere.

Responding to Drought

Until recently, little effort had been devoted to drought preparedness in the United States. Governments instead relied on the traditional reactive or crisis management approach. Strategies for responding to and preparing for drought are numerous and range from household or community level to regional or national level. These strategies take many forms. At the local level, people and communities possess detailed knowledge of the likely occurrence of drought and its effects and have developed a broad range of strategies over the years to help them reduce its effects and promote recovery once the rains have returned.

At the state or national level, governments may respond to drought in three ways: pre-drought mitigation programs for impact reduction; post-drought relief programs to provide emergency assistance to victims; and preparedness or contingency planning to develop institutional capacity to respond in a more timely and effective manner and reduce impacts (Parry & Carter, 1987). Mitigation is defined as short- and long-term actions or programs implemented before drought that reduce the degree of economic, social, and environmental risk to people, property, and productive capacity. These activities are long-term preventive or anticipatory measures. Examples of pre-drought mitigation programs include development of an early warning system, augmentation of water supplies, demand reduction, and crop insurance. Post-drought interventions refer to those reactive programs or tactics implemented by government and others in response to drought. This includes a wide range of emergency measures such as low-interest loans, transportation subsidies for livestock and livestock feed, water transport, and drilling wells for irrigation and public water supplies. This reactive, crisis management approach has been criticized by scientists, government officials, and many relief recipients as inefficient, ineffective, and untimely (GAO, 1979; Wilhite *et al.*, 1986; Riebsame *et al.*, 1990; Wilhite, 1993a). More recently, the provision of emergency assistance or relief has been criticized as being a disincentive to the sustainable use of natural resources because it does not promote self-reliance (White *et al.* 1993; Bruwer, 1993). In fact, this approach may increase vulnerability to drought. Preparedness actions refer to the development of policies and plans that can be useful in addressing future episodes of drought. These policies and plans build institutional capacity to anticipate potential problems, coordinate information flow and decision making, and implement programs to reduce impacts.

Deficiencies that have been noted in the reactive or crisis management approach in the United States and elsewhere include (Wilhite, 1992a):

- the lack of a monitoring system to provide early warning of developing drought conditions;
- inadequate tools and methodologies for early estimation of impacts in various sectors;
- insufficient information flow within and between levels of government on drought severity, impacts, and appropriate policy responses;
- inappropriate or untimely emergency assistance programs;
- poorly targeted emergency assistance programs that do not reach vulnerable population groups and economic sectors;
- lack of emphasis on proactive mitigation programs aimed at reducing vulnerability to drought;
- institutional deficiencies that inhibit effective emergency response; and
- lack of coordination of policies and programs within and between levels of government.

Drought policies and plans must address each of these issues if progress is to be made in drought management during the remainder of this decade and beyond.

State-level Drought Planning in the United States

The number of states with drought plans has grown from 3 in 1982 to 28 in 1996. In addition, 5 states (Oklahoma, Alabama, Louisiana, Texas, and New Mexico) are now pursuing plan development (Figure 3). Several of these states are pursuing plan development as a direct result of the severe drought that occurred in 1996. Two states, California and Florida, have delegated drought planning authority to local or basin authorities.

The basic goal of state drought plans is to improve the effectiveness of state response efforts by enhancing monitoring; impact assessment; and preparedness, response, and recovery programs. These plans are also directed at improving coordination within agencies of state governments and between state and federal government. The growth in the number of states with drought plans suggests an increased concern about the potential impact of extended water shortages and the attempts of state governments to address these needs. In the United States, state governments have clearly become the policy innovators for drought management (Wilhite, 1991). Although the increase in the number of state drought plans is an

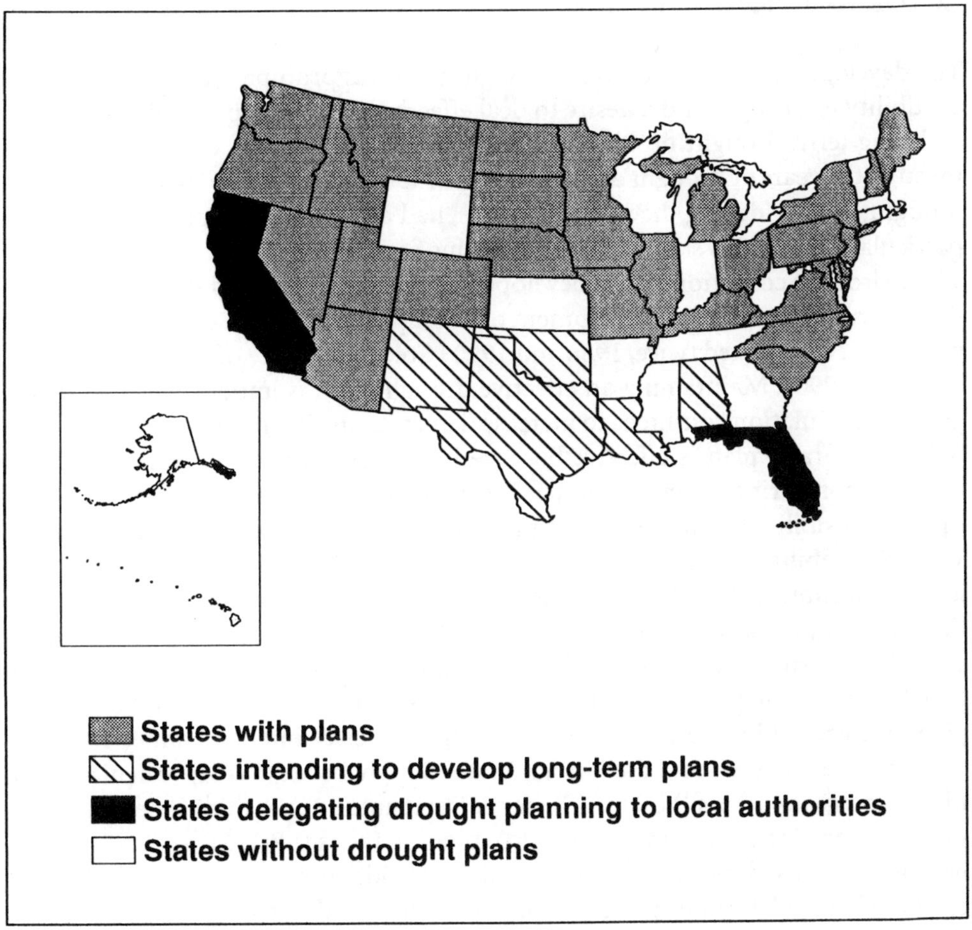

Figure 3. Status of state drought plans, August 1996.

extremely positive sign, plans are still largely reactive, treating drought in an emergency response mode.

State drought plans take many forms. Some concentrate largely on impacts in one principal sector (*e.g.*, agriculture, municipal water supply), while others attempt to address a full range of impacts within the state. One of the first states to develop a drought plan was Colorado. This plan was developed in 1981 at the request of the governor. Since development, this plan has undergone revisions to improve institutional capacity to deal with extended periods of water shortage. The Colorado drought plan is described below as an example of one approach being taken to respond to the impacts of drought.

COLORADO DROUGHT PLAN

The development of the Colorado drought plan was prompted by the state's susceptibility to drought and a desire to deal effectively and systematically with short and long term drought problems. Like its neighboring states, Colorado suffered through the years of drought during both the 1930s and 1950s. In 1976–77, a severe drought, mainly during the winter months, had an immense impact on the state, particularly the state's skiing industry, causing severe economic impacts statewide. When drought conditions again developed during the spring and summer of 1981, the governor initiated the development of a comprehensive state drought plan. The plan has been updated twice, 1986 and 1990 (Colorado Office of Emergency Management, 1990). Maintaining an effective drought plan is important, as a rapidly growing population continues to increase demands on Colorado's water resources.

The Colorado plan is effective because it incorporates three primary components: a monitoring system, an impact assessment system, and a response system. The responsibility for monitoring the availability of water resources is given to the Water Availability Task Force. This task force makes monthly assessments and projections of snowpack, soil water levels, reservoir and ground water levels, precipitation, temperatures, and stream flow collected by numerous state and federal agencies. This information can provide 'early warning' of developing drought conditions to help the state prepare for a potential drought situation. Activation of the drought plan is triggered by the values of two drought severity indices: Palmer Drought Severity Index (PDSI) and the Surface Water Supply Index (SWSI)[2] calculated for nine river basins across the state (Shafer & Dezman, 1982). A third index based on precipitation alone, the Standardized Precipitation Index (SPI) (McKee *et al.*, 1993), has also been used recently as part of this statewide monitoring system. The SPI can be calculated for multiple time scales. It is calculated by taking the difference of the precipitation from the mean of the particular time scale (*e.g.* 3-, 6-, 12-months) and then dividing it by the standard deviation. This index has proved useful in communicating information on the spatial and temporal dimensions of drought to a diverse audience.

The assessment system of the drought plan comprises eight different impact task forces covering the following water-related areas: municipal water, wildfire protection, agricultural industry, tourism, wildlife, the economy, energy loss, and health. The goal of each task force is to identify existing and potential drought-related problems and assess possible impacts on society. Each task force is activated based on criteria specifically identified within the plan. Members of the task force are representatives from agencies directly involved with the issue. A final task force, called the Review and Reporting Task Force, is responsible for coordinating all assessments from the impact task forces and reporting this information to policy makers, media, and others.

Finally, the response system is designed to deal with the unmet needs identified by the specific task forces. Local responses are initially encouraged. State action is taken, however, when local capabilities are exceeded. The agency most closely associated with the response is assigned to take action and enlist the cooperation of other agencies. These lead agencies are identified in the plan. For complex emergency responses, an Interagency Coordinating Group is established.

Colorado has determined that it is important to have specific criteria established and responses clearly identified so that as a drought begins, the state can immediately begin to cope with the situation. Colorado's drought plan also calls for a post-drought evaluation. Suggestions made in these evaluations can be incorporated into the plan to help mitigate the impacts from future drought.

State Drought Mitigation Tools

A recent study reviewed and evaluated ongoing and developing federal, interstate, and state drought mitigation technologies, programs, and policies in the United States (Wilhite, 1993b). This study was based on the assumption that the roles of federal and state governments in drought mitigation needed to be reexamined, given the severity of drought that the United States had experienced since 1986; the economic, social, and environmental costs associated with these droughts; and the mitigation actions and policy efforts underway at all levels of government. One of the goals of the study was to identify opportunities to improve the effectiveness of drought mitigation efforts by the Natural Resources Conservation Service of the U.S. Department of Agriculture and other federal agencies. It was believed that the nation's ability to cope with and manage water shortages resulting from drought would only be improved if an integrated approach within and between levels of government, involving regional organizations and the private sector where appropriate, were adopted.

One of the objectives of this study was to identify emerging drought prediction, assessment, and mitigation technologies that were being employed by various levels of government to lessen the effects of future severe droughts. Increasing demand for water throughout the United States and the widespread occurrence of drought resulted in the introduction of numerous innovative institutional technologies to manage water more effectively and efficiently during periods of shortage. The goal of this objective was to identify the lessons learned from these institutional adjustments and the specific constraints that exist to the implementation of

TABLE 1. Drought-related mitigative actions taken by states during recent droughts

Category	*Specific action*
Assessment Programs	– Developed criteria or triggers for drought-related actions – Develop early warning system, monitoring program – Conduct inventories of data availability – Established new data collection networks – Monitored vulnerable public water suppliers
Legislation/Public Policy	– Prepared position papers on public policy issues – Examined water rights statues for possible modification during water shortages – Passed legislation to protect instream flows
Water Supply Augmentation/ Development of New Supplies	– Issued emergency permits for water use – Provide pumps and pipes for distribution – Proposed and implemented program to rehabilitate reservoirs to operate at design capacity – Undertook water supply vulnerability assessments – Inventoried self-supplied industrial water users for possible use of their supplies for emergency public water supplies – Inventoried and reviewed reservoir operation plans
Public Awareness/ Education Programs	– Organized drought information meetings for the public and the media – Implemented water conservation awareness programs – Published and distributed pamphlets to individuals, businesses, and municipalities on water conservation techniques and agricultural drought management strategies – Organized workshops on special drought-related topics – Prepared sample ordinances on water conservation for municipalities and domestic rural supplies
Technical Assistance on Water Conservation	– Provide advice on potential new sources of water – Evaluated water quantity and quality from new sources – Advised water suppliers on assessing vulnerability of existing supply system – Recommended the adoption of water conservation measures to suppliers
Demand Reduction/Water Conservation Programs	– Established stronger economic incentives for private investment in water conservation – Encouraged voluntary water conservation – Improved water use and conveyance efficiencies – Implemented water metering and leak detection programs
Emergency Response Programs	– Established alert procedures for water quality problems – Stockpiled supplies of pumps, pipes, water filters, and other equipment – Establish water hauling programs for livestock from reservoirs and other sources – Compiled list of locations for livestock watering (cont.)

TABLE 1 continued

Category	*Specific action*
	– Established hay hotline – Provided funds for improvement of water systems, developing new systems, and digging of wells – Provide funds for recovery programs for drought and other natural disasters – Lowered well intakes on reservoirs for rural water supplies – Extended boat ramps and docks in recreational areas
Water Use Conflict Resolution	– Acted to resolve emerging water use conflicts – Negotiated with irrigators to gain voluntary restrictions on irrigation in areas where domestic wells were likely to be affected – Establish a water banking program – Clarified state law regarding sale of water – Clarified state law on changes in water rights – Suspended water use permits in watershed with low water levels – Investigated complaints of irrigation wells interfering with domestic wells
Drought Contingency Plans	– Recommended to water suppliers the development of drought plans – Establish state-wide contingency plan – Evaluated worst-case drought scenarios for possible further actions

other drought mitigation programs, including the development of drought preparedness plans.

Data on drought mitigation actions were collected through a survey of states and key federal agencies with responsibility in the management of water and other natural resources during periods of drought. The survey was directed at specific actions taken or programs adopted during the period from 1986 to 1992, a particularly drought-affected period (see Figures 2 and 3).

Mitigative actions were purposely not defined in the survey instrument referred to above—states were given the flexibility to define mitigation by including those actions or activities that they felt were appropriate. Those mitigation activities identified by states and/or local municipalities during recent droughts were diverse, reflecting regional differences in impacts, legal and institutional constraints, and characteristics of contingency plans. The diversity in responses was also related to the wide range of state agencies with principal authority for planning and mitigative actions (*e.g.*, agriculture, natural resources, water resources, emergency or disaster management).

State mitigation actions used to address issues during recent droughts are clustered into nine primary areas in Table 1. These actions represent the full range of possible mitigative actions, from monitoring and assessment programs to the development of drought contingency plans. Some of the actions included were adopted by many states, while others may have been adopted only in a single case. One of the most innovative and successful mitigation actions implemented during recent droughts was the California Drought Water Bank Program (California Department of Water Resources, 1992). This program was created in 1991 and could acquire water in three ways: (1) by purchasing water from farmers who did not irrigate, choosing to allow water to flow past their farms; (2) by purchasing surplus water from local water districts; and (3) by paying farmers or water districts to use ground water instead of surface water.

Many of the mitigative programs implemented by states during recent droughts can be characterized as emergency or short-term actions taken to alleviate the crisis at hand, although these actions can be successful, especially if they are part of a preparedness or mitigation plan. Other activities, such as legislative actions, drought plan development, and the development of water conservation and other public awareness programs, are considered actions with a longer-term impact. As states gain more experience assessing and responding to drought, future actions will undoubtedly become more timely and effective and less reactive. Mitigative actions taken by states in response to recent drought conditions are numerous when considered in total, but most individual state actions were quite narrow. In the future, state drought contingency plans need to address a broader range of potential mitigative actions, including provisions for expanding the level of intergovernmental coordination. Improved coordination will require a greater commitment by federal agencies to work together and with states to promote an integrated approach to drought planning. Coordination at the federal level will likely require the establishment of an interagency task force, as recommended by the Congressional Office of Technology Assessment (1993).

Federal Drought Response and Planning

The federal government became the principal player in the provision of drought relief in response to the economic depression and droughts of the 1930s (Wilhite, 1983). Before that, assistance was provided primarily by the private sector (*e.g.* churches, Red Cross). The level of assistance required during the 1930s far exceeded the capacity of this sector to respond. The federal government was the principal provider of drought assistance during subsequent drought events, most notably the 1950s in the southwest, southern plains, and midwestern states; the 1960s in the northeast; the mid-1970s in the midwest and western states; and the recent series of drought years beginning in 1986. Until recently, state government and the private sector assumed a relatively passive role in drought management. Drought relief remains largely a federal responsibility.

Post-drought evaluations or audits are not routinely completed in the United States. However, the General Accounting Office (GAO) characterized the assessment and response programs implemented in the mid-1970s as largely ineffective, poorly coordinated within and between levels of government, and untimely. GAO (1979) found that assistance provided by federal agencies to farmers, communities, businesses, and water user organizations was available too late to lessen the effects of drought. Wilhite *et al.* (1986) confirmed these findings and also concluded that the decision-making process was seriously flawed. For example, the designation and revocation process for determining eligibility for the more than $5 billion of disaster relief expended in 1976–77 was confusing and was not based on consistent, established criteria. In total, 16 federal agencies administered 40 separate assistance programs in 1976–77. GAO (1979) recommended a national drought plan be developed to provide 'future assistance in a more timely, consistent, and equitable manner.' GAO further recommended that the plan address the following issues: (1) the identification of respective roles of agencies involved to avoid overlap and duplication, (2) the need for legislation to more clearly define those roles, and (3) the need for standby legislation to permit more timely response to drought-related problems. GAO suggested that effectively implementing a national plan required establishing uniform criteria for determining 'priorities for the type of projects to be constructed; eligibility of applicants; and interest rates, terms, and repayment requirements for loans.'

Wilhite *et al.* (1986) concluded that four basic requirements were necessary for federal government to improve response to drought. These were (1) reliable and timely information and dissemination plans, (2) objective and reliable impact assessment procedures, (3) objective and timely designation (and revocation) procedures, and (4) appropriate disaster programs and efficient program administration and delivery systems. It was concluded that a national drought plan would help coordinate the activities of the federal government in mitigating the effects of future droughts. It was also suggested that state government and regional organizations should play a more active role in drought management and that those activities be coordinated between levels of government.

Following the severe droughts of 1976–77 and the demonstrated inability of the federal government to adequately cope with the problems that emerged, scientists and policy makers expressed considerable concern about the inefficiencies of this effort and repeatedly issued 'calls for action' for the development of drought contingency plans, including the development of a national drought plan. These calls included recommendations from the Western Governors Policy Office (1978), General Accounting Office (1979), National Academy of Sciences (1986), Great Lakes Commission (1990), American Meteorological Society (Orville, 1990), and Interstate Council on Water Policy (1987; 1991). In light of a possible increase in the frequency and severity of extreme events in association with changes in climate, an Environmental Protection Agency report (Smith & Tirpak, 1989) called for the development of a national drought policy to coordinate federal response to drought. The intent of a national drought policy is to move the country away from a reactive, crisis management approach toward a proactive, risk management approach.

To date, no progress has been made at the federal level on the development of a national plan. Federal efforts to respond to the series of droughts between 1987 and 1989 proved to be only slightly, if any, better than previous attempts (Riebsame *et al.*, 1990). As a direct result of the droughts of the late 1980s and early 1990s, a few federal agencies initiated changes in policies and programs, but coordination between agencies continues to be largely nonexistent. The Corps of Engineers' National Water Management During Drought Study and the Bureau of Reclamation's new strategic plan are excellent examples of individual agency efforts to improve drought response.

Drought Policy: A Plan of Action

Drought policy in the United States has not been stated explicitly by the federal government. What has evolved since the 1930s has been a *de facto* policy, one of reacting to, rather than preparing for, periods of water shortage. This crisis management approach, as discussed previously, has been ineffective. Unfortunately, the decision on whether to provide drought relief has been based more often on political rather than economic reasoning. Without a clearly stated drought policy, it is unlikely that significant improvement will occur in federal response efforts in the United States. Drought relief or assistance, if necessary, must only be provided if it is consistent with stated national policy. This policy would provide the framework for a national drought plan; incentives could be given to states to make their plans consistent and compatible with national goals.

Although governments in the United States have shown some progress in responding to recent droughts, the dilemma facing government, particularly the federal government, is whether to continue with the approach of the past (*i.e.* crisis management) or seek a new direction. Given that previous attempts to mitigate drought in the United States have been largely unsuccessful, it seems clear that fundamental and sweeping program and policy changes must occur in order for the nation to more adequately address the drought management problems that exist today and will exist in the future. As a nation we can no longer afford to tinker with a system that is seriously flawed.

A wide range of risk reduction measures exist that the federal government could implement to lessen vulnerability to drought in the United States (Wilhite, 1992b; U.S. Congress, Office of Technology Assessment, 1993). These measures are divided into five categories in Table 2. Improved assessment programs cover a broad range of initiatives that seek to improve the capacity of governments to monitor

TABLE 2. Recommended risk management and risk minimization measures that the federal government in the United States should consider to lessen the effects of drought

Category	*Specific action*
Assessment Programs	– Develop a comprehensive, integrated national drought watch system (NDWS) – Inventory data availability in support of a NDWS – Improve existing and establish new data collection networks (e.g., soil moisture, AVHRR) – Develop new indexes to assist in the early estimation of impacts in various sectors – Establish objective 'triggers' for the phase-in and phase-out of relief/assistance programs
Legislation/ Public Policy	– Develop a national drought policy and plan – Examine federal land use policies to ensure appropriate management of natural resources and consistency with national drought policy – Promote a reexamination of state water rights statutes for possible modification – Review all federal drought relief/assistance programs, federal crop insurance program, and other agricultural/water policies for consistency with national drought policy
Water Supply Augmentation/ Demand Reduction	– Promote vulnerability assessments of public water supply systems – Review reservoir operation plans for improved management of water supplies during drought – Improve operational flexibility of water allocations between users – Establish stronger economic incentives for private investment in water conservation – Promote improved water use and conveyance efficiencies – Promote the implementation of municipal water metering and leak detection programs
Public Awareness Programs	– Establish a national drought mitigation center to provide information to the public and private sectors – Improve data/information products and delivery systems to provide timely and reliable information to users – Develop and implement water conservation awareness programs
Drought Preparedness Planning	– Promote the establishment of comprehensive state drought plans – Promote intergovernmental cooperation and coordination on drought planning – Evaluate worse-case scenarios for drought management – Evaluate the potential effects of climate change on regional hydrology and its implication on federal and state water policies – Promote the establishment of drought plans by public water suppliers – Conduct post-drought audits of federal drought assessment and response efforts

water availability. Current monitoring programs are fragmented between federal agencies and levels of government, inconsistent from region to region, and poorly coordinated. In addition, near-real time assessments of soil moisture conditions, a critical variable in early estimation of crop yield potential and other evaluations, are piecemeal at best. A comprehensive, integrated national drought (or climate) assessment system would rectify these problems.

In the domain of public policy, a national drought policy and plan could profoundly enhance governmental capacity to lessen the risk of future events. The development of a policy would require a reexamination of federal land use policies and current federal drought assistance programs, including crop insurance, and other agricultural and water policies for consistency. A national policy could also promote a thorough reexamination of water rights through the provision of incentives to states. Tackling the issue of water rights is critical to the adoption of a proactive approach to drought management. Many states have initiated changes in water rights, a process that must continue if the nation is to adjust to current and future water use trends and deal more effectively with conflicts between users for an increasingly scarce resource.

From a management perspective, there are many opportunities to increase water availability through demand management. Significant savings can be achieved through well-designed water conservation programs, improved conveyance efficiencies, metering and leak detection programs, and modification of reservoir operation plans. States should also conduct vulnerability assessments of public water supply systems to determine the sensitivity of these systems to extended periods of water shortage and promote programs to reduce that sensitivity.

One of the deficiencies noted by many decision makers has been the lack of or untimeliness of relevant information on drought conditions and appropriate response options or mitigation measures that could be employed (Wilhite & Wood, 1994). Public awareness programs are useful in this regard, but there is also a continuing need to consolidate and centralize information to facilitate availability and distribution. In 1995, the National Drought Mitigation Center was established at the University of Nebraska-Lincoln with one of its principal goals to develop a comprehensive drought information clearinghouse.[3]

The rapid increase in the number of state drought plans has improved the ability of states to detect emerging drought conditions, estimate potential impacts, and formulate a variety of response measures to reduce hardships and impacts. Drought preparedness can be fostered further through the adoption of mitigation measures that are directed at reducing impacts, such as those programs or measures included in Table 1. Post-drought audits are critical in identifying the successes and failures of previous response efforts. Such audits are rare in the United States at either the state or federal level. Without a systematic post-drought evaluation effort, the deficiencies of previous responses are often repeated. It is also important to coordinate drought plans between levels of government. For example, municipal drought or water shortage plans should be coordinated with state plans; river basin or reservoir operation plans should be coordinated with state-level plans that exist within the basin. Coordination of drought planning at multiple administrative levels provides a strong rationale for the adoption of a national drought policy or philosophy that establishes the principles for the sustainable use of natural resources.

Conclusions and Recommendations

During the past decade in the United States, widespread and severe drought has resulted in an increased awareness of the nation's continuing vulnerability to this creeping natural hazard. This experience has resulted in numerous initiatives by state and federal governments to improve the timeliness and effectiveness of response efforts. Although some progress has been made, much remains to be done. For the most part, governments continue to deal with drought in a reactive, rather than proactive, mode. The growth in the number of states with drought plans is one positive sign that greater emphasis is now being placed on drought preparedness, although most state responses continue to stress emergency assistance. States have developed and implemented a wide range of mitigation measures but the shift from crisis management to risk management continues to be a difficult transition for governments.

For this transition to be successful, the deficiencies of previous drought response attempts noted in this chapter must be addressed in a systematic way. Creating a federal interagency task force with the authority to develop and implement a national drought policy and plan would represent an important first step. The task force must develop the objectives of an integrated national policy in concert with extensive public involvement. This policy should promote the concept of risk management, although the policy cannot ignore the need for government assistance during extended periods of severe drought. The policy should promote self-reliance while at the same time protecting the natural and agricultural resource base. This interagency task force should coordinate the drought-related activities of the federal government (*i.e.* forecasting, monitoring, impact assessment, response and recovery, and planning), linking these activities to state and regional planning activities and plans. This national policy should also incorporate incentives for all drought-prone states to develop plans that promote a more proactive, anticipatory approach to drought management. Lessons learned from previous drought response attempts need to be documented, evaluated, and shared with all levels of government.

Drought inflicts considerable pain and hardship on society. The impacts of contemporary droughts in the United States have demonstrated this fact again and again over the past several decades. Drought illustrates in innumerable ways the vulnerability of economic, social, political, and environmental systems to a variable climate. It also illustrates the dependencies that exist between systems, reinforcing the need for improved coordination within and between levels of government.

Extended periods of normal or benign weather conceal the vulnerability of societies to climate variability, while drought exposes these sensitivities. Projected changes in climate because of increased concentrations of carbon dioxide and other atmospheric trace gases suggest a possible increase in the frequency and intensity of severe droughts in the future. Coupled with the world's increasing population and associated rise in demand for water and other shared natural resources, there is a sense of urgency to reduce drought-inflicted personal hardships, as well as economic and environmental impacts of drought.

Drought is a good analog for climate change because it helps identify societal strengths and weaknesses in coping with extreme climatic events today, thus allowing us to bolster the strengths and sharply reduce or eliminate the weaknesses in the future. Stated another way, drought first exposes vulnerabilities to climate variability and then provides us with the opportunity to adjust or adapt present systems to be more in harmony with the stresses imposed by that variability. Drought planning is the tool to accomplish this goal.

Notes

1. The PDSI compares quantitatively the actual amount of precipitation received during a specific period with the normal amount expected during that same period. The PDSI is based on a procedure of hydrologic or water-balance accounting, by which excesses or deficiencies in moisture are determined in relation to average climatic values. It is the most commonly used index in the United States.
2. Designed to complement the PDSI, the SWSI is an indicator of surface water conditions in which mountain snowpack is a major component. The SWSI incorporates four hydrological and meteorological inputs: snowpack, streamflow, precipitation, and reservoir storage.
3. For more information on the National Drought Mitigation Center, consult the NDMC's home page (http://enso.unl.edu/ndmc) on the World Wide Web.

References

Blaikie, P., T. Cannon, I. Davis, and B. Wisner (1994) *At Risk: Natural Hazards, People's Vulnerability and Disasters*. Routledge Publishers, London.

Bruwer, J.J. (1993) Drought policy in the Republic of South Africa. *In* D.A. Wilhite (ed.) *Drought Assessment, Management, and Planning: Theory and Case Studies*. Kluwer Academic Publishers, Boston.

California Department of Water Resources (1992) Report to the California Water Commission. Sacramento, CA, May 1.

Domeisen, N. (1995) Disasters: Threat to social development. *STOP Disasters: The IDNDR Magazine* 23 (Winter) IDNDR Secretariat, Geneva, Switzerland.

General Accounting Office (1979) Federal response to the 1976–77 drought: What should be done next? Report to the Comptroller General, Washington, D.C.

Great Lakes Commission (1990) *A Guidebook to Drought Planning, Management and Water Level Changes in the Great Lakes.* Great Lakes Commission, 61 pp.

Houghton, J.T., G.J. Jenkins, and J.J. Ephraums (eds.) (1990) Climate change. The IPCC Scientific Assessment. Report prepared for the IPCC by Working Group I. Cambridge University Press, Cambridge.

International Federation of the Red Cross and Red Crescent Societies. (1995) *World Disasters Report 1995.* Geneva, Switzerland

Interstate Council on Water Policy. (1987) *Statement of Policy 1986–87.* Interstate Conference on Water Policy, Washington, D.C.

Interstate Council on Water Policy (1991) *Statement of Policy 1991–92.* Interstate Conference on Water Policy, Washington, D.C.

McKee, T.B., N.J. Doeskin, and J. Kleist (1993) Drought monitoring with multiple time scales. *American Meteorological Society Preprints,* Eighth Conference on Applied Climatology, 17–22 January, Anaheim, CA, pp. 179–184.

National Academy of Sciences (1986) *The National Climate Program: Early Achievements and Future Directions.* Washington, D.C.

Colorado Office of Emergency Management (1990) *The Colorado Drought Response Plan.* Denver, Colorado.

Orville, H.D. (1990) AMS statement on meteorological drought. *Bull. of the Amer. Meteorol. Soc.* 71:1021–23.

Palmer, W.C. (1965) Meteorological drought. *Research Paper No. 45.* U.S. Weather Bureau, Washington, D.C.

Parry, M.L. and T.R. Carter (1987) Climate impact assessment: A review of some approaches. *In* D.A. Wilhite and W.E. Easterling (eds.) *Planning for Drought: Toward a Reduction of Societal Vulnerability.* Westview Press, Boulder, CO, Chapter 13.

Riebsame, W.E., S.A. Changnon, Jr., and T.R. Karl (1990) *Drought and Natural Resources Management in the United States: Impacts and Implications of the 1987–89 Drought.* Westview Press.

Shafer, B.A. and L.E. Dezman (1982) Development of a surface water supply index (SWSI) to assess the severity of drought conditions in snowpack runoff areas. *Proceedings of the Western Snow Conference,* pp. 164–175.

Smith, J.B. and D. Tirpak (eds.) (1989) *The Potential Effects of Global Climate Change on the United States.* Environmental Protection Agency (EPA-230-05-89-050).

U.S. Congress, Office of Technology Assessment (1993) *Preparing for an Uncertain Climate,* Volume 1. OTA-0-567. U.S. Government Printing Office, Washington, D.C.

Western Governors Policy Office (WESTPO) (1978) *Managing Resource Scarcity: Lessons from the Mid-Seventies Drought.* Institute for Policy Research.

White, D.H., D. Collins, and M. Howden (1993) Drought in Australia: Prediction, monitoring, management, and policy. *In* D.A. Wilhite (ed.) *Drought Assessment, Management, and Planning: Theory and Case Studies.* Kluwer Academic Publishers, Boston, MA.

Wilhite, D.A. (1983) Government response to drought in the United States: With particular reference to the Great Plains. *J. of Climate Appl. Meteor.* 22:40–50.

Wilhite, D.A., N.J. Rosenberg, and M.H. Glantz (1986) Improving federal response to drought. *J. of Climate Appl. Meteor.* 25:332–342.

Wilhite, D.A. (1991) Drought planning: A process for state government. *Water Res. Bull.* 27:29–38.

Wilhite, D.A. (1992a) *Preparing for Drought: A Guidebook for Developing Countries*. Climate Unit, United Nations Environment Programme, Nairobi.

Wilhite, D.A. (1992b) Drought management and climate change. Contractors' report prepared for the Office of Technology Assessment, Washington, D.C.

Wilhite, D.A. (ed.) (1993a) *Drought Assessment, Management, and Planning: Theory and Case Studies*. Kluwer Academic Press, Boston, MA.

Wilhite, D.A. (1993b) An assessment of drought mitigation technologies in the United States. Final Report to the Soil Conservation Service/USDA. IDIC Technical Report 93–1.

Wilhite, D.A. and D.A. Wood (1994) *Drought Management in a Changing West: New Directions for Water Policy* (Proceedings of a conference). IDIC Technical Report Series, 94–1. University of Nebraska-Lincoln.

ABOUT THE AUTHORS

Donald A. Wilhite is professor of agricultural climatology in the Department of Agricultural Meteorology at the University of Nebraska-Lincoln. He is also the Director of the National Drought Mitigation Center and the International Drought Information Center. He specializes in studies of the impact of climate on society and societal response to climatic events, particularly drought. Dr. Wilhite is co-editor of *Planning for Drought: Toward a Reduction of Societal Vulnerability*, published by Westview Press in 1987. He has authored a guidebook on drought preparedness for developing countries under the sponsorship of the U.N. Environment Program and is editor of the book *Drought Assessment, Management, and Planning: Theory and Case Studies*, published by Kluwer Academic Press in 1993. Dr. Wilhite is editing a book on drought to be published by Routledge Publishers as part of a series on natural hazards and disasters.

Prof. Donald A. Wilhite
Director
National Drought Mitigation Center,
University of Nebraska
Lincoln, Nebraska 68583-0749, U.S.A.

Michael J. Hayes is assistant professor of agricultural climatology in the Department of Agricultural Meteorology at the University of Nebraska-Lincoln. He is also a climate impact specialist in the National Drought Mitigation Center (NDMC). Dr. Hayes joined the NDMC staff in July 1995, and his responsibilities include conducting research on the impact of drought and the development of new monitoring and impact assessment methodologies. He interacts with users on drought mitigation and preparedness technologies, available at the NDMC's home page on the World Wide Web. Dr. Hayes received his Ph.D. degree in 1994 from the University of Missouri-Columbia in agricultural meteorology.

Prof. Michael J. Hayes
Department of Agricultural Meteorology and also
National Drought Mitigation Center
University of Nebraska
Lincoln, Nebraska 68583-0749, U.S.A.

Chapter 3

Interregional Environmental Policy in Australia's Murray-Darling Basin

Joseph M. Powell

Introduction

The Murray-Darling Basin encompasses several 'old' settlement frontiers in Australia's semi-arid, sub-humid and temperate zones, roughly equivalent in area to France and Spain combined. Running from subtropical areas of the state of Queensland through New South Wales and Victoria to 'Mediterranean' South Australia, and taking in cool and wet high country and great plains fringing the interior deserts, this enormous region is not too badly described as Australia's heartland. Certainly, its story goes to the very center of a national experience in which communities of immigrants have engaged in a massive geographical experiment. Increasingly, the territorial gains of the pioneering years were threatened by environmental degradation, principally but not only in the drier sectors, until decisions were made to promote a 'bioregional' perspective in which the newest frontier is, in fact, the very concept of the natural unit itself. Bioregionalism offers a timely and motivating challenge for a young nation currently confronting the multifaceted challenge of ecological sustainability in the driest of the inhabited continents.

Australia is a young, multicultural nation in which many of the demands of modern citizenship were taken comparatively lightly until recent decades. Some of the soul-searching critiques associated with the much vaunted centenary of Australian federation in 2001 suggest a need to move towards a more secure coming-to-terms. Interpretations of demanding environments, and of the distressing legacy of ecological disruption resulting from the wanton or unwitting mismanagement of natural resources, advise that the gathering debates be made to incorporate a recognition of the defining centrality of 'environmental responsibility', both individual and collective. An inspection of the transformations wrought in the vast Murray-Darling Basin and of the changing relationships of Australia's immigrant communities with that region, including novel programs designed to improve those

H. J. Bruins and H. Lithwick (eds.), The Arid Frontier, 55–73.

relationships, should offer a timely focus. The example tests the associations between modern federalism and the management of a critical and imperiled natural unit which transcends political boundaries.

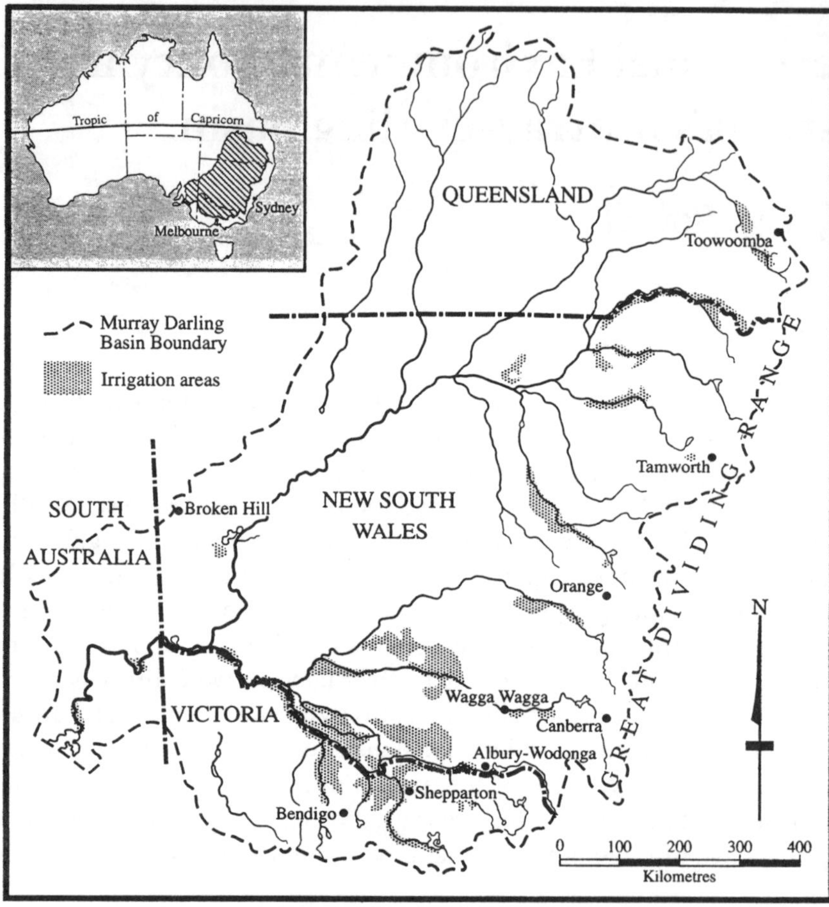

Figure 1. The Murray Darling Basin: Location.

The Old and New Frontiers

Encompassing approximately one-seventh of Australia, the Murray-Darling Basin includes all of the Australian Capital Territory, 75% of New South Wales, 56% of Victoria, 15% of Queensland and 8% of South Australia. This deceptively simple geographical delineation is therefore politically charged, given the states' original (1901) constitutionally-guaranteed jurisdictions over natural resources. Despite the gradual intensification of federal involvement during the twentieth century, in many respects the huge state units continue to function as the country's most powerful regional divisions. Until recently that was emphatically so in the case of the twin fields of environmental appraisal and natural resource management, yet that is precisely where improved federal-state dialogue has delivered important national reviews which insist on a re-ordering of priorities.

Another characteristically Australian feature is the predominance of flat to rolling topography, seldom rising to more than 200 meters (about 650 feet) above sea level. On the other hand, the Basin also includes Australia's highest peaks (over 2,000 meters, or 6,500 feet) on its southern and eastern fringes. Interlinked land and water resources have been pivotal to life in the prehistoric and historic eras, and it is difficult to ignore the significance of the complex *interaction* of these primary elements. On the other hand, the relentless, monotonal relief tends to concentrate the eye and the mind on the sinuous lines of the rivers, which provided much of the initial stimulus for European incursions. Spurred on by tales of 'inland seas' and navigable waterways leading into and through the continent's mysterious arid center, the early white explorers wrote the Murray-Darling region into the national psyche. Their journals and official reports, and the countless fictional and non-fictional works they inspired, bulk large in Australia's literary heritage (Powell, 1993).

As for the present climatic regime, the broad picture is now confirmed. To the east, maritime influences are inhibited by the presence of the Great Dividing Range; for the most part, the 'inland' characteristics of sub-humidity therefore prevail. The other main factors are the large latitudinal range, from the subtropical north to the temperate south, and subregional altitudinal effects producing cool and humid conditions in the fringing uplands. Serious droughts and floods have affected every part of the Basin since the commencement of written records. The accepted or working totals for annual average rainfall range from more than 1,400 mm (55 inches) in the higher country to less than 300 mm (11 inches) in the northwest. Annual variability is generally high, increasing from east to west. Average summer temperature maxima are highest in the northwest (over 30°C), and in the uplands winter maxima frequently fall below 0°C. Evaporation rates follow this trend line, far exceeding the rainfall totals over most of the Basin, except for the uplands.

If this points to nature's parsimony, then other data can be produced to illustrate sheer caprice. For example, searching adjustments are required to deal with the remarkable variations in the seasonal flows of the Basin's major rivers. Expressed as proportions (%) of the established (*i.e.* historic) average discharges, extreme flows commonly range from below 10% to above 500%, and in the most demanding cases from zero or virtually zero to above 900%. Over the past 200 years, spectacular cessations of flow have stranded paddle-steamers on prodigiously remote stretches of the Darling, while the Murray itself, reduced to a mere trickle, has occasionally failed to reach the sea. Detailed records for Menindie (New South Wales) show that the Darling ceased to flow 48 times between 1885 and 1960.

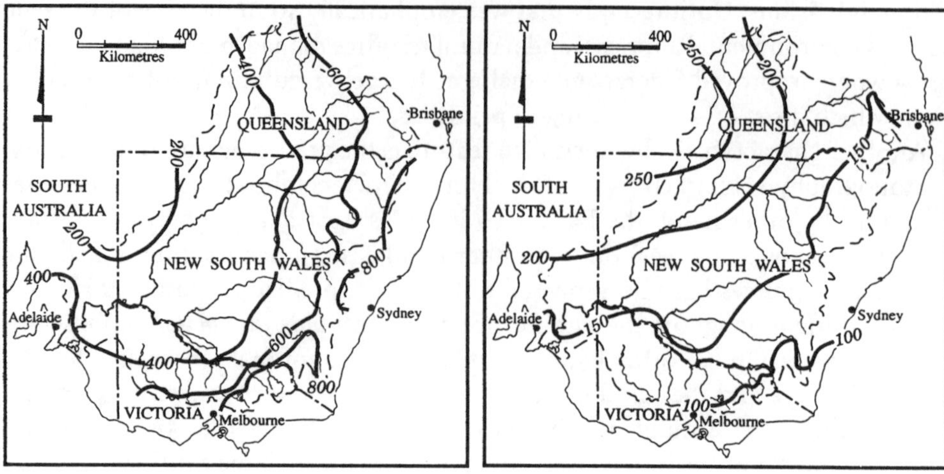

Figure 2. Median annual precipitation in millimeters (left); mean annual evaporation in centimeters (right). From the *Atlas of Australian Resources.* Division of National Mapping, Canberra.

From the early nineteenth century the westward, northward and southerly expansion of (1) the pastoral frontier across the Basin, combined with the subsequent expansion and consolidation of (2) the wheat belt and (3) government-sponsored irrigation settlements, identified the coarser environmental parameters and established a gigantic heritage. It is now the very hearth of rural production in Australia (Heathcote, 1965, 1977; Camm, 1972; Robinson, 1976; Powell, 1976, 1991a, 1993). From the early twentieth century, senior bureaucrats and politicians advocated interregional policies to safeguard this critical heritage. Recently, their successors have been galvanized by the economic and environmental potentials of a far more inclusive integration. The new vision calls for the adoption of the most generous of spatial conceptualizations for interregional policy-making: nothing less than the Murray-Darling Basin itself, in its stupendous entirety.

The Basin accounts for approximately one-third of the nation's total output from industries based directly on renewable natural resources—supporting a little under half of Australia's sheep and lambs, over 40% of its wheat-growing country, perhaps a quarter of the cattle and dairy farms, almost three-quarters of the irrigated land, and the overwhelming proportion of rice, cotton and grape production (Table 1). Annual output values exceed $A10 billion. In terms of over-riding human values, it should be added that the Basin included 40% of the population resident on all Australian holdings in 1991 (Murray-Darling Basin Ministerial Council, 1987).

Over the past decade, academic writings and media coverage have alerted the Australian public to the fact that the devastation caused by environmental degradation has been undermining national investment in the Basin (Table 2). With

Table 1. National economic significance of the Murray-Darling Basin

(MDB shares as % of selected national totals, 1991)	
Total no. rural production units	42
Production of wheat for grain	44
Wool production	40
Rice production	97
Grape production	84
Area of irrigation, crops and pastures	74
Area under cotton	92
Area under orchard trees	59
Total dairy cattle	31
Total meat cattle	28
Total sheep and lambs	46

Source: Australian Bureau of Statistics and Murray-Darling Basin Commission; these statistics are merely broad estimates, and will vary between seasons.

Table 2. Land degradation costs in the Murray-Darling Basin

Type	*$A million p.a.*
Largely non-reversible and cumulative	
Wind erosion, cropland	0.1
Water erosion, cropland	5.0
Reversible but requiring expensive infrastructure	
Shallow watertables, irrigation areas	39.0
Dryland salinization, cropland	0.4
Reversible through appropriate on-farm management	
Soil acidification, cropland	25.0
Soil structure decline, cropland	145.0
Total Costs	215.0

Source: Murray-Darling Basin Commission.

productivity as the supreme goal, comprehensive transformations were introduced over most of the area taken from the Aboriginal stewards, and with inadequate knowledge of the subtle ecological matrices that had maintained life and livelihood. Extensive tracts are affected by continuing desertification, as land is still being taken out of production by salinity and soil erosion. Polluted rivers—some with minuscule volumes by world standards—run a violent blue-green over hundreds of kilometers. 'Interventionist' remedies are costed in the billions of dollars, while the whispered alternative is outright 'retirement' of land—and of people, with no discount for any 'good will' accrued in the pioneering years (Barr & Cary, 1992; Lawrence *et al.*, 1992).

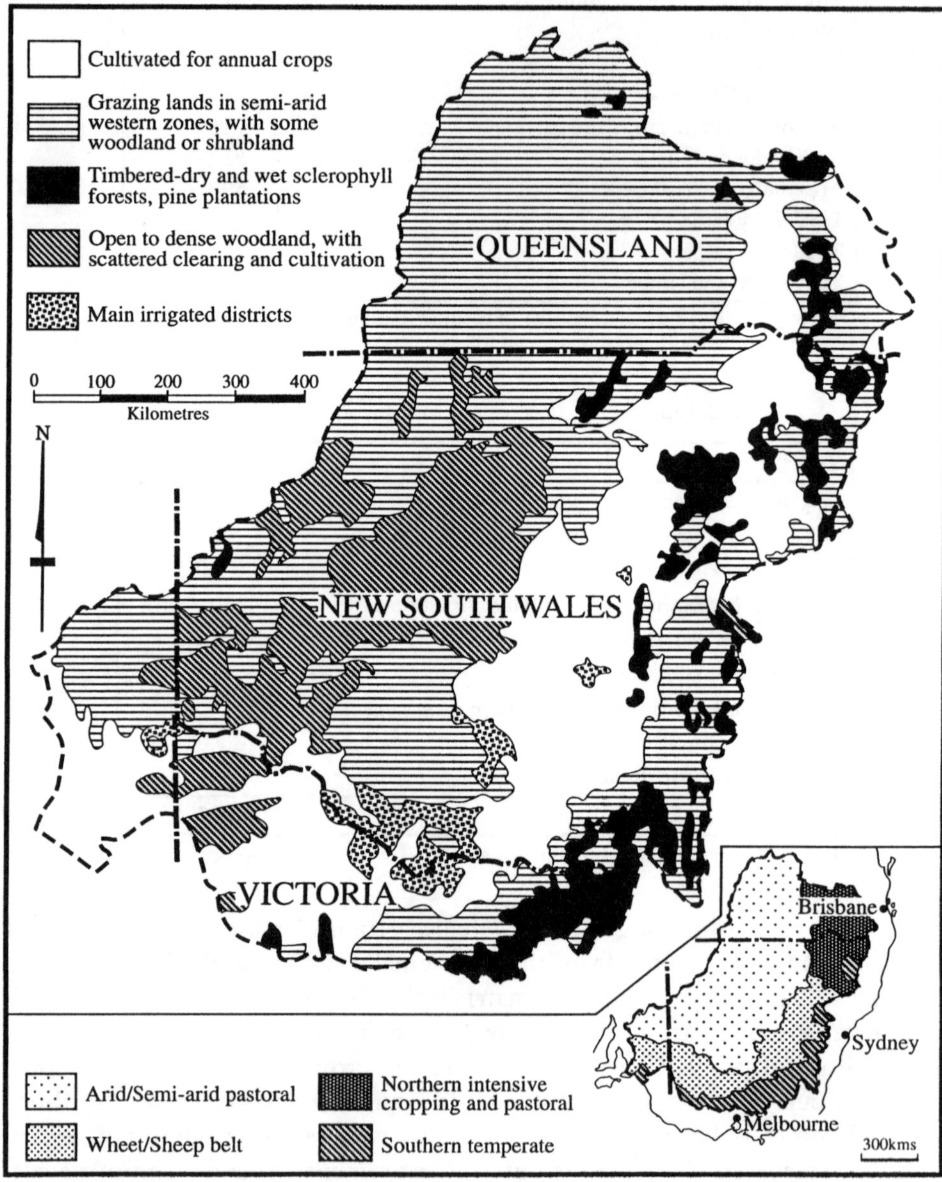

Figure 3. Generalized distribution of land uses in the Murray Darling Basin. Inset: Simplification of the main categories. Source: Murray Darling Basin Commission.

Region and Community

After decades of bitter, often indulgent disputation, in 1915 the three southeastern states and the young federal government reached an agreement on the multiple use of the River Murray, which dominates the southern portion of the Basin (Figure 1; Clark, 1982, 1983). There were two unfortunate outcomes to this otherwise progressive decision. First, there was heavy concentration on the southern sector. Secondly, the constrained activities of the resultant 'River Murray Commission' elevated a style of 'line-based' (that is, river-based) management and research above the wider demands of the Basin *as a whole.* This situation prevailed, essentially, until the 1970s and 1980s, but there were several important initiatives in the inter-war period.

REGIONALISM AND BIOREGIONALISM

Between the 1880s and the opening years of the twentieth century, Victoria's governments tried to jettison English riparian law, on the grounds that it was unsuited to the development of dry regions (Powell, 1989). Gradually, they moved to abolish private rights over water frontages and to establish government control over all the natural sources of water supply. These innovations were variously imitated and adapted throughout Australia. In 1905 the Victorian government set up a single 'State Rivers and Water Supply Commission' with full control of key water management affairs over most of the non-metropolitan area of the state. Thus, it enjoyed an important measure of influence over most aspects of *regional development,* involving an area which was almost as large as the entire United Kingdom (Powell, 1989).

But Australia's technocrats were also intensely interested in supplementing these initiatives with gleanings from overseas practice. These included British elaborations, in the 1930s and 1940s, of an administrative framework for water management based on river basins and on devolution of responsibility to local bodies representing each of those natural units (Powell, 1976:127–31; Powell, 1989). They were also influenced by related U.S. experience—initially by John Wesley Powell's visionary design for a 'technological democracy' in the American West founded on the supremacy of river basin divisions (Worster, 1985), and above all by the New Deal administration's partial resurrection of Powell's design, in the promotion of 'multiple-purpose' development during the 1930s. Some Australians argued that the most lauded creation, America's Tennessee Valley Authority, had been anticipated by reforms in their own country, most notably in the case of Victoria's State Rivers and Water Supply Commission. There was a grain of truth in that, although the scales of operation were poles apart in terms of the financial investments and the sizes of the respective communities, and by any reasonable measure of organizational complexity.

Yet between the 1930s and 1950s, recurrent campaigns throughout Australia for the creation of new states induced an improved 'sense of place', which was either the base of community attachment or the source of disaffection, or both. Two of the more vigorous campaigns were located in the Murray-Darling Basin (Powell, 1991a). Similarly, wartime exigencies permitted any number of government-led tinkering with administrative frameworks. The logic of the 'natural region' gained in the process a certain cachet, in the compilation of economic and environmental statistics and in the formulation of planning programs which favored 'grass-roots' participation.

The concept of bioregional planning declares an interest in those environmental problems and issues related to the development of natural resources which transcend political and other artificial boundaries. Accordingly, Australia—with so few political boundaries, such a small population, and such extensive and generally loosely-defined natural units—could scarcely be expected to contribute as much to the emerging discourse as the United States. Another distinguishing characteristic of the Australian context is the absence of media-massaged catalytic contexts on the scale of the 'Dust Bowl' devastations in the United States before World War II (Worster, 1970). Again, in the realm of philosophical concerns, which would eventually infuse modern bioregional thinking with a type of biocentricism, Australia boasted no renowned environmental thinker to rival U.S. forester-ecologist Aldo Leopold—and no well-publicized 'land-ethic' beyond some vaguely understood conceptions of the drastically diminished stewardship exercised by one tiny fraction of the national population, the traditional Aboriginal communities.

Australians were not easily diverted from perceived mainstream issues. Preoccupied with a revivified development imperative, and thereby with a post-war immigration boom which was coupled with the aggressive use of tariffs to stimulate manufacturing industries in the urban centers of the country's periphery, most of them knew little about the environmental problems of the most productive part of the interior which had bolstered their trading economy for more than a century. Generally, when they focused on any aspect of resource management in the Murray-Darling Basin they became absorbed by the leading federal 'project' of their day, the Snowy Mountains Scheme in the southeastern section of the Basin. Officially launched in 1949, the work was not declared complete until 1974. It was a mammoth, multiple-purpose enterprise (hydro-electricity, irrigation, water supply), by the international standards of the time. The resultant sudden invasion of the (comparatively water-rich) 'high country' by teams of scientists and technocrats, although nurtured by nationalistic emphases on development, was bound to improve the modest quantum of environmental knowledge. Inevitably, established land and water uses were brought under scrutiny.

The foothills and high plains of the Basin had been managed by graziers, miners and timber companies since the mid-nineteenth century. Exhaustive reviews of the catastrophic, drought-fed bushfires of 1939 led to a concentrated debate on these modes of exploitation in Victoria and New South Wales. The steady growth of Canberra, the new national capital to the immediate north, ensured the involvement of the federal and New South Wales governments in high country water resources. The prestigious Snowy Mountains Authority was obliged to enter the debate because the international literature had established the connections between upland erosion and the siltation of expensive reservoirs. So the tide was moving strongly at the official or government level towards a conservationist critique. The broadening of the foundations of that critique—at the level of community involvement—had commenced a few years earlier with well-publicized investigations into predisposing conditions and actual triggers of the big bushfires, and with a subsequent closely related inquiry into the environmental effects of poorly regulated grazing enterprises in the upland forests. Judge L.E.B. Stretton, who spearheaded both exercises, produced a number of eloquent statements on the need to safeguard the *Inseparable Trinity* of forests, soil and water. The catch-cry was astutely adopted by a promising 'Save the Forests Campaign', which drew its support from the urban middle classes (Powell, 1993).

The outline thus far suggests an increasingly refined appreciation of the phenomena and processes of environmental degradation in its various manifestations, moving from a focus on individual facets to an appreciation of the significance of ecological composites or networks. Preoccupations with soil erosion had not only emphasized an association with clearing activities and with the hazardous behaviors of miners, farmers and graziers. It had also begun to draw expert and popular interest away from the Murray *per se*, and towards a consideration of the significance of the land-water-community theme across a much wider territory. Yet the story is incomplete without a brief encounter with the most didactic local example of desertification, 'the salt problem'.

Even before the end of the nineteenth century, suspicions had been raised in Victoria's northern and northwestern irrigation districts about careless 'overwatering' and natural and incidental or unplanned 'waterlogging'. In one of these areas, the water table had risen from between eight and ten meters below the surface in 1880 to less than four meters in 1890. The first big outbreaks of salt were reported in the same places in the early 1920s, and by the next decade several hundreds of thousands of hectares were judged to be badly degraded. The affected terrain, like so much of the Basin's riverine plains, was underlain by the deposits of ancient seas: once more, nature's own intrinsic dynamic was being asserted, with a little help from immigrant resource managers. All too often, management decisions reflected highly localized political factors and the poor contemporary understanding of the intricate relationship between geographical variations in groundwater pressures and sub-surface palaeohydrological patterns (Crabb, 1993; Kellow, 1992; Powell, 1993).

Some of the least satisfactory responses came from irrigators who successfully lobbied their parliamentary representatives for more water: irrigation was sent to fix what irrigation had caused. Certainly, they exhibited political constraints and inadequate environmental knowledge yet again, but by the early 1970s there were signs that it was becoming more difficult to quarantine local environmental problems. *Environmentalism* was beginning to redefine the people's patrimony. A little later, state and federal governments seemed to become intoxicated with 'corporate management' models and the 'managers', whether high or low, were expected to improve their performance. At one level the pace of reform was dramatic, but the sub-text exhibits more evolution than revolution.

INSTITUTIONAL REFORM AND THE BIOREGIONAL ALTERNATIVE

Environmental issues were forced onto the political agenda after the late 1960s and early 1970s (Mercer, 1991; Powell 1991a). Lively disputes between the new radicals and the old development interests attracted a wide following. The environmental movement's vital emotional and intellectual core was well tended by a series of detailed surveys ranging from specific environmental impact statements to landmark national reports on valued 'heritage' sites and landscapes, and by provocative inventories underlining the lamentable condition of fundamental natural resources. Included amongst these reports was a very influential document, deceptively entitled *A Basis for Soil Conservation Policy in Australia* (Department of Environment, Housing and Community Development, 1978). Its substantive message was inescapable: about 50% of the nation's agricultural and pastoral lands needed urgent 'remedial' treatment, and at 1975 prices the total cost of the required construction works was put at nearly $A700 million, with another large and presumably unspecifiable amount for improvements in management systems. Discussions of its detailed tabulations communicated the complete degradation profile, ranging from acidification and wind and water erosion to various types of salinization (Woods, 1983; McTainsh & Boughton, 1993). The combined threat posed to the Murray-Darling Basin was made quite clear in the original document.

The ensconced export-orientation of Australia's rural economy has meant that farmers and graziers are highly specialized by world standards. The kinds of production they have favored are generally capital-intensive operations, with low labor requirements and an unusual dependence on large areas; the concomitant risks are both environmental and economic. During the post-war era the tendency towards fewer working units of larger size, with a reduced workforce, has been no more significant than an increasing preference for cropping systems instead of emphasizing livestock production for meat or wool. There has also been a related intensification of cultivation practices in the climatically favored districts and an expansion of cropping into previously 'marginal' zones characterized by lower and more variable rainfall (Powell, 1991a, 1991b; Cocks, 1992). In this fashion, pockets of subhumid Australia have been converted from extensive grazing districts, with related environmental problems and demands, into more specialized cropping or mixed farming districts with different environmental demands and problems.

In the Murray-Darling Basin the synthesis of entrenched structural characteristics and the recent expansions aggravated the pressures on fragile and depleted soils. Cropping intensities increased everywhere except on the driest and wettest fringes—far western New South Wales and southwest Queensland, and the southern high country (Cocks, 1992). Government researchers and agricultural extension officers could usually keep abreast of the main processes, which were being driven by the kinds of short-term economic motivations which meshed more comfortably with the short-term goal-setting of the political systems. But scientists and technologists were not well equipped to cope with the essence of the modern challenge: the priming of public opinion, while twisting the arms of political paymasters who had been bizarrely out of touch with problems on the land that went directly to the heart of Australia's long-term security as a nation. Environmentalists did much better. They showed that the disturbing profile offered in *A Basis for Soil Conservation Policy* could not be side-stepped with impunity. It could hardly be marketed as Dust Bowl fare, but its derivatives fueled animated debates in schools, colleges and universities.

For the public at large salinization dwarfed the rest of the list in its intimations of sinister forces. The grim spectacle of 'salting' could be interpreted as a warning of primeval natural purifications, a violent purging of presumptuous intruders. The title 'land degradation' had been intended to schedule a busy confessional, but in itself the descriptor was too prosaic to prick fatigued consciences, let alone rouse jaded politicians. 'Salinization', in contrast, represented an inquisitorial rebuke of the very tenure of Australia by Europeans: figuratively, of course, it demanded that they 'turn back'—to first principles. Basin communities were no less deeply riven over the reasons for the rapid deterioration than by the vexed question of culpability. The important *Murray Valley Salinity Investigation* (Gutteridge *et al.*, 1970) was resolutely studied. For the most endangered districts it had predicted that the amount of territory underlain by very shallow water tables would be tripled by the year 2020; over the same period production losses would increase six-fold without further remedial action.

The essential parameters of this problem were, of course, no less social and political than environmental. By the early 1980s the management of land degradation was becoming more open and properly 'inter-disciplinary' in scope, and displayed greater commitment to holistic bioregional interpretations. Well and good, but the next two comments amount to an important caveat. First, the 1970s had brought particularly wet seasons to the southern part of the Basin, and sharp rises in water-tables resulted in dramatic outbreaks of salt in northern Victoria. By the early 1980s the total amount of seriously affected land amounted to 100,000 ha (about 250,000 acres). Secondly, during the droughts of 1982 and 1983—despite all the progress in scientific, technological and extension services—amazing rates of erosion of up to 100 tons of soil per hectare (about 40 tons to the acre) were reported after severe rain-storms on the worst of the fire-devastated slopes. Wind erosion affected over half of the cropland in Victoria's semi-arid Mallee country, with average losses of 60 tons per hectare (around 24 tons/acre) in a number of places (Berndt & Stephens, 1987; Powell, 1991b).

And yet how slowly some pennies drop. Environmental processes in the Basin continued to follow the ancient dynamic of nature beneath all of these trends, while the management institutions seemed fixed on expedient denials of nature's transcendent importance. The very work of the old River Murray Commission—fine though so much of it was, undoubtedly—exacerbated the dangers in those denials by its southerly preference, thereby projecting the image of a curiously lop-sided network, in which the River Murray was crudely excised from its systemic context. That situation was increasingly at odds with contemporary social, political and institutional changes. The gradual resolution of this confusion included the hesitant introduction of fully-fledged bioregionalism in the Basin.

A reformist federal Labor government introduced a new era in Australian public administration in the mid-1970s. The main objective was the achievement of 'accountability' in public sector management (Beringer *et al.*, 1985). British and American precedents were again influential, and the driving paradigm was 'comprehensive corporate management'. It had revolutionized much of the administrative culture before the early 1990s, when it was beginning to show signs of having run its course. In the water sector (intriguingly and repeatedly referred to by some of the reformers as the water *industry*), the trend was away from the heroic or 'developmental' orientation towards a more 'mature' phase which de-emphasized big construction.

The new evangelists said that they wanted to create more 'adaptive' administrative structures and to make water management systems more 'open'—that is, more responsive—so that the community and its elected representatives would be enabled to make direct contributions to the determination and implementation of policy. In addition, there seemed to be an unwritten and largely unspoken assumption that the activities of the huge resource management agencies of the state administrations had been allowed to develop rather too independently of government. Increasingly, it was argued that the predominantly uni-disciplinary engineering cultures of some of those agencies had become incongruous with the burgeoning ideals of multi-skilling and 'inter-disciplinarity' (Kellow, 1985; Beringer *et al.*, 1986). Neither of these charges stands up to historical scrutiny, but they stuck.

The effect was to expose all existing approaches to the reformers' searing critiques. The water management sector, ever the realm of opportunity for would-be pioneers, presented a range of prime targets.

As Kellow (1985, 1986, 1991) has explained very well, the River Murray Commission was patently vulnerable, but the public angst over land degradation, and particularly over salinity, had laid bare its primary weaknesses in advance of the surge for institutional reforms. In truth, the old agency had performed very well in facilitating the development of irrigation in the Basin, but it had not been empowered to cope with *system-wide* problems. Since 1917, its responsibilities had been progressively increased by formalized amendments to the original agreement and, just as importantly, by informal extensions of practice. Even so, the simmering potential for inter- and intra-state conflict was occasionally realized in fiery exchanges, such as those over dam construction in the late 1960s and early 1970s (Powell, 1989). Suggested strategies for dealing with salinity threatened to reinforce the political divisions at the expense of efficient 'comprehensive' planning. There were objections that the residents of New South Wales were paying too much for mitigation measures which would chiefly benefit South Australia. The South Australians tried, in return, to block further water withdrawals for New South Wales irrigators. In the 1980s it was still a muddle. Upstream states retained a right of veto. None of the states was required to notify the River Murray Commission of development plans which might affect water quality. And, as the furor intensified, the Commission itself was actually denied access to important tributaries to conduct essential water quality tests.

It is tempting to accentuate the historic conjuncture of the early 1980s, when Labor administrations were in power in the three southern states as well as in Canberra. Even so, 'political will'— assisted by the maneuvers of a few percipient senior technocrats who were committed to fostering the interpenetration of key agencies across state boundaries— entered the equation with a vengeance. In November, 1985, a meeting of twelve federal and state ministers holding land, water and environmental portfolios called for a 'benchmark' environmental study of the whole Basin. The meeting also spelled out the need for a powerful Ministerial Council (representing state and federal bodies) —'to promote and coordinate effective planning and management for the equitable, efficient and sustainable use of the land, water and environmental resources of the Murray-Darling Basin.' That Council met for the first time in 1986.

The research purview of the old River Murray Commission ('RMC') had become more ambitious over the years, and its original engineering mandate had been offset, to some extent, by its direct and indirect involvement in the accumulation of applied ecological expertise. The most obvious departures from standard practice were in the envisaged geographical scale of management operations and in the required institutional reforms. Under the *Murray-Darling Basin Agreement,* the 'RMC' was replaced by the Murray-Darling Basin Commission, which was to advise the Ministerial Council on Basin-wide environmental management and to take on the existing statutory responsibilities of the RMC. The *Agreement* was signed by the four participating governments (federal, New South Wales, Victoria, South Australia) on 30 October 1987 and—with transparent nods to posterity—it came into force at the beginning of 1988, Australia's bicentennial year. Gradually, 'Murray-Darling Basin' became quite regularly communicated as *MDB*, just as the Tennessee Valley Authority had become celebrated as the 'TVA' in the United States. That signaled a large aspiration.

The Queensland government kept its distance for a few years and its absence handicapped the early sorties in regional integration (Powell, 1991b). From the outset, however, the new Commission seemed determined on a course of rolling change, openly canvassing the links between environmental concern and civic responsibility with the Basin's rural and urban residents. A Community Advisory Committee was formed to engage grass-roots interest and support, and this produced a serious effort to identify, foster, coordinate and harness the Basin's loose network of community organizations, regardless of their special interests and geographical coverage. There was no plain sailing. Whereas in the nineteenth century the central issue had been the disharmony between macro-level political interests in navigation and irrigation, in the late twentieth century inter-state differences in water allocation remain, but there is probably far more potential for quarrels at the scale of local neighborhoods.

The process of community engagement was assisted in two main ways. Firstly, the Commission monitored and sometimes co-opted the initiatives taken in the states to ensure broad community involvement in the battle against land degradation. Each state benefited from the federal government's very public sponsorship of a national *Landcare* movement which was aimed at the rehabilitation and reformulation of endangered properties, partly by a battery of land-use and land-shaping (contour banking, etc.) advice and by selective tree planting. New South Wales and Victoria had launched their own programs for the enlistment of community support in dealing with the key problems of soil erosion and salinity, and Victoria's *Salt Action: Joint Action* proved highly influential insofar as it introduced a thorough ten-year program, based on coordinated sets of activities linking the government agencies and the local communities (Government of Victoria, 1987). In that sense the salinization theme, so divisive in the past, began to point to a potential for intra-Basin bonding.

Elsewhere, the high national profile of the Landcare program, coupled with the intense media coverage of droughts and floods in the interior and of alarming recurrences of blue-green algae in the Darling (in one instance the scum or 'bloom' stretched for 1,000 km), helped to re-assert a more genuine Basin-wide perspective. At least it has thrown down another challenge to the inherited southerly orientation. So has the rising economic significance of the Upper Darling and its tributaries, mainly in Queensland. Over the past two decades there has been a marked swing in that state towards the production of cotton, oilseeds and summer grains, partly connected with increases in livestock carrying capacities. These trends have been supported by greater investments in clearing operations, irrigation and fertilization. Potentials for environmental degradation and conflict with downstream (New South Wales) water users have escalated commensurately. If it is to be fair and abiding, the great Basin game needs powerful umpires as well as special rules.

Many deeply involved commentators had stressed to earlier generations that negotiated adaptations are also crucial to the play (Powell, 1993). And yet, as Aldo Leopold (1949) and other champions of the *land ethic* had explained very well in the United States, there is an ecological dimension to this moral principle which puts limits on our freedom of action. A moment's reflection on that observation suggests that, whatever the ruling management system, entrenched values and attitudes concerning freehold and quasi-freehold tenures are bound to raise large obstacles to fundamental adaptation across the Basin. More familiar with the ethics of person-to-person relationships, Australians must now learn the ethics of society-environment relationships. Too often, private property rights have been allowed to prevail over the public interest. That judgment is succinctly captured in one early exposé of the relentless production of 'small paradises and very large deserts' (Tepper, 1895:390). Clearly, Australia's indigenous people could teach a thing or two about a communal 'stewardship' philosophy. Environmental change in the MDB mocks the folly of attempting to maintain any semblance of insulated exclusivity. Unarguably, ownership carries permanent social and ecological mortgages. What, then, of the

TABLE 3. Selected natural resources management policies in the Murray-Darling Basin

1.	The declaration of a moratorium on further clearing of native vegetation.
2.	Selective replanting of trees where that promises an effective retardation of the spread of dryland salinization.
3.	The introduction of transferable water rights to break the legalistic land/water tie.
4.	The imposition of full and realistic water charges—*i.e.* incorporating the costs of dams, channels and other related works.
5.	Construction of networks of evaporation basins.
6.	The rejection of ideas for the location of entirely new cities in the Basin and a more cautious approach to the expansion of existing regional centers.
7.	The re-purchase, where possible, of cropping rights in designated 'marginal' districts.
8.	Development of compulsory erosion-reduction cultivation regimes.
9.	Conservation of native flora and fauna and the protection of the cultural heritage of the Basin.
10.	Guarantees of environmental impact statements for proposed new industrial projects.
11.	Mechanisms for co-ordinating the efforts of communities, individuals, and federal and state agencies.

Source: Murray-Darling Basin Commission, 1989.

prospect for modifications to the individualistic interpretations favored by modern Australian society?

Historical geographers have shown that the universal tension between environment and development in Australia is best understood within the context of official-popular interaction, with all of its nuances and frustrations, and against the backdrop of chronic if scarcely unrelieved uncertainty (Heathcote, 1965; Powell, 1970, 1991a). One of the recent products of that interaction in the MDB is the severance of the hallowed knot joining land and water rights.

Officially-sanctioned sales of 'transferable' water rights now permit drastic reductions of water use in proven hazard zones (or in less productive localities), and increased applications of water in areas considered more secure, environmentally and economically. That is not inserted at this late stage to return our discussion to the narrow, focusing threat of salinization. Quite the reverse: the purpose is to point to a maturing attention to multi-functionality and multi-disciplinarity now infusing the agenda of the Murray-Darling Basin Commission (Table 3).

Prospect: Regionalism and Regionalization

In applying myself to the task of crafting a short conclusion to this chapter, I found that my geographer's instincts turned me at once to some lines from Stephen Spender: 'Different living is not living in different places / But creating in the mind a map'. One of the more honorable quests of my own nation of immigrants (daily augmented by wave after wave from so many troubled parts of the world) is indeed to find ways of making places better by living differently in them. The MDB story shows one large example of how that objective is at last being approached. However, returning to Spender, it is always wise to be wary of the guides set before us. One should not confuse the planners' 'regionalizations', which may confine us to a choice between rough expedients, with 'regionalism'—which is surely a sense of belonging that cannot be conjured and needs, most urgently in the case of the MDB, to be cultivated.

Bioregionalism addresses such tasks and dilemmas through its efforts to link experts, institutions, and rural and urban communities in a mutual endeavor. In this Australian case the prefix in 'bioregionalism' is richly inclusive: it calls for a knowledge of environmental processes as well as a definite emphasis on the hopes, attitudes and actions of the people within and outside a great natural unit. In those lights the informing spirit of this enterprise is entirely laudable, but what is to be made of the practical approaches, and what of the outcomes? It is too early for answers, but I am inclined to say that the quest is much less quixotic than the baffling, awesome scale may suggest to some non-Australians. Regional thinking cannot be imposed from 'above', but it is possible to nurture it by the clear identification and diffusion of shared concerns and expectations. Moreover, an exploration of the past serves to declare a common and reasonably accessible frame of reference. Today's Murray Darling Basin Commission is, therefore, at once a regional and a regionalizing institution: that is, part of its *raison d'être* is the production and maintenance of regional consciousness. This logic insists that, with the establishment of an improved 'sense of belonging', the collective conceptualization must influence the decisions and behaviors of individuals. And thus the region will truly live—no longer a perplexing abstraction, it will have real meaning in the affairs of the communities residing within and beyond it.

I have pointed to the inheritance of a blend of local and international discourse. Allow me now to return to my earlier note on the boost given to bioregionalism by Franklin D. Roosevelt's New Deal planners by their favoring of river basin units. Given the unusual extent of the MDB it might be pertinent to place one of Roosevelt's more careful qualifications on the record. Over-zealous commitments to the river basin ideal were to be avoided. Flexibility, he advised, was essential—whatever the designated boundaries they should suit the appointed task. Where the demands of a social and economic program were paramount, the selected planning region might differ markedly from one in which environmental issues dictated the choice of a river basin, or indeed any other natural unit. The Murray-Darling Basin Commission has decided on themes and strategies which are to be managed at a mixture of scales—basin-wide and sub-regional (the latter focused on a number of minor contributory catchments). I believe that shows a welcome acceptance of the importance of locating the fine balance between pragmatism and idealism. And the documented evidence attests that the most significant of the Commission's current policies (Table 3) could be said to amount to an appeal for a new kind of civic responsibility informed by environmental history. I hope that observation does not rest too much on my passion for applied historical-geographical analyses. Admittedly, I have become easily irritated by the careless neglect of the inherent contradictions in that done-to-death term, 'global village'. Permitting only an ersatz intimacy with people and places, it seems to induce alienation and passivity, and that does nothing good for civics and the promotion of environmental responsibility. Patently, young nations like my own cannot neglect the need to promote a sustainable and sustaining sense of place, and the mission signaled by the term 'MDB' is curiously appropriate for the uncertainties which abound in the present era.

In some respects, it is more noble and more ambitious than that prescribed for the original 'TVA'. It is not simply an organizational device, then; and it is more than just another step in the Australians' continuing experiment. As a 'timely' project it is bound to be compared with the great Snowy project (Wigmore, 1968), but the challenges it offers are far greater. Conceptually and in its sheer scale, of course, it outranks the Snowy scheme. Furthermore, whereas the philosophy of the Snowy behemoth was essentially one of 'wise use' with 'development' assumed, the MDB mission puts such a premium on 'sustainability' that some reversals of earlier developments must be considered likely. It is consciously framed to take this vital region into the 21st century and beyond, which is indeed a timely objective as the nation prepares for the centenary of federation. Founded on a review and revision of the Australian engagement with Nature, it could represent one of our boldest throws of the dice.

References

Barr, N. and J. Cary (1992) *Greening a Brown Land. The Australian Search for Sustainable Land Use.* Macmillan, Melbourne.

Beringer, I., G. Chomiak, and H. Russell (1986) *Corporate Management. The Australian Public Sector.* Hale and Iremonger, Sydney.

Berndt, R.D. and R.W. Stephens (1987) Soil conservation in Queensland. *Queensland Agricultural Journal* 113:311–16.

Camm, J.C.R. (1972) Development of Rural Landscape and Economy of the Darling Downs, Queensland, 1885–1915. Ph.D. Thesis, University of Newcastle, NSW.

Clark, S.D. (1982) The River Murray Waters Agreement: Down the drain or up the creek? *Civil Engineering Transactions* 24:201–8.

Clark, S.D. (1983) The River Murray Waters Agreement: Peace in our time? *Adelaide Law Review* 9:108–41.

Cocks, D. (1992) *Use with Care. Managing Australia's Natural Resources in the Twenty First Century.* New South Wales University Press, Kensington.

Crabb, P. (1993) *The Murray-Darling Basin: A Resource at Risk.* Longman Cheshire, Melbourne.

Department of Environment, Housing and Community Development (1978) *A Basis for Soil Conservation Policy in Australia.* Australian Government Publishing Service, Canberra.

Government of Victoria (1987) *Salt Action: Joint Action. Victoria's Strategy for Managing the Salinity of Land and Water Resources.* Government Printer, Melbourne.

Gutteridge, Haskins, and Davey, Ltd. (1970) *Murray Valley Salinity Investigation.* 2 vols., River Murray Commission, Canberra.

Heathcote, R.L. (1965) *Back of Bourke. A Study of Land Appraisal and Settlement in Semi-arid Australia.* Melbourne University Press, Melbourne.

Heathcote, R.L. (1977) Pastoral Australia. *In* D.N. Jeans (ed.) *Australia: A Geography.* Sydney University Press, Sydney, pp. 252–88.

Kellow, A.J. (1985) Managing an ecological system: 2. The politics and administration. *Australian Quarterly* 57:107–27.

Kellow, A.J. (1986) Federalism, development and the environment. *Regional Journal of Social Issues* 18:75–84.

Kellow, A.J. (1991) The Murray-Darling Basin. *In* B. Galligan, O. Hughes, and C. Walsh (eds.) *Intergovernmental Relations and Public Policy.* Allen and Unwin, Sydney, pp. 129–45.

Kellow, A.J. (1992) *Saline Solutions. Policy Dynamics in the Murray-Darling Basin.* Centre for Applied Social Research, Deakin University, Geelong.

Lawrence, G., F. Vanclay, and B. Furze (eds.) (1992) *Agriculture, Environment and Society. The Australian Experience.* Macmillan, Melbourne.

Leopold, A. (1949) *A Sand County Almanac.* Oxford University Press, New York.

McTainsh, G.H. and W.C. Boughton (1993) *Land Degradation Processes in Australia.* Longman Cheshire, Melbourne.

Mercer, D.C. (1991) *A Question of Balance. Natural Resources Conflict Issues in Australia.* Federation Press, Sydney.

Murray-Darling Basin Commission (1989) *Murray-Darling Basin Natural Resources Management Strategy.* MDB Ministerial Council, Canberra.

Murray-Darling Basin Ministerial Council (1987) *Murray-Darling Basin Environmental Resources Study.* MDB Ministerial Council, Canberra.

Powell, J.M. (1990) *The Public Lands of Australia Felix.* Oxford University Press, Melbourne.

Powell, J.M. (1976) *Environmental Management in Australia, 1788–1914. Guardians, Improvers and Profit: An Introductory Survey.* Oxford University Press, Melbourne.

Powell, J.M. (1989) *Watering the Garden State. Water, Land and Community in Victoria, 1834–1988.* Allen and Unwin, Sydney.

Powell, J.M. (1991a) *An Historical Geography of Modern Australia: The Restive Fringe.* Cambridge University Press, Cambridge.

Powell, J.M. (1991b) *Plains of Promise, Rivers of Destiny. Water Management and the Development of Queensland 1824–1991.* Boolarong Press, Brisbane.

Powell, J.M. (1993) *'MDB'. The Emergence of Bioregionalism in the Murray-Darling Basin.* Murray-Darling Basin Commission, Canberra.

Robinson, M.E. (1976) *The New South Wales Wheat Frontier 1851 to 1911.* Research School of Pacific Studies, Australian National University Publication HG10, Canberra.

Tepper, J.G.O. (1895) Remarks on weeds: The soil and its fertility. *Agricultural Gazette of NSW* 113:385–90.

Wigmore, L.G. (1968) *Struggle for the Snowy.* Oxford University Press, Melbourne.

Woods, L.E. (1983) *Land Degradation in Australia.* Department of Home Affairs and Environment, Australian Government Publishing Service, Canberra.

Worster, D. (1979) *Dust Bowl.* Oxford University Press, New York.

Worster, D. (1985) *Rivers of Empire. Water, Aridity and the Growth of the American West.* Pantheon, New York.

ABOUT THE AUTHOR

J.M. Powell holds a personal chair in geography in the Department of Geography and Environmental Science, Monash University, Melbourne, Australia. A graduate of both Liverpool University, England, and of Monash, he is a past president of the Institute of Australian Geographers and a former editor of Australian Geographical Studies. He is the author of several books and numerous articles on the geography and history of resource appraisal and environmental management, principally in Australia, but with sorties into New Zealand and North America. His most recent publications include *An Historical Geography of Modern Australia* (Cambridge, 1988, 1991), *'MDB': The Emergence of Bioregionalism in the Murray Darling Basin* (Canberra, 1993), and *Griffith Taylor and 'Australia Unlimited'* (Brisbane, 1993).

Prof. J.M. Powell
Monash University
Department of Geography and Environmental Science
Clayton
Melbourne, Victoria 3168
Australia

Chapter 4

Environmental Law and Management in the Gulf of Aqaba

Deborah Sandler

Introduction

Sustainable development of the Gulf of Aqaba is dependent upon individual state efforts as well as regional cooperation. Although the Gulf is bordered by diverse states, each with its own political, social and cultural traditions, the area can be regarded as a single unified environmental system.

The Gulf of Aqaba is one of the world's most rare and beautiful ecosystems, bordered by Egypt, Israel, Jordan and Saudi Arabia (Figure 1).[1] The Gulf, 180 km long and 5–26 km wide, contains an astonishing diversity of coastal lagoons, sandy and rocky outcrops, seagrass beds and coral reefs. It is the habitat for approximately 1,000 species of fish and over 200 species of coral. To put this in its proper global perspective, the richness of life in the Gulf can be compared to the waters surrounding Hawaii, comprising an estimated 45 coral species, or to the waters of Bermuda with an estimated 15 species of corals (Alawneh, 1993; ISPAN Irrigation Support Project for Asia and the Near East, 1992).

The climate over the Gulf of Aqaba and the surrounding land region is hyper-arid and one of the driest in the world. The meager annual rainfall averages below 25 mm, while evaporation is very high, ca. 3,500 mm per year. This is due to the extreme summer temperatures which rise to well over 40°C, as well as the low humidity in the air. Indeed, the entire Gulf of Aqaba region can be regarded as an arid frontier zone.

Development of the Gulf of Aqaba is essential for its coastal states—most significantly for Egypt, Israel and Jordan. Aqaba is Jordan's only outlet to the sea; Eilat is an important port for Israel. The Gulf's attraction as an international tourist center for most of the riparian states has resulted in the investment of millions of dollars, rapid coastal development and large numbers of year round visitors. Egypt, for example is pressing ahead with plans to turn the Sinai coastline into a Middle East Riviera.

H. J. Bruins and H. Lithwick (eds.), The Arid Frontier, 75–93.

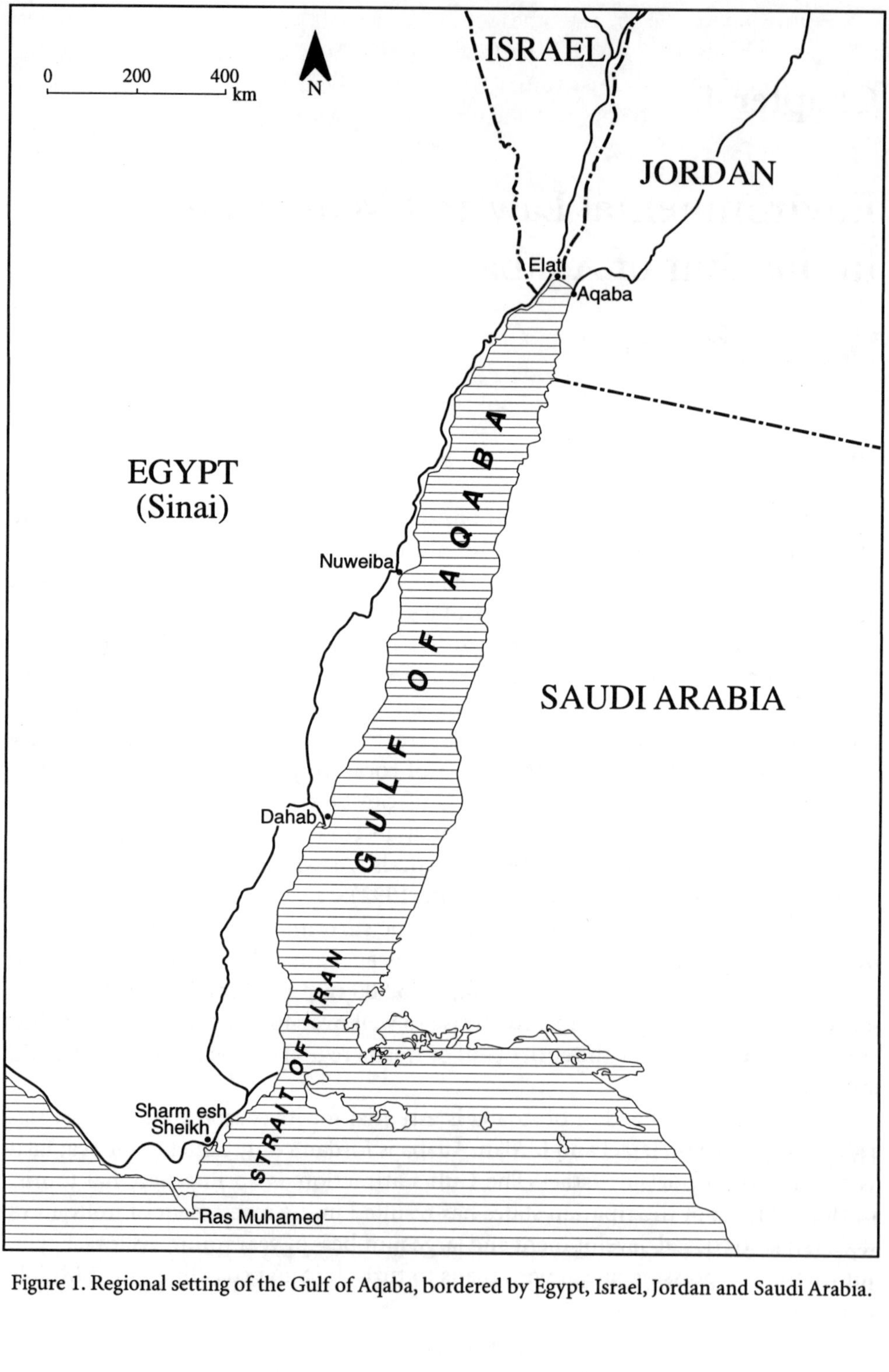

Figure 1. Regional setting of the Gulf of Aqaba, bordered by Egypt, Israel, Jordan and Saudi Arabia.

In consequence of these and other activities, the Gulf's delicate ecosystem is threatened by oil spills, unregulated coastal development, municipal sewage, air borne pollutants and uncontrolled tourist activity. Today, if just one tanker sailing up these waters were to spill its cargo of oil, all the countries could be faced with ecological and economic catastrophe. Moreover, the task of assessing and apportioning responsibility for the damage would be a nightmare.

Agreement between the Gulf of Aqaba states is, therefore, of paramount importance in the creation and maintenance of an effective framework for pollution prevention and environmental protection of the sea and its surroundings. Concurrently, the success of any regional effort ultimately relies upon each state's willingness and ability to administer environmental policy. Even if there is international law or a regional agreement which can be applied to the Gulf of Aqaba, compliance will depend largely upon the enforcement strategies of each individual state.

This chapter provides an overview of some of the more significant state and regional policies which impact upon the Gulf of Aqaba. It then looks at possible models for establishing a comprehensive framework for sustainable development of the area. The first section identifies the environmental risks faced by each state and evaluates the measures taken to deal with them. The aim of the section is to provide a fuller understanding of specific environmental difficulties and the legal and managerial mechanisms which have been employed in practice. The second section looks at regional efforts to protect the Gulf of Aqaba, especially in light of the Middle East peace process. The third and final section examines existing international agreements which are relevant to the Gulf of Aqaba. The elements of those treaties which may be useful to the Gulf of Aqaba states in constructing a comprehensive regime of environmental protection are highlighted.

Environmental Risks and State Responses

This section seeks to identify the environmental risks facing the Gulf countries and to evaluate the state responses to these dangers. Unfortunately, the policies and practices of Saudi Arabia will be excluded from this discussion since at present there is no public access to the relevant information.[2]

JORDAN

The port city of Aqaba has experienced dramatic growth in recent years. According to Jordanian census data, the 1979 population of Aqaba was 27,000 (Alawneh, 1993:98).[3] In 1990 it was reportedly 55,000 and by the year 2005 the population is expected to reach 178,000. These numbers reflect expected increases in commerce, industry and tourism. The major environmental threats to the Gulf from the Jordanian side include municipal expansion, tourism, industry, including air-borne pollution and vessel-based pollution. Jordanian policy and practice for dealing with these hazards is explored below.

Vessel-based Pollution

As Jordan's only port, Aqaba plays a major role in the country's regional and international trade. Between 1972 and 1989 the volume of goods passing through Aqaba increased from 1.2 million tons to 20 million tons (Alawneh, 1993:100). The number of ships calling annually at the port during this time rose from 327 to approximately 2,446, with the risk of a shipping collision or accident rising as well (Al-Khoshman, 1993:217). While the peace process has facilitated the development of both domestic and regional arrangements to deal with small to medium sized spills, Jordan has yet fully to implement training plans, appropriate technology and emergency response strategies. Like the other Gulf countries, Jordan is still lacking a regional emergency response plan as well as the capacity to cope with a large and significant spill.[4]

Despite the magnitude of its shipping activity, Jordan is not a Contracting Party to the International Convention for the Prevention of Pollution from Ships 73/78 (MARPOL, 1973/78).[5] This important global convention, discussed below, regulates shipping and port practices world wide. One consequence of the Jordanian position is that in contravention of MARPOL, Jordanian law permits the release of bilge waters into the sea. These discharges contain processed oils which accumulate with time and are damaging to the ecosystem as well as to human health.

Industry

Industrial pollution of the Gulf has been of significant concern to Jordan. Many of the industries in Aqaba need large amounts of cooling water and are therefore located near the sea. Since the marine organisms found within the Gulf live within a few degrees of their upper thermal limits, a sustained increase in the temperature of the sea could adversely affect their ability to survive (Wahbeh, 1993:30).

The environmental impact of Aqaba's important and lucrative phosphate industry is also problematic. More than eight million tons of phosphates are exported annually through Aqaba, with increased exports to be expected. When phosphates are loaded onto ships a large amount of dust is released into the atmosphere, settling on the water. The most serious consequence of this process is 'phosphate poisoning' which prevents the calcification of corals, mollusks and other organisms. In addition, the dust also increases sedimentation in the water and reduces the penetration of sunlight, an essential element in the survival of the coral reefs. This issue has recently been addressed through the installation of a choke feeder system, which aims at reducing the amount of dust released into the air as the minerals are loaded onto the ships. Although this initiative by the local authorities represents a positive first step in dealing with phosphate pollution, it must be noted that Jordan still lacks legislation regarding the control and supervision of these activities.

Other sources of industrial pollution, such as chlorine, an anti-fouling chemical used in the city's industrial and power plants, also pose a threat to the Gulf. When this chemical mixes with seawater a toxic derivative results, which is harmful to marine life (Wahbeh, 1993:30–31).

Municipal Growth and the Discharge of Sewage

One common consequence of rapid municipal growth is the discharge of sewage into the sea and the resulting health risks from coliform bacteria as well as other threatening agents. Moreover, the influx of sewage leads to a rise in nutrient matter in the sea, which in turn increases algae growth and blocks sunlight penetration. Until 1987 the municipality of Aqaba discharged its untreated sewage directly into the Gulf. In June of that year, the municipality began treating its sewage at an inland plant and utilizing the waters for irrigation (Wahbeh, 1993:29).

Jordanian management of Aqaba's municipal sewage should be applauded and looked to as a model for other cities around the Gulf. At the same time, however, it must be pointed out that sewage continues to enter the sea from recreational, industrial and domestic source not integrated into the municipal infrastructure of Aqaba. In 1992 it was reported that the municipal sewage system served only about half of the town population (ISPAN Irrigation Support Project for Asia and the Near East, 1992).[6]

Tourism

The activities of tourists add to the unsettling of the sea, the littering of the beaches and the destruction of the reefs either through intentional ransacking of the corals and shells or accidental damage caused by swimmers, divers and boaters. In addition, intense construction along the coastal strip causes substantial ecological disturbance, adding to the sedimentation threatening the coral reefs. During 1989, it was reported that 430,000 tourists arrived in Aqaba by sea and 40,000 by air. (Wahbeh, 1993:29) That number is expected to increase significantly as plans for regional tourism are implemented. In order to cope with this rapid growth, the Jordanian coastline has been divided into four development zones: the town of Aqaba, the port area, a proposed tourist zone and an industrial area (ISPAN Irrigation Support Project for Asia and the Near East, 1992:12). It is intended that these divisions will result in planned and regulated development of the Aqaba region and the marine area.

Environmental Law and Administration

Perhaps the most fundamental difficulty that Jordan faces in addressing its environmental concerns regarding the Gulf of Aqaba is the lack of a central administrative body which is responsible for and capable of supervising the legislation, administration and implementation of Jordanian policy. In general, jurisdiction over policies and programs affecting the environment is shared by several national ministries and sub-ministries, including the Ministry of Municipal, Rural Affairs and the Environment, the Ministry of Health, the Ministry of Water and Irrigation, the Ministry of Industry and Trade, the Ministry of Agriculture and the Ministry of Tourism. National legislation which impacts upon the Gulf is found in different types of laws, each aiming to regulate an area of environmental administration. For instance, while there are regulations in place which govern the discharge of wastewater into the sea, there is another set of laws, often in the form of circulars which regulate shipping and port activities.

Moreover, close examination reveals that even this division of jurisdiction is complex. Ziad Jaber Alawneh, an Extension Officer at the Jordanian Society for the Control of Environmental Pollution, reports that under the Water Authority Law, No. 18 of 1988, overall responsibility for regulating water pollution lies with the Jordanian Water Authority (Alawneh, 1993:107–8). However, the authority for regulating water pollution from industrial sources predates this law and emanates from the Military Ordinance No. 6 of 1980, which is implemented by the Jordanian Department of Standards and Specifications, within the Ministry of Industry and Trade. In 1982, this department issued Jordanian Standard Specifications No. 202 to set limits on industrial discharges into the water. Enforcement jurisdiction for these standards was never clearly allocated and as a result is shared by the Water Authority, the Ministry of Industry and Trade, the Ministry of Health, and the Department of the Environment.

The lack of cohesion on the national level is complicated by the administrative organization on the local level where there are three main agencies responsible for the environmental protection of the Gulf of Aqaba region: the Aqaba Region Authority, the Aqaba Municipality and the Aqaba Port Corporation (Alawneh, 1993:107–8). The Aqaba Region Authority was established by the Government of Jordan to oversee the economic and social development of the area. Its mandate covers responsibility for coastal zone management, including environmental monitoring of all major construction activities. The Aqaba Municipality is the main administrative agency for the city of Aqaba and is responsible, among other things, for the collection and disposal of the city's solid wastes, the licensing of various commercial establishments, the oversight of conservation activities around the city and the prevention of environmental nuisances. The Aqaba Port Corporation has jurisdiction over the operation and maintenance of Aqaba port, including supervision of loading and unloading of all vessels calling at the port. The Port Corporation is also responsible for providing services for the vessels in port. While it does not have a specific environmental mandate, the Port Corporation has nonetheless taken the initiative for the installation and maintenance of the choke feeder system discussed earlier. Other governmental and non-governmental organizations also influence the environmental protection activities of the Gulf of Aqaba. Most significant of these is the Aqaba Marine Science Station, operated jointly by the Universities of Jordan and Yarmouk. The Station is responsible for environmental monitoring and scientific research on the Gulf.

The overlapping jurisdictions regarding the environment, as well as the competing national and regional interests represented by the various authorities, inevitably hinder the effectiveness of national as well as regional policy. Perhaps because of the pressing difficulties Jordan faces in protecting its environment, there seems to be a growing awareness of the need for significant reform. Currently, the Jordanian Cabinet is considering a draft law which proposes the creation of a Higher Council for Environmental Protection. This body would oversee, coordinate and implement Jordanian environmental policies, including many of those affecting the marine and coastal activities discussed above.

EGYPT

While Jordan clearly leads the Gulf countries with its commercial activities in the port of Aqaba, it is Egypt which may soon emerge in the forefront of the Gulf of Aqaba tourist industry. The Egyptian coastline, 200 km in length, dominates the western portion of the Gulf of Aqaba, and is the object of intense and massive development. As the Sinai peninsula has remained virtually unchanged and unpolluted since biblical times, its protection is of intense international concern and interest.[7]

Tourism is a vital component of the Egyptian economy. Three million tourists visit Egypt annually and it is estimated that by the year 2000 there will be 5 million visitors a year (Sultan, 1993:178).[8] In order to accommodate this growth, it is essential that Egypt diversify its tourist industry away from the traditional antiquities in the Cairo region. According to the former Egyptian Minister of Tourism, His Excellency Fouad Sultan, the Cairo region has reached its absorption capacity and the opportunities for real expansion now lie with developing beach and sea resorts, especially those endowed with a combination of antiquities, sea shore, coral reefs, mountains, deserts, religious shrines and a warm sunny climate. In other words, the Gulf of Aqaba.

In 1990, the Government of Egypt began to act on this view and formulated a national priority plan for the development of tourism, designating South Sinai and the Red Sea coast as high priority areas. The entire length of the Egyptian coastline was divided into five planning sectors, each 30–70 km long: Taba, Nuweiba, Dahab, Wadi Kid and Sharm-el-Sheikh. Each sector is designed to share major infrastructure facilities and will encompass an established community as well as several resort centers.

A brief examination of plans for the Taba sector will illuminate this development concept. Currently there is only a small community in Taba, but a holding company has been created in order to proceed with its expansion. Planning for Taba includes the construction of 40,000 beds by the year 2005. It is estimated that 20,000 new jobs will be generated, with the majority of workers being housed in a new village to be established at the northern end of the area. The development costs for the Taba sector are projected at $555 million, including $142 million for infrastructure (Sultan, 1993:188–190).

Allocation of specific lands for development is under the jurisdiction of the Egyptian Ministry of Tourism. Project selection is based on free market criteria with the goal of attracting foreign capital that will in turn, draw in masses of foreign visitors. Because of the high priority given to developing the region and the large expanses of land available, there are lower costs, fewer legal issues and reduced bureaucratic delays.

The potential threats from the tourist industry, including swimmers, snorklers, divers and boaters, will be very serious. The anticipated growth in towns and resort centers will lead to a dramatic increase in construction waste and other land-based pollution such as sewage. At present the infrastructure along the Red Sea coast is poorly developed. While some of the major towns treat their sewage in oxidation ponds, many of the newer hotels are far from these sites and are left alone to develop their own sewage treatment units. These units are, however, costly to build, maintain and inspect. As a result, sewage, as well as recreation and construction wastes, often flow directly into the sea (Sultan, 1993:184). In order to protect the Gulf region and the most vulnerable areas of the Sinai coast, environmentally oriented rules need to be established and implemented.

Environmental Law and Administration

The potential environmental hazards in the Gulf of Aqaba are a concern of the Egyptian Environmental Affairs Agency (EEAA) which was created in 1985 to act in an advisory capacity regarding environmental matters. With time, the EEAA's powers and influence have increased and a draft law has been under consideration by the Egyptian government which would award the agency new and sweeping powers (ISPAN Irrigation Support Project for Asia and the Near East, 1992:17). The Egyptian Ministry of Tourism is, as discussed above, the other and probably most significant government organ involved in the development and protection of the Sinai coast.

During the early planning stages of the Gulf of Aqaba Tourism Development Zone, the EEAA and the Ministry of Tourism collaborated in preparing a survey of the entire coastline with the aim of identifying areas to be protected and those to be developed. Large tracts of coastline were designated as protected areas. The Ras Muhammed Park, at the tip of the Sinai is one such protected reserve and certain steps have been taken to reduce damage to the area. Mooring buoys have been placed at popular diving spots to minimize improper anchoring. In addition, the EEAA and the Ministry of Tourism have the authority to regulate the tourist flow to the reserve in order to limit damage to the reef. However, as officials of the Egyptian ministries acknowledge, even preserving a protected area can be a difficult task. Damage is difficult to control in practice, and reports indicate deterioration of the coral reefs due to excessive diving and boating. The mangroves in Ras Muhammed are subject to chronic oil pollution (Sultan, 1993:179).

In order to regulate the planning and building process the Ministry of Tourism requires environmental impact assessments to be submitted by developers to the Tourist Development Authority of the Ministry of Tourism for review. Project compliance with the Ministry's rules and problems of enforcement in an area so distant from the center of the country are beyond the scope of this chapter, but are of obvious concern.

ISRAEL

The Israeli interest in the Gulf of Aqaba is intense and varied, although the Israeli coastline is only 14 km long. The city of Eilat, which dominates the northwestern part of the Gulf, has thousands of hotel rooms (Sandler, 1993:84),[9] numerous entertainment and eating facilities, an oil terminal, a pleasure boat marina, a naval base, a commercial harbor, a research laboratory, an underwater observatory, a coral reef nature reserve and a dolphinarium. On average, there are more than 6,000 daily visitors to the city (Sandler, 1993:84). During the holiday seasons of Sukkot and Passover the city hosts between 80–100,000 people. Many of these visitors engage in Gulf-related activities.

Vessel-based Pollution

As Israel's point of entry to the Red Sea and Indian Ocean, Eilat is also a strategically important port city. Although the volume of goods transported through Eilat is much smaller than that which passes through Aqaba, the threat of an oil spill is ominously present. For example, the Israel Ministry of Environment reports that in 1991 there were 14 reported spills in the Israeli waters (Ministry of the Environment, 1992:317). Five were from large commercial vessels, two from pleasure boats, one from a naval vessel, two from land-based sources and four from unknown sources—which often means other territorial seas. Fortunately, the spills were contained in all cases and the damage was minimal.

The facilities in Eilat which most endanger the environment are the port of Eilat and the Eilat-Ashkelon pipeline terminal, comprised of two fuel jetties, located just north of the coral reef reserve (Tal, 1993:259). Indeed, the pipeline terminal is the only location in the Gulf of Aqaba through which oil is transferred on a regular basis; in 1994 approximately 5 million tons of oil passed through the oil terminal, almost all of it transported to Eilat by tanker.

In addition to oil spills there are threats from the release of ballast and bilge waters into the sea. Israel has, however, taken significant steps to reduce the threats to the Gulf from these facilities. Israel is a Contracting Party to MARPOL 1973/78 and has, in compliance with the Convention, constructed reception facilities in the port of Eilat for the disposal of ballast waters from visiting ships (Tal, 1993:259). There are also safety devices which have been put into place at the fuel jetties in order to prevent spills during the unloading of oil. This is supported by an aggressive and thorough inspection system which checks every tanker that enters Israeli waters to ensure, among other things, that its oil record books are in order. Nonetheless, as long as the pipeline continues to operate, the danger of a large and uncontrollable oil spill exists.

Land-based Pollution

Along with the dangers of pollution from ships, there are also continuous threats from land. Despite the concentration of hotels and harbors along Israel's short Gulf coastline, the potential for still more coastal development is considerable. Conflicts over the use of resources and the impact of development activities are intense (Brachya and Marinov, 1993:195). In order to mediate these conflicting interests successfully, Israel's Planning & Building Law, 5725 (1965) requires that any major project be subject to an environmental impact assessment before receiving project approval (Brachya, 1984; Ministry of the Interior, 1981; Ministry of the Environment, 1992). The recommendations which emerge as part of the final environmental impact statement are then incorporated into the planning and building of the project as legally binding obligations incumbent upon the project developer (Brachya & Marinov, 1993:196). Application of the environmental impact statement has provided in practice a necessary and beneficial tool for the planning and construction of ports, airports, power stations, sewage treatment plants and marinas.

Until recently, Eilat discharged its municipal sewage into the sea, amounting to about 3 million cubic meters of waste water annually, bringing increased sedimentation and coli bacteria. On a number of occasions the amount of coli surpassed acceptable levels forcing the Eilat authorities to close the beaches. Moreover, the discharge of sewage has marred the aesthetic beauty of the area and provided a breeding ground for mosquitoes.

The Israel law for the Prevention of Marine Pollution from Land-Based Sources, 1988 states that 'no person shall dump or flow into the sea waste or sewage from a land-based source in order to dispose of them in the sea, directly or indirectly, save under a permit pursuant to this law...' Following its evaluation of the situation, the Permit Committee established under the Land-Based Sources Law ordered the Municipality to cease discharging sewage into the sea by the beginning of 1992. After three years of legal action and negotiations between the Ministry of Environment and the Eilat Municipality, this issue has been resolved. As of June 1995 the sewage of the Eilat municipality is transported 35 km north to be treated and used in the irrigation of agricultural lands.[10]

Mariculture

Eilat is host to two general types of mariculture stations. One is landbased, where fish are grown in pools or ponds and aquariums. These facilities are regulated by the Law on Land-based Sources. The other type of station is marine-based, where fish are grown in cages. This industry, which is gaining in importance, provides an average 400–500 tons of fish a year. The pollution problems in both types of stations are similar. Uneaten food and fish wastes result in increased sedimentation and nutrients, which may adversely affect the Gulf. In order to regulate the marine-based facilities the Ministry of Environment is currently drafting environmental regulations stipulating the conditions which must be met and adhered to in order for station operators to receive their business licenses.[11]

Tourism

Israel has addressed the problem of tourist damage to the coral reef through the creation of a Coral Reef Nature Reserve. The Reserve, which is managed by the Israel Nature Reserves Authority under the direction of the Law of National Parks, Nature Reserves, National Sites and Memorial Sites (1992),[12] aims at encouraging visits to the reef through a program of strict regulation. For example, the Reserve area along the shore is fenced and guarded. An entrance fee is required of visitors and snorkeling is permitted only in clearly designated places. The Reserve is completely off limits to boaters, and violators of these rules are subject to fines. But perhaps of greatest significance is the Underwater Observatory which is located in the heart of the Reserve. In this partnership between development and environmental protection, thousands of visitors can view marine life in the Reserve annually, without causing any damage to the reef environment (Sandler, 1993:88).

Biological diversity has received heightened attention in Israel recently, with the 1994 ratification of two international conventions dealing with nature conservation: The Convention on Biological Diversity and the Convention on Wetlands of International Importance (the Ramsar Convention). Under the Ramsar Convention coral reefs and coastal lagoons are considered wetlands. Both of these conventions have the potential of influencing development and protection of the Gulf of Aqaba region.

Environmental Law and Administration

With the establishment of the Israel Ministry of Environment in 1988,[13] authority over environmental matters in the country was significantly consolidated under one central organ.[14] The Ministry's Marine and Coastal Environment Division (MCED) oversees the environmental protection of the Mediterranean and Red Seas.

In implementing its mandate the MCED has designed an extensive environmental management program aimed at addressing the particular needs of the Gulf of Aqaba region. The program includes a broad approach towards community education, involving corporate establishments, commercial and government industries, the public schools and the public at large. The MCED operates a visible inspection program aimed at deterring would-be violators and sanctioning those who disobey the law. These activities are coordinated by the staff of the local Marine Pollution Control Unit who are authorized to carry out marine and coastal surveillance, inspect in-coming vessels, locate sources of pollution, impose fines and assist in the preparation of legal claims and criminal indictments. The staff of the MCED are housed in the Eilat Marine Pollution Control Station, which is located between the oil terminal and the Coral Reef Reserve. The Station's primary tasks are the prevention, treatment and abatement of oil spills. It operates 24 hours a day and is fully equipped to deal with small to medium sized oil spills with maximum speed and efficiency (Sandler, 1993:74).[15] The activities of the station are funded by the Marine Pollution Prevention Fund, which largely draws its sources from fees levied on ships calling at Israeli oil terminals and ports and from fines imposed upon polluters. This implementation of the 'polluter pays' principle is a good example of complementary management of development and environmental protection.

Regional Efforts to Protect the Gulf of Aqaba

Even before the Madrid Peace Conference in October 1991, representatives from Egypt, Jordan and Israel were meeting secretly to explore ways to preserve the environmental quality of the Gulf and to establish strategies for regional cooperation on preventing marine pollution (Maranz, 1993). These discussions identified several pressing issues, such as emergency response to marine incidents, which were then developed during the peace process following the Madrid Summit in 1991. On 8 May 1995, in accordance with the Treaty of Peace between Israel and Jordan,[16] the two countries initialed a general Agreement on Cooperation in Environmental Protection and Nature Conservation, intended to be ratified within a year. In Article 3 of the Agreement the Parties agree to promote cooperation relating to the protection of the marine environment and to coastal resource management. Bi-monthly meetings between representatives of the Parties will review current and future activities and be responsible for assessing the results of cooperation to date (Article 10).

More specifically, the countries have put forward a number of plans designed to develop the commercial potential of the Gulf of Aqaba and to protect its natural resources as well. In preparation for the Second Economic Conference, held in Amman, Jordan in October 1995 representatives from Egypt, Israel and Jordan worked to draft plans for coastal zone management, potential tourist ventures and environmental parks. One Israel Ministry of Environment official cited a joint Israeli-Jordanian proposal for a shared Gulf of Aqaba Underwater Reserve as reflective of the type of sustainable development which is most desirable.[17]

Despite the apparent enthusiasm for the idea of sustainable development, to date the only plans which have been realized are those regarding emergency response in the event of an accidental spill in the Gulf waters. Egypt, Israel and Jordan have concluded an agreement for a joint action program based on the establishment of three centers to combat oil spills (Israel Environment Bulletin, 1995:10; 1994:16). The centers—in Aqaba, Nueiba and Eilat—will be equipped to deal with small to medium sized spills; acting together they will have the capacity to combat larger spills.[18] The project formulation was completed in 1994 and is to be followed by the establishment and equipment of the individual centers and the implementation of a regional contingency plan.[19] These centers are expected to be operational in 1996 and will serve as a model of cooperation for other projects in the Gulf region and elsewhere.

A Comprehensive Agreement for the Gulf of Aqaba: Models to Learn From

Any comprehensive arrangement regarding the Gulf of Aqaba should consider similar international contexts in which environmental agreements have been made. At the same time the historical relations between the countries as well as the unique features of the Gulf must be taken into account. This section examines several multilateral agreements which could provide useful models for a Gulf of Aqaba accord (Marinov & Sandler, 1995).

THE MEDITERRANEAN ACTION PLAN

In 1975, 18 states bordering the Mediterranean agreed to work together within the framework of the Mediterranean Action Plan (MAP), a wide-ranging program aimed at preventing, controlling and abating pollution in the Mediterranean Sea (Raftopoulos, 1993:315). The work within MAP entails on-going information sharing, team planning and participation in joint research projects. Some of the important issues include integrated coastal zone planning, the control of marine pollution and waste disposal. A program for specially protected areas and the Blue Plan, a 21st century forecast of the situation in the Mediterranean, underscores the comprehensive and long-term vision to which the parties have committed themselves (UNEP, 1984, 1988).

The Contracting Parties to the Barcelona Convention (1976) and Protocols, the legal component of MAP, include Israel, Algeria, Egypt, Lebanon, Libya, Morocco,

Syria and Tunisia.[20] The legal instruments aim at regulating almost all types of pollution affecting the Mediterranean.[21] It was within the framework of the Barcelona Convention that Israel and its Arab neighbors first agreed to cooperate on regional issues (Marinov &Sandler, 1995).

MAP has been extremely successful in bringing the diverse Contracting Parties together on many levels, from the highly political to the professional and technical, with little conflict (Haas, 1990).[22] The consistent emphasis on benefits available through cooperation rather than conflict has laid a foundation for dealing with the environmental hazards threatening the Mediterranean Sea.

There are many lessons which can be drawn from the experience in the Mediterranean and applied to the Gulf of Aqaba. A non-adversarial approach to environmental protection can create a strong basis for cooperation among the Gulf states. The thrust of such an effort should recognize the need of all the countries to develop their interest in the Gulf in a manner which will safeguard its long term value. Encouragement of individual state activities through educational, technical and financial assistance would be a useful vehicle for fostering harmonization of various programs such as emergency response, coastal zone management, and the creation of protected areas. Employment of the polluter pays principle can provide a sensible and substantial source of regional income, as illustrated by Israeli incorporation of this mechanism.

It has been suggested elsewhere that following the example provided by MAP, coordination of Gulf-wide environmental protection efforts would be best served by a Gulf of Aqaba Regional Activity Center (Marinov &Sandler, 1995).[23] The Center would be responsible, among other things, for the creation of a scientific agenda for the Gulf, including, monitoring, research and information sharing. The Center would serve as a database for legislative efforts, act as the headquarters for regional marine pollution prevention and control strategies, and provide a consultative and educational resource for the entire area.

At the same time, the shortcomings of MAP should be carefully considered by the Gulf of Aqaba states. For example, the emphasis on cooperation within the MAP framework has tended, until recently, to shift the focus away from Contracting Party compliance with their legal obligations. As a result, the states have amended the Convention and Protocols in order to emphasize the need for implementation and to integrate developments which have occurred in the region and internationally.[24]

Drawing upon this experience, the Gulf of Aqaba states may find it useful to address issues of compliance and enforcement at the outset. Regional monitoring of the Gulf can provide the foundation for measuring and encouraging compliance. Public assessment of monitoring results and community involvement in Gulf-wide activities will help safeguard the reef and beach areas as well as focus national attention on the need for continuous vigilance. While enforcement policies are to be determined by the individual states, regional strategies in the event of non-compliance can encourage state action. Arbitration, mediation and resolution of disputes in a non-litigious manner would probably be the most likely to succeed in the Gulf of Aqaba.

THE INTERNATIONAL CONVENTION FOR THE PREVENTION OF POLLUTION FROM SHIPS

The MARPOL 1973/78 Convention is recognized the world over as the most important international agreement dealing with vessel-based pollution. MARPOL 1973/78 has been ratified by many states, including Egypt and Israel; the Contracting Parties to this Convention comprise ownership of 80–85% of the world's shipping tonnage. The Mediterranean states, for example, have chosen not to prepare a separate Protocol on vessel-based pollution in the Mediterranean, but instead to defer to the rules of MARPOL 1973/78. Indeed, all the MAP countries, with the exception of Albania, are Contracting Parties to MARPOL 1973/78.

MARPOL 1973/78 regulates pollution by oil, chemicals, other harmful substances, garbage and sewage in five specific and detailed annexes. MARPOL 1973/78 has designated the Red Sea as a special area in which discharges of oil are completely banned and the establishment of reception facilities by port states are required. Flag states must ensure that their ships comply with very precise environment construction standards. In addition, MARPOL 1973/78 sets forth a comprehensive and thorough enforcement scheme, based on cooperative efforts and communication between the flag, coastal and port states.

Given the potential benefits from this Convention, it is unfortunate that Jordan, the leading Gulf of Aqaba state in shipping and port activity, is not a Contracting Party. Jordan's participation in the treaty regime would be an important step forward in achieving harmonization of standards in the region and in preventing pollution in the Gulf from operational as well as accidental spills of oil and other substances.

REGIONAL CONVENTION FOR THE CONSERVATION OF THE RED SEA AND OF THE GULF OF ADEN, 1982

This regional agreement was modeled after the Barcelona Convention and aims at balancing environmental conservation with economic development interests. The Convention calls for the prevention, abatement and combating of pollution of the sea as a result of accidental or intentional discharges from ships, dumping of wastes from ships and aircraft, land-based sources and other human activities. Its Protocol Concerning Regional Cooperation in Cases of Emergency prescribes procedures for the coordination of national plans in the event of an emergency spill.

Israel is the only Gulf-bordering state that is not a Contracting Party to this Convention and despite the peace treaties with Jordan and Egypt, Israel has not yet been invited to join. Nonetheless, it is instructive to examine the treaty in order to ascertain the other countries approaches to the protection of the Red Sea. In particular, the Saudi Arabian position is worth noting since this influential coastal power has not entered into the present regional discussions regarding the protection of the Gulf of Aqaba.[25] The standards to which the Contracting Parties to the Jeddah Convention have concurred can provide an important starting point for agreement on the Gulf of Aqaba.[26]

Conclusions

While the flurry of attention devoted to the Gulf of Aqaba is encouraging, it is also important to bear in mind the formidable task ahead. The success of any regional project lies in each nation's commitment to strengthening its own environmental management program. Issues of implementation, must be addressed by both national institutions and regional authorities. Educational programs, monitoring networks, coastal zone planning with the use of environmental impact statements are some of the tools which can be applied across the region in order to facilitate constructive and sustainable development.

Consideration of established models of cooperation which promote a regional approach towards environmental protection should be examined closely. Of particular relevance is the Mediterranean Action Plan, which has established a comprehensive and wide-ranging regime for the eighteen politically and culturally diverse states of the Mediterranean. At the same time, any comparisons must take into account the differences in the respective programs. In particular, the unique features of the Gulf and the individual challenges facing each of the countries must be addressed. Regarding Jordan, for example, air pollution from the loading of phosphates is still a serious concern. Likewise, there is a question as to how much of Aqaba's sewage is diverted into the sea, despite official policy not to discharge municipal wastes. And finally, as the busiest port state in the Gulf, Jordan should be encouraged to become a Contracting Party to the MARPOL 1973/78 Convention.

Concerning the Israeli coast, a major source of contention was eased when the Municipality of Eilat stopped discharging its sewage into the sea. Coastal zone management remains an issue, although strict environmental regulations seek to regulate coastal development. The largest threat to the Gulf from Eilat is the Eilat-Ashkelon pipeline, which necessitates the continuous loading and unloading of oil. Now that peace has been established with Egypt and Jordan, it would be wise to consider declaring the Gulf of Aqaba a 'tanker free zone' and provide for the shipment of oil to Israel through the Suez canal, or overland across Egypt or Jordan, depending upon the point of departure.

And finally, while Egypt does not yet have major municipal problems along the Aqaba coastline, it may soon face such difficulties unless it incorporates environmental considerations into its development plans. Unchecked and unregulated growth will surely spoil the beautiful and rare Sinai coastline. Implementation of environmental impact assessments and vigilant supervision of protected areas is essential in order to protect the area and the sea.

Notes

1. The Jordanian coastline is 27 km long, the Israeli coast 14 km, the Saudi Arabian coast 130 km and the Egyptian coast 200 km, tracing the Sinai peninsula (see Figure 1).
2. See *Irrigation Support Project for Asia and the Near East* for a brief overview of Saudi Arabian Coastal Development (US Agency for International Development, 1992)
3. Alawneh, 1993:98. This article is the source of statistical data presented in this section.
4. See below regarding the agreement between Egypt, Jordan and Israel concerning oil spills in the Gulf of Aqaba.
5. Jordan is however a Contracting Party to the 1954 International Convention for the Prevention of Pollution of the Sea by Oil (1954). This convention preceded MARPOL 1973/78 and also designated the Red Sea as a special area, prohibiting oil discharges and requiring the construction of reception facilities for incoming ships to deposit their oily wastes.
6. Irrigation Support Development Project for Asia and the Near East, (US Agency for International Development, 1992). It was also reported that the municipality had drawn up plans to connect premises to the system within the next five years, if the funds were available.
7. Under Israeli occupation of the Sinai, for example, development of the entire coastline was strictly regulated and for the most part, forbidden.
8. Sultan, 1993:178. This section relies heavily on information provided in the article by Sultan.
9. In 1995 there were reportedly 8,000 hotel rooms in Eilat.
10. Interview with Erik Adler, Division Chief, Marine and Coastal Environment Division, Israel Ministry of the Environment, August 1, 1995, by telephone.
11. See Licensing of Business Law, 5728, 1968. This law provides the Ministry of the Environment officials with powerful mechanisms for preventing and regulating certain types of pollution. For instance, if the environmental conditions are not met by the mariculture station operators, the station will not be allowed to function (Adler Interview).
12. This law replaces the old Nature Reserve Law of 1963. See also the Wild Animals Protection Law, 5716, 1955.
13. As early as 1973, the Government of Israel established an environmental agency, the Environmental Protection Service (EPS). The EPS was comprised of an interdisciplinary group of professionals who served in an advisory capacity. In the years that followed, it became clear that a central environmental authority was necessary to deal with the growing environmental issues in the country. Prior to the creation of the Ministry of the Environment authority over environmental matters was shared by several ministries, sometimes rather haphazardly, and often resulting in confusion and duplication. (Sandler, 1993:71).

14. Although it is important to note that there are still essential environmental matters which remain under the jurisdiction of other ministries. For example, the Ministry of the Interior has authority over the building and planning process and the Ministry of Transport is responsible for regulating ships. Nonetheless, on environmental matters these ministries are obliged by law to work together with the Ministry of the Environment on the planning and implementation of policy.
15. The station has, for example, a speed boat, skimmers, pumps, booms, absorption materials, oil separators and containers (Sandler, 1993:74).
16. The Treaty was signed on October 26, 1994.
17. Interview with Esther Korn-Ancri, Foreign Affairs Coordinator, Israel Ministry of the Environment, August 2, 1995 by telephone.
18. The Jordanian center is to be funded by Japan; the Egyptian center by the European Community; and the Israeli center will be upgraded to an even higher standard.
19. There is an Emergency Response Steering Committee which meets twice a year to oversee this project. (Adler interview).
20. The other members of MAP include Albania, Cyprus, France, Greece, Italy, Malta, Monaco, Spain, Turkey and former Yugoslavia.
21. An exception to this is ship-based pollution, which is regulated by the MARPOL Convention.
22. For an analysis of the impact of epistemic communities and their impact on the environmental protection efforts of the Mediterranean, see Haas, 1990.
23. In Marinov and Sandler (1995), the authors propose the creation of a Middle East Environmental Action Plan (MEEAP) in order to oversee the establishment and implementation of regional environmental policy. MEEAP is similar in concept to MAP, however adapted to meet the specific environmental, economic and political conditions of the Middle East. The authors propose the establishment of regional activity centers to address specific issues. In addition to a Gulf of Aqaba Regional Center, centers have been proposed for the protection and rehabilitation of the Jordan River. These centers would be coordinated by a Secretariat for Sustainable Development—a regional headquarters for environmental protection efforts, modeled after the MAP headquarters in Athens. The amendments to these instruments can be found in UNEP (OCA)/MED IG.5/16, 8 June 1995.
24. While the Saudis were involved in the multilateral talks on the Middle East, including those on the Gulf of Aqaba, they are not involved in the planning process which is taking place between Jordan, Israel and Egypt.
25. Alternatively, in the event that Israel does become a Contracting Party to the Jeddah Convention, it could provide the basis for comprehensive protection efforts of the Gulf.

References

Alawneh, Z.J. (1993) Jordanian environmental laws, institutions and treaties affecting the Gulf of Aqaba. *In* D. Sandler *et al.* (eds.) *Protecting the Gulf of Aqaba: A Regional Environmental Challenge.* ELI, Washington, D.C., pp. 97–116.

Al-Khoshman, M.A. (1993) Oil spill prevention, cleanup, and emergency response in the Gulf of Aqaba: A technical analysis. *In* D. Sandler *et al.* (eds.) *Protecting the Gulf of Aqaba: A Regional Environmental Challenge.* ELI, Washington, D.C., pp. 217–237.

Barcelona Convention for the Protection of the Mediterranean Sea Against Pollution, February 16, 1976, 15 I.L.M, p. 306.

Brachya, V. (1984) Integrated Planning and Management of Coastal Zones—Israeli Report to RAP/PAP, United Nations Environment Program (UNEP), 1984.

Brachya, V. and U. Marinov (1993) Coastal zone management in Israel and prospects for regional cooperation. *In* D. Sandler *et al.* (eds.) *Protecting the Gulf of Aqaba: A Regional Environmental Challenge.* ELI, Washington, D.C., pp. 195–211.

Haas, P. (1990) *Saving the Mediterranean.* Columbia University Press, New York.

ISPAN Irrigation Support Project for Asia and the Near East (1992) *Gulf of Aqaba Environmental Data Survey.* U.S. Agency for International Development, Washington, D.C.

Maranz, F. (1993) Bridging the Gulf. *The Jerusalem Report,* April 22, p. 28.

Marinov, U. and D. Sandler (1995) A proposed model for environmental protection in the Middle East. *In* S. Spiegel and D. Pervine (eds.) *Practical Peacemaking in the Middle East Vol. II: Environment, Water, Refugees, and Economic Cooperation and Development.* Garland, New York.

MARPOL 1973/78 International Convention for the Prevention of Pollution from Ships (12 I.L.M. 1319) and the Protocol of 1978 Relating to the International Convention for the Prevention of Pollution from Ships (17 I.L.M. 546).

Ministry of the Environment, Israel (1992) *Annual Report, 1992* (Hebrew).

Ministry of the Environment, Israel (1994) *Israel Environment Bulletin* 17(3), Autumn, 5755, 1994.

Ministry of the Environment, Israel (1995) *Israel Environment Bulletin* 18(1), Winter, 5755, 1995.

Ministry of the Environment, Israel (1992) National Masterplan for the Mediterranean Coast, TMA 13, Part C (Hebrew).

Ministry of the Interior, Israel (1981) National Masterplan for the Mediterranean Coast (Hebrew).

Raftopoulos, E. (1993) The Mediterranean Action Plan: Appraisal of a model for regional cooperation. *In* D. Sandler *et al.* (eds.) *Protecting the Gulf of Aqaba: A Regional Environmental Challenge.* ELI, Washington, D.C., pp. 315–345.

Sandler, D. (1993) Environmental law and policy for the Gulf of Aqaba: An Israeli perspective. *In* D. Sandler *et al.* (eds.) *Protecting the Gulf of Aqaba: A Regional Environmental Challenge.* ELI, Washington, D.C., pp. 69–96.

Sultan, F. (1993) Tourism development along the Aqaba coast: An Egyptian perspective. *In* D. Sandler *et al.* (eds.) *Protecting the Gulf of Aqaba: A Regional Environmental Challenge.* ELI, Washington, D.C., pp. 177–194.

Tal, A. (1993) Preventing pollution from ports and maritime Activity in the Gulf of Aqaba: Current practices in Israel. *In* D. Sandler *et al.* (eds.) *Protecting the Gulf of Aqaba: A Regional Environmental Challenge.* ELI, Washington, D.C., pp. 259–282.

UNEP (1984) Integrated planning and management of Mediterranean Coastal zones MAP/PAP, UNEP.

UNEP (1988) The blue plan: futures of the Mediterranean Basin Environment/Development 2000–2025, executive summary and suggestions for action, UNEP, BP/RAC.

Wahbeh, M. (1993) An agenda for scientific research. *In* D. Sandler *et al.* (eds.) *Protecting the Gulf of Aqaba: A Regional Environmental Challenge.* ELI, Washington, D.C., pp. 25–40.

ABOUT THE AUTHOR

Deborah Sandler is currently a doctoral candidate in international law and a Junior Research Fellow at Wolfson College, Oxford, England. A member of the Israel and California bar associations, she has specialized in the practice of international environmental law. She has written on various related topics and co-edited *Protecting the Gulf of Aqaba: A Regional Environmental Challenge.* Ms. Sandler is a member of Kibbutz Ein Gedi, Israel.

Deborah Sandler
University of Oxford
Wolfson College
Oxford OX2 6UD
England

Part Three

Desertification and Climatic Variability in Arid Environments

Chapter 5

Bioclimatic Aridity, Climatic Variability, Drought and Desertification: Definitions and Management Options

Hendrik J. Bruins and Pedro R. Berliner

Introduction

Bioclimatic aridity is the principal constraint for habitation and development of arid regions. Moreover, climatic variability further increases the vulnerability of human activities in drylands, as rainfall variations are large and unpredictable from year to year, negatively affecting food and water security. When the rains fail, drought gradually develops as a natural part of climatic variability. Drought is the most ominous, natural dryland hazard, which can develop in a disaster. Desertification is defined (UNCED, 1992) as land degradation in drylands resulting from various factors, including climatic variations and human activities.

Drylands are usually classified on the basis of bioclimatic aridity. Four classes are distinguished: hyper-arid, arid, semi-arid and sub-humid zones (UNESCO, 1979). Rainfed agriculture, based on direct precipitation falling on the field, is generally possible in the sub-humid and semi-arid zones, albeit with risks due to drought and rainfall variability. The arid and hyper-arid zones, being too dry for rainfed agriculture, are often used for livestock raising. Cattle requires pasture in the wetter parts of the arid or semi-arid zone, while goats and camels are also suited for the hyper-arid zone.

Humankind may partly overcome the constraints of aridity through irrigation development or intensified resource management such as rainwater harvesting. The spatial position of a climatic arid frontier may move gradually, for better or worse, through climatic change, which may go unnoticed on a human timescale. Societies living in drought-prone areas have usually developed traditional coping mechanisms, partly based on the formation of reserves and shared risk distribution through social relations and networks.

H. J. Bruins and H. Lithwick (eds.), The Arid Frontier, 97–116.

Both developing and developed countries are vulnerable to the impact of drought. The severe drought in the Sahel in the 1970s and 1980s (Warren, Chapter 6) had a considerable impact on environment and society through complex biological, socio-economic and political mechanisms. Large portions of the United States of America are known to be sensitive to drought, which frequently results in costs and losses exceeding $1 billion (Wilhite & Hayes, Chapter 2; Roos, Chapter 14). Although drought is a recurring, natural part of climatic variability, we cannot predict in time and space when and where drought will strike. However, its frequency of occurrence and possible impact in a given area can be studied. Such knowledge has to be translated into action through proactive planning and interactive management (Bruins & Lithwick, Chapter 1).

The dramatic consequences of severe multi-annual drought in the Sahel, which began in 1968, led to the first United Nations Conference on Desertification (UNCOD), held in Nairobi in 1977, where a Plan of Action to Combat Desertification (PACD) was adopted. Combating, controlling and preventing desertification is, therefore, very much related to landscape stabilization and improvement. Water and soil conservation by means of rainwater harvesting, for example, are capable of doing just that, if related correctly to society in technical, socio-economic, land-right and political terms. Management options for irrigation development are presented in detail by Shanan (Chapter 13). Options for dryland management are highlighted in Figure 1, with particular focus on intensified resource management of agro-pastoral systems.

In this chapter we make specific recommendations for a standardized system to express bioclimatic aridity uniformly on a world-wide basis. We also advocate the inclusion of the hyper-arid zone in the desertification concept. Options to increase preparedness for the typical dryland hazards of drought and desertification, through disaster planning and management, must include designs for environmental refugee camps with built-in elements of self-sufficiency, productivity, and education aimed at development, rehabilitation and resettlement.

Defining Bioclimatic Aridity and Boundaries between Arid Zones

Several scientific systems to define dry climatic areas have been developed this century. Botanists carried out pioneering research to study the various vegetation zones on Earth, which are invariably related to climate. They had already accomplished world-wide vegetation surveys before an increasing number of meteorological stations began to deliver their data. Hence global vegetation maps predate world climate maps. The former provided in fact the basis for the latter (Heathcote, 1983). It was Wladimir Köppen (1931), initially trained as a botanist, who developed a well-known global climatic classification system. He defined two arid zones: the Steppe

(BS) and Desert (BW). Further subdivisions, such as BWk'w, indicate the average annual temperature (k' means <18°C) and rainfall distribution (w means >70% of the rain falls in the six summer months). However, the fact that climatic zones are represented by such letter codes, as the above example, largely based on temperature and rainfall, but lacking data on evaporation and transpiration, makes it difficult to define and compare aridity around the globe in a numerical way. However, Köppen had not much choice, as detailed meteorological data on a world-wide basis were not available in his time. His method defined 12.0% of the land surface of the Earth as desert (BW) and 14.3% as steppe (BS), giving a total of 26% drylands.

Thornthwaite (1948) developed a climate classification system based on moisture input (Precipitation) and the potential moisture loss to the atmosphere from both abiotic evaporation and plant transpiration combined (Potential Evapotranspiration), which is indeed the most straightforward and physically correct way to define climate anywhere around the globe numerically in terms of a Moisture Index. However, the calculation of potential evapotranspiration is complicated, as several approaches are possible, some more practical based on few available data, and others more precise, but difficult to implement due to lack of sufficient data on a world-wide basis. Thornthwaite (1948) provided a formula to calculate potential evapotranspiration on the basis of latitude and temperature data alone. This was a considerable break-through at the time, providing the basis for his climatic classification system, which differentiates the drylands between Arid (15.3%) and Semi-arid (15.2%), giving a total area of 31% drylands around the globe.

Peveril Meigs (1953) compiled the first global map of arid lands, using Thornthwaite's method to calculate the potential evapotranspiration. The moisture index of Meigs (1953) differentiates, in his map, between Extremely Arid (-57 and below), Arid (-40 to -56), and Semi-arid (-20 to -40). These three arid zones cover 4.3%, 16.2% and 15.8% of land, respectively, giving a total area of 36% drylands on Earth. Meigs excluded from his map areas too cold for crop growth, based on Köppen's criteria. The extremely arid zone must, on his definition, also be characterized by years without any rainfall and by the lack of seasonal precipitation. The latter criteria unnecessarily complicate the differentiation between arid and extremely arid regions, which is a weak point in an otherwise well established system.

UNESCO (1979) produced a map of the world distribution of arid regions, in the framework of the Programme on Man and the Biosphere (MAB). This map is based on a better and more straightforward approach than the Thornthwaite-Meigs methodology. The moisture index is based on the P/ETP ratio (average annual precipitation, divided by average annual potential evapotranspiration). It is explained that the P/ETP ratio is to be preferred to a P minus ETP subtraction. The latter method might give a figure of -50, which can result from 200 mm (P) - 250 mm (ETP), but also from 800)mm - 850 mm, just indicating potential lack of annual moisture in quite different climatic regions. The ratio P/ETP expresses the degree of aridity better in arid and semi-arid regions, "because it gives the same value for all climates in which the potential water loss is proportionally the same in relation to rainfall. Aridity increases as values of this ratio decline. In addition, this ratio is biologically accurate in climates with highly contrasted seasons, since it represents well the ratio ETR/ETM (ETR = real evapotranspiration of a soil-plant system; ETM = maximum evapotranspiration in the absence of a water constraint), which largely determines vegetative dry-matter production. In these areas, annual precipitation (P) is a satisfactory proxy for ETR, and potential evapotranspiration (ETP) is very close to ETM, being its upper limit" (UNESCO, 1979).

The P/ETP ratios for the map were calculated by D. Henning of the Meteorological Institute of the University of Bonn, using Penman's formula for the calculation of ETP, which requires data on solar radiation, atmospheric humidity and wind. "The great advantage of Penman's formula is that it has been used in numerous biological and physical studies of climate, the results of which have been widely diffused. In addition, it is today considered more satisfactory than the formula used in the Meigs map" (UNESCO, 1979). The Penman method is indeed widely used to determine crop water requirements (Doorenbos & Pruitt, 1984), essential in agricultural planning and management.

Four main zones of aridity were delimited by UNESCO (1979), presented in Table 1 of Chapter 1 (Bruins & Lithwick, this volume). The Sub-humid zone, not shown on the Meigs (1953) map, is included, "because desertification as a result of soil and vegetation degradation also occurs in this zone" (UNESCO, 1979:11). There is another important reason to include the sub-humid zone: its sensitivity to recurring drought years, in which the degree of aridity increases dramatically, causing severe problems to grazing systems and rainfed agriculture. The hyper-arid zone covers 5.8 % of the land surface of the Earth, the arid zone 13.7 % and the semi-arid zone 13.3 %. This gives a global dryland area of 33%, excluding the sub-humid zone.

The most recent classification of arid zones was published by the United Nations Environment Programme (UNEP, 1992), as a basis for the world atlas of desertification. The map excludes the hyper-arid zone, because it was not included in the latest definition of desertification. The UNEP (1992) approach is a step backward, because in their methodology to determine the P/ETP ratio's, potential evapotranspiration (ETP) calculations are based on the Thornthwaite (1948) method, *which is less accurate* than the Penman (1948) method. The reason given by UNEP (1992) is pragmatic: a lack of detailed data required for ETP calculations with Penman's method on a world-wide basis. The UNEP map of arid regions, therefore, cannot be directly compared with the UNESCO (1979) map, as the latter is based on the Penman potential evapotranspiration method.

Moreover, the UNEP (1992) map changed the numerical boundaries between some arid zones, based on the P/ETP ratio, for the following reasons: "The

Thornthwaite method is known systematically to underestimate PET (= ETP) for dry conditions and to overestimate values for moist and cold environments. Consequently an empirical adjustment factor was ... applied to the data to bring the values closely in line with those of the Penman method" (UNEP, 1992:3). The P/ETP boundary in the UNEP map between the *hyper-arid* and *arid* zone is, therefore, changed from 0.03 (in the UNESCO map) to 0.05. Likewise the upper boundary of the *sub-humid* zone is changed from 0.75 (in the UNESCO map) to 0.65. The other boundaries have not been changed, as shown in Table 1.

Table 1. Comparison between the UNESCO(1979) and UNEP(1992) definitions for bioclimatic aridity and the boundaries between the various arid and sub-humid zones

Climatic zone	*P/ETP Ratio (Penman method) UNESCO (1979)*	*P/ETP Ratio (Thornthwaite method) UNEP (1992)*	*Interannual rainfall variability*
Hyper-arid	< 0.03	< 0.05	~ 100%
Arid	0.03–0.20	0.05–0.20	50-100%
Semi-arid	0.20–0.50	0.20 –0.50	25–50-%
Sub-humid	0.50–0.75	0.50–0.65	< 25%

We clearly prefer to retain the boundaries as established by UNESCO (1979), basing P/ETP ratio's on Penman-related methods only. Both UNESCO (1979) and UNEP (1992) agree on the usefulness of the P/ETP ratio to express the degree of dryness. Precipitation (P) data are usually not a problem, but the estimation of water losses (ETP) constitutes the main difficulty. UNESCO (1979) decided to use the sophisticated Penman (1948) approach, while UNEP (1992) opted for the simplified Thornthwaite (1948) method, as mentioned above. The approach by Penman is physically sound, as he used both turbulent transfer and energy balance in the development of his formula. Thornthwaite, however, derived an empirical relation between temperature and ETP (Sellers, 1965).

The equation of Penman (1948) defines the *Potential Evaporation* from an open body of water (evaporation pan). A large number of experimental trials showed a close correlation between the evaporation rates determined with Penman's formula and those obtained when crops were grown under conditions in which water was not a limiting factor for plant growth. Penman's formula has therefore been used for estimating *Potential Evapotranspiration*. The practical disadvantage of Penman's sophisticated formula is that data on solar radiation, temperature, relative humidity and wind speed are needed. The establishment of a rather dense network of meteorological stations throughout the world during the last decades makes it possible today to obtain sufficiently long records for all parameters except radiation. This problem led several authors to propose formulae in which radiation does not explicitly appear. However, the role of solar radiation is crucial in water evaporation. Hence the lack of a radiative term in a formula implies that some other parameter (usually temperature) is correlated with solar radiation. Such site specific correlations have been carried out for annual periods (Budyko, 1956), but the correlations tend to break down for shorter periods.

It is preferable, therefore, to approximate the monthly radiation balance averages on the ground from known extraterrestrial radiation fluxes, cloudiness, temperature and relative humidity, even if some assumptions have to be made. The assumptions can always be tested against at least one ground station in which solar radiation is measured (Brutsaert, 1982). Progress in remote sensing allow us to estimate reflected short wave radiation from satellite data. If the albedo of the area under consideration is known, or estimated with the aid of remotely sensed data, the incoming solar radiation can be estimated for large areas. This approach will make radiation data available in the near future for remote locations and therefore only equations of the Penman type should be used.

However, for establishing a standardized and accurate system to express ETP on a world-wide basis, we recommend the use of Class-A evaporation pans [1]. It appears that data from Class-A evaporation pans are an accurate proxy for Penman's ETP. A very good correlation between both Penman formula calculations, based on direct measurements of all the metorological parameters, and simultaneous Class-A evaporation data, was established for several stations in the Sonoran Desert (Berliner *et al.*, in preparation). The correlation is linear with a slope being close to one. Class-A evaporation pan data are widely available from all over the world. We, therefore, suggest to base globally comparable P/ETP ratio calculations in the future on Class-A evaporation pan data as a standardized system to define bioclimatic aridity throughout the world.

Climatic Variability and Drought

Besides the general constraint of aridity, drylands are characterized by significant climatic variability, particularly in terms of precipitation. Interannual and intra-annual rainfall variations in sub-humid and semi-arid zones constitute a great predicament for farmers (Dietz, Put & Subbiah, Chapter 8), who have to manage their affairs in the face of irregularly recurring droughts as well as floods. Rainfall variability is even larger in arid and hyper-arid regions, causing problems for pastoralists, who traditionally responded through flexible management and mobility, *i.e.* nomadism. Drought is a recurring, natural part of climatic variability in drylands. *Drought can be defined in spatial and temporal terms as a dry period characterized by significantly less precipitation than the average amount.* The time period and spatial extent covered by the average amount must be indicated of course.

Such a general definition of drought is necessarily vague and non-specific. It is, therefore, not very helpful for policy makers, planners and managers. Wilhite & Glantz (1985) correctly recommend to define drought in specific physical and social terms, which can only be done for a particular region. Dietz, Put & Subbiah (Chapter 8) make such a specific drought risk assessment for a semi-arid part of Andhra Pradesh in India. It is obvious that drought affects rainfed agricultural crops and pastures negatively in a rather direct way, but its indirect impact on farmers, pastoralists, related industries and society is much more complex. It is possible to differentiate droughts according to causes or disciplinary perspectives, such as meteorological, agricultural, hydrological, and socio-economic drought (Wilhite & Glantz, 1985), to which pastural drought and man-made calamity drought may be added (Bruins, 1996).

If the climatic average itself is changing, one may speak of climatic change. However, the time period chosen for the average, *i.e.* 30 years, 1000 years or much more, may also determine the semantic switch from climatic variability to climatic change. The continuation of the biosphere on planet Earth during the Phanerozoic, following the explosion of biodiversity in the Cambrium period, is not just remarkable, it is extraordinary. The biosphere was able to continue because *temperature variability was relatively minor* during such a long stretch of time. Planet Earth neither entered into a state of complete glaciation nor into excessive heat. The probability of this long existence of the biosphere is extremely low from a climatic-astronomic viewpoint, according to Budyko, Golitsyn & Izrael (1988).

Natural climatic changes caused deserts to expand and contract repeatedly, while grazing animals followed in a planetary transhumance. Grazing must have been much more attractive in parts of the Sahara and the Negev some 5000 years ago during a considerably wetter climatic period (Nicholson & Flohn, 1980; Lamb, 1982; Bruins, 1994; Issar *et al.*, 1995).

Making the time-scale smaller again helps us to appreciate possible warming in the next century, as a result of human-made increase of so-called greenhouse gases like carbon-dioxide (CO_2). This aspect of climate change and its effect on regional climate is being investigated and debated in many circles. The greenhouse effect itself is of course not human-made. Part of the infra-red radiation emitted from the Earth's surface is absorbed by the greenhouse gases (water vapour, carbon dioxide, chloro-fluorocarbons, methane, nitrouse oxide and others), producing an approximate temperature increase of 35°C (Williams & Balling, 1996). Without this effect, the Earth, long ago, would have become a white, glaciated planet without a biosphere, having an average planetary temperature of about -20°C.

Following the Industrial Revolution, humankind is burning fossil fuels in large amounts, which caused a rise in CO_2 from approximately 280 ppm (parts per million) in 1800 to about 350 ppm by 1990. Climatic models have been developed to study the effect of 600 ppm CO_2 on temperature and other climatic parameters. There is general agreement that the projected temperature increase, estimated to range in the order of 1.6°C to 5°C, will be least in the tropics and most in the high latitudes. The effect on rainfall and drought patterns in the arid zones is complicated, as the models are generally considered uncertain at regional geographic scales (Williams and Balling, 1996). Possible effects of a warmer climate on deserts are discussed by Goudie (1994).

Desertification

The severe drought in the Sahel and other parts of Africa, that began in 1968 and continued with fluctuating intensity through much of the 1970s and 1980s, became an environmental landmark in global human perception. It was the main factor that led to the United Nations Conference on Desertification (UNCOD), held in Nairobi in 1977, where a Plan of Action to Combat Desertification (PACD) was adopted. The international institutional development of desertification during the 1970s and 1980s is discussed in detail by Verstraete (1986) and by Thomas & Middleton (1994). These activities continued in the 1990s. The United Nations Conference on Environment and Development (UNCED), held in 1992 at Rio de Janeiro, called on the United Nations General Assembly to establish an Intergovernmental Negotiating Committee (INCD) to prepare a Convention to Combat Desertification in Those Countries Experiencing Serious Drought and/or Desertification, Particularly in Africa. The General Assembly responded positively and in December 1992 agreed by adopting Resolution 47/188. Subsequently, the United Nations Convention to Combat Desertification (UNCCD) was adopted in Paris on June 17, 1994, and opened for signature by individual nations from October 14, 1994. The convention would enter into force 90 days after its ratification by 50 countries, which occurred on December 26, 1996. The Convention adopted the 1992 UNCED definition of desertification as land degradation in arid, semi-arid and dry sub-humid areas resulting from various factors, including climatic variations and human activities (UNCCD, 1995).

It is wrong, in our opinion, to exclude the hyper-arid zone from the desertification concept, because landscape degradation is not limited to the sub-humid, semi-arid and arid zone, but also occurs in the hyper-arid zone. The exclusion of the hyper-arid zone possibly stems from a semantic misconception: desertification will not make the landscape of the hyper-arid zone more desert-like, as it already represents the "inner" desert. This logic does not hold, because the arid zone also contains genuine deserts in many parts of the world. If the arid zone is embraced in the desertification concept, then the hyper-arid zone must be included as well. Living in the Negev desert close to the hyper-arid boundary, we notice that landscape degradation can occur in both the hyper-arid southern Negev and the arid northern Negev. Moreover, the term desertification is a useful and established term to signify land degradation in drylands, including the sub-humid zone. Taking the term desertification too literally at face value is wrong, because typical desert land forms in the geomorphic sense do not usually form in sub-humid or semi-arid zones as a result of landscape degradation.

The remainder of this section on desertification is partly based on an encyclopedia entry by the first author (Bruins, 1997, in press).

The phenomenon of desertification is probably as old as the earth and its climatic fluctuations. Geologists, geographers and paleoclimatologists are aware of both expansion and contraction of deserts in the past, due to climatic changes. The involvement of humankind is more recent in relative terms, yet dating back to antiquity. Ancient writers were aware that human beings can be responsible for landscape degradation, leading to more desert-like or desolate conditions. The Bible contains several portions describing land degradation related to human activities, for example in Jeremiah 12:10-11: "Many shepherds will ruin my vineyard and trample down my field; they will turn my pleasant field into a desolate desert. It will be made a wasteland, parched and desolate before me; the whole land will be laid waste because there is no one who cares". The classical author Columella discussed agricultural degradation in ancient Italy during the first century AD in his treatise De Re Rustica (I.1.18ff). He emphasized human factors, like carelessness, vis-à-vis natural factors concerning "the bad state of agriculture in contemporary Italy. But he refuses to side with the pessimists who blame the climate or the exhaustion of the soil instead of the carelessness, ignorance and lack of skill of those who till it" (White, 1970:34).

This tension of who is to blame, climate, humankind, or both, is still very much debated in the modern perception and interpretation of the concept of desertification. Land degradation in the Mediterranean lands during historical times was viewed by Huntington (1924) as the result of adverse climatic changes, in the framework of his hypothesis of climatic determinism. The human factor as a cause for desertification (as well as the reverse process of land reclamation) was eloquently emphasized by Lowdermilk (1944:1): "By neglect, ignorance and suicidal agriculture, peoples have bequeathed to their descendants "man-made deserts" of sterile, rocky and gullied lands".

The term desertification was apparently first publicized by the Frenchman A. Aubréville in his monograph *Climats, Forêts et Désertification de l'Afrique Tropicale*, published in Paris in 1949. Phrases like man-made deserts (Lowdermilk, 1935; Stebbing, 1938) and desert encroachment (Klintworth, 1948) were already in use earlier this century. Aubréville, who was Inspecteur Général des Eaux et Forêts des Colonies, distinguished two major causes of desertification:

(1) Desertification through destruction or degradation of the soil.
(2) Desertification through the action of climatic factors.

In relation to the first cause, he considered man the main culprit for the degradation of African forests into savannah. Thus he used the phrase savannization as a form of desertification. Aubréville also included climatic variations as a cause for desertification. His expression "désertification climatique" (p. 140) is quite clear in this respect. Aubréville coined the term desertification to describe deterioration of vegetation and soil conditions in a rather wide sense, not just restricted to arid zones.

Documented examples of human influence on desertification in Near Eastern history are presented by Reifenberg (1955), in his classical work "*The Struggle Between the Desert and the Sown*", as well as by Whyte (1961). Climatic changes influenced valley changes and desertification in northern Sinai (Bruins, 1990; Tsoar, 1995). A more restricted term 'desertization' has been suggested since 1959 by Henri N. Le Houérou to describe "the extension of desert landforms and landscapes to areas where they did not occur in the recent past" (Le Houérou, 1977:364). In his opinion, "The term desertification has been misused for decades to describe the degradation of vegetation and soil conditions in arid, semi-arid and even humid zones. In many cases these processes have nothing to do, scientifically, with the desert. The regression of tropical forest into savanna and that of Mediterranean forest into steppe are examples of this. Neither climatically, geomorphologically nor biologically can savanna and steppe be classified as deserts." However, as noted before, desertification has become a useful and established term to describe land degradation in drylands. It does not mean, neither was it proposed as such by Aubréville, to signify the formation of real desert landforms.

Many publications about desertification began to appear since the UNCOD conference: Glantz (1977), Dregne (1983), El-Baz and Hassan (1986), to name but a few. Almost 4000 annotated publications are listed in the World Desertification Bibliography (Summers et al., 1991). The concept and definitions of desertification were reviewed by Glantz and Orlovsky (1984), Verstraete (1986), Warren and Agnew (1987) and by Thomas and Middleton (1994). Land degradation may be irreversible in the short run, but reversible over a longer time period. This time factor was raised by Mainguet (1991; Chapter 7, this volume), who suggested using the term desertification only when land degradation cannot be reversed within one human generation. Scientific views of desertification usually comprise human causes on the one hand and climatic causes on the other. Both should indeed be included, as done in the UNCED (1992) conference and in the 1994 United Nations Convention to Combat Desertification (UNCCD). Basic human perception of land degradation has not really changed since the days of Columella.

Management Options

The chief principle in managing the constraints of bioclimatic aridity, climatic variability and drought is the formation of reserves during the wet periods for use during times of drought. Such reserves may consist of water, food grains, grazing animals, and money (Bruins, Evenari & Rogel, 1987). Pastoralists tend to build large herds during good years as a reserve for the lean years, in addition to flexible and mobile management systems (Dahl & Hjort, 1979; Sandford, 1983; Behnke, Scoones & Kerven, 1993). Traditional management is becoming more and more difficult in modern nation states, due to boundaries, changing land rights and decreasing size of grazing lands. In a secure money-based economy, farmers and pastoralists may form capital reserves for the lean years, while the government may also give drought subsidies in the form of financial compensation (Wilhite, 1986). Grain storage has long been an important strategy in proactive drought planning, as recorded for example in the book of Genesis. Joseph stored grains in ancient Egypt during the seven fat years as a reserve for the seven lean years (Schneider, 1976; Bruins, 1993).

Water reserves are of most direct importance for agriculture and pastoralism in day-to-day management. Dependable surface and groundwater reserves are used in irrigated agriculture, which has become very important, producing one third of all the food in the world. Irrigation and related environmental management are discussed in detail by Shanan (Chapter 13). The various aspects of large-scale drought and water management in California are presented by Roos (Chapter 14), while the costs and benefits of mega-water projects in India and China are analyzed by El-Bihbety & Lithwick (Chapter 15).

Water management with regard to rainfed agriculture, runoff agriculture and pastoralism must be aimed at storing the maximum or desirable amount of water during each rainfall event. Different strategies can be implemented to prevent losses to the atmosphere by direct evaporation, depending on the objectives, the infrastructure and available capital. Concentrating runoff water in large ponds to recharge artificial aquifers or storing the water in closed tanks are viable options to manage drinking water availability for both human beings and animals (Orev, 1986). However, this approach may be too expensive for crop production. Direct storage of runoff in the soil profile of agricultural and pastoral lands is usually cheaper. Runoff water is passively or actively guided to suitable fields which have check-dams to retain the water, allowing for its infiltration in the soil. No pumping devices are needed as the plant roots fulfill this function. The "buffering capacity" of the system is determined by the water-holding capacity of the soil and its depth. Rainwater-harvesting systems are described by Evenari, Shanan & Tadmor (1982), Bruins, Evenari & Nessler (1986), Orev (1986), Reij, Mulder & Begemann (1988), and Ben-Asher & Berliner (1994).

Concentrating runoff in so-called liman fields, partly surrounded by dikes, for agro-forestry and fuelwood production, is presented in Chapter 18 (Sauerhaft, Berliner & Thurow). If the soil is deep enough, large amounts of water can be stored. A crop with a deep root system is needed to retrieve the water. Trees can easily obtain water from a depth of several meters, while water uptake from a depth of 10-12 m is not uncommon (Adar *et al.*,1995). Trees can be intercropped with an annual crop in order to provide grains for human consumption (Lövenstein *et al.*, 1991). Competition for water, solar radiation and nutrients can be minimized and the overall productivity per unit water maximized by properly managing both communities, trees and intercrop. Runoff Agroforestry is a resilient system. It may not have the highest water use efficiency but production usually continues even under conditions of severe drought.

Arid zones, too dry for viable rainfed agriculture, are the classical grazing regions since time immemorial. However, traditional extensive pastoralism in arid zones is faced with profound difficulties and change in many parts of Africa and Asia. The area available for grazing and the range of pastoral movement have diminished in the present century, due to the establishment and development of nation states, increasing population pressure and agricultural expansion. The viability of extensive pastoralism has suffered, while the impact of drought has increased as a result. Disturbance of the traditional environmental equilibrium in rangeland management by various outside changes may lead to desertification (Meir and Tsoar, 1996). Extensive, subsistence pastoralism is often in a state of combined socio-economic and environmental crisis (Dietz, 1986). Its successful development and adaptation require both a participatory and functionalist approach. Perevolotsky (1995) presents examples that the effect of various grazing systems on the environment in the Near East is less than the effect of natural climatic fluctuations. This is a very important point in the debate on grazing, nature conservation and desertification.

Proper development of pastoralism in drylands (Orev, 1986) is by no means an easy matter. There is often a tendency toward reductionism, in research as well as in development projects, by which the pastoral system as a whole is reduced to the sum of individual parts. Development of the system may then be related to the development of a single item. However meaningful the single-part contribution may be, it usually does not succeed in developing the system as a whole. We certainly have to search for new elements that can be added to the system, in addition to the development of existing parts. The use of runoff water, soil fertility enhancement, solar energy, solar-powered water pumps, groundwater, brackish water, treated wastewater, duckweed, algae, food production and food reserves form important elements to be enhanced or added to the system. However, these elements need to be integrated into the entire system at the user level through participatory development. Thus a new, viable and mutually-adjusted system equilibrium for agro-pas-

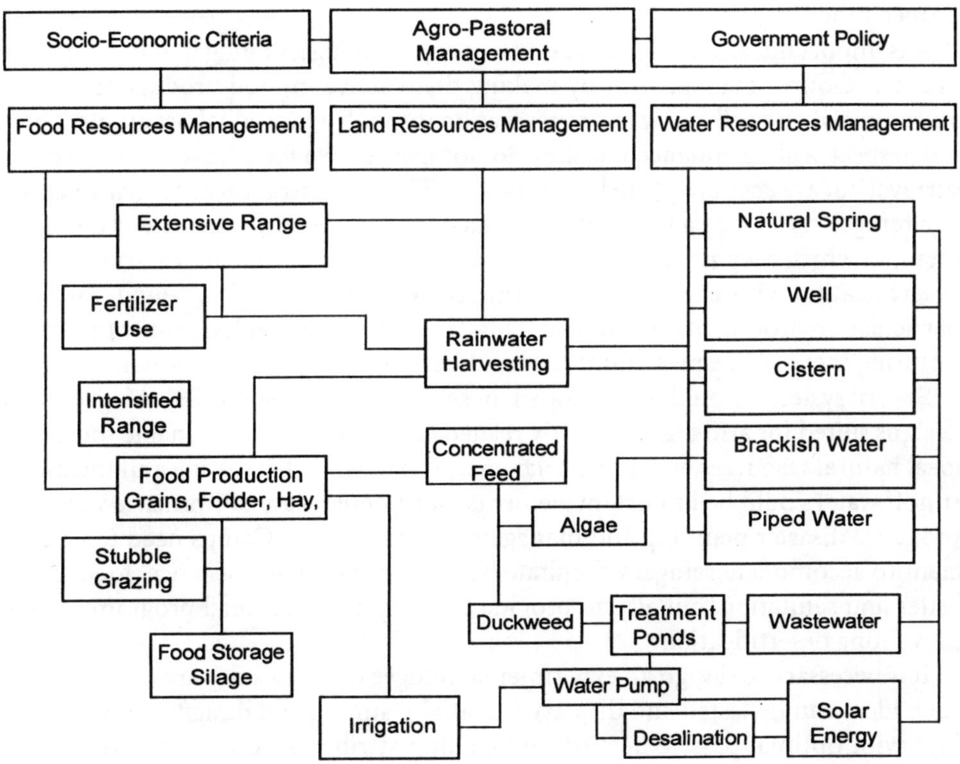

Figure 1. Development and management options for agro-pastoralism in arid zones.

toralism may be achieved through intensified resource management. Extensive range grazing increases plant biodiversity up to a certain limit of grazing pressure. Possible development relationships of existing as well as new constituent elements are displayed in Figure 1.

Dryland Disasters and Refugees

Drought and desertification constitute hazards for humankind living in arid regions which sometimes result in disasters. Management options of such dryland disasters have to include the absorption of refugees. There are various cases of migration in recent years in relation to drought and desertification, such as 10 million displaced persons in the Sahel in the 1980s (Warren, Chapter 6).

Although the reason for migration is often more complex than a single environmental factor, the term environmental refugees is sometimes used to distinguish them from political refugees. The concentration of large numbers of refugees in a certain place, whatever the reason for their migration, may also lead to environmental degradation and desertification, if their food and energy needs are not properly met.

It is important, also for social reasons, to engage refugees living in refugee camps in local resource management by making them active in food production (crops and animals) and soil and water conservation through educational programs. Their self-respect will be enhanced, if they do not just receive food hand-outs for mere survival but are encouraged to be productive. The concentration of so many people in a refugee camp may also be viewed as an opportunity for participatory development, in which they receive education about new options to manage drought and desertification with new possibilities (intermediate technologies) suited for their particular environment and culture. This can make a tremendous contribution to restoring their self-respect and fostering their rehabilitation or resettlement.

Special systems should be developed for self-sufficient (Meir, 1989) refugee camp designs suited for arid regions, partly related to Figure 1, that will enable the use of local natural resources, including solar energy. Rainwater harvesting and the use of runoff water could be important elements. Such designs are of immense value for proactive disaster planning and management in drylands. Camps need to be efficient to accomodate refugees, facilitate basic humanitarian assistance, food, clean water and sanitation, but also to provide training in educational programs, while preventing desertification.

It is necessary to design an experimental refugee camp in a dry area, in which all such ideas can be tested and displayed: spatial planning and design, simple dwellings with optimal passive heating and cooling attributes, runoff collection from roofs and natural catchments, water storage, food production and storage, solar cookers, photovoltaic power for essential electricity supply, water pumps, use of groundwater, and inventive sanitation systems.

Conclusions

1. Values for potential evapotranspiration in P/ETP ratio calculations to express bioclimatic aridity, an essential element in dryland planning and management, should be based on Class-A evaporation pan data as a comparable, standardized system throughout the world.
2. For the P/ETP boundaries of climatic zones in drylands, we recommend the values as published by UNESCO (1979) in the map of the world distribution of arid regions. These values relate to ETP determinations based on Penman (1948), which are compatible with Class-A evaporation pan data (Berliner *et al.*, in preparation). The hyper-arid zone has P/ETP values < 0.03; the arid zone from 0.03-0.20; the semi-arid zone from 0.20-0.50; and the sub-humid zone from 0.50-0.75.
3. The definition of desertification should include the hyper-arid zone, besides the arid, semi-arid and sub-humid zones.
4. Drought and desertification are serious hazards in drylands. Projected climatic change in relation to the so-called greenhouse effect may increase the risk of these hazards in certain areas. Governments usually respond through ad-hoc crisis management rather than proactive disaster planning and management. Crisis management is always inferior and more costly in economical and environmental terms than carefully-formed response and mitigation plans, developed ahead of time in joint cooperation between academic experts and decision makers (Wilhite, 1996; Chapter 2). Proactive planning and interactive management are essential to develop successful preparedness and mitigation schemes to be activated in the event of drought and desertification (Bruins and Lithwick, Chapter 1).
5. Environmental refugee camps should be designed for arid regions to cope with disasters and to prevent desertification. Self-sufficient productive elements, education, training and development programs must be integral components of such camps, in order to promote participatory development and rehabilition of the refugees.

Notes

1. The standard way of measuring evaporation is by means of an American Class-A evaporation pan, which is made of galvanized iron, circular in shape, 1.21 m in diameter, 25 cm deep. The

Class-A pan has to be mounted on a wooden open-frame platform 15 cm above the ground in an accurately horizontal position. The pan is filled with water to a depth of 20 cm. Each morning at the same time (usually 8 AM) the water level is measured in centimeters by making a reading from the fixed point or hook gauge. Water is added due to evaporation or extracted after heavy rainfall to keep it at a depth of about 20 cm. Allowance must be made for these additions, artificially or by rainfall, or extractions, to obtain the correct evaporation amount in mm for a period of 24 hours each time (Stern, 1979).

References

Adar, E.M., I. Gev, P.R. Berliner and A.S. Issar (1995) The effect of forestation on a shallow groundwater reservoir in an arid sand dune terrain. *Journal of Arid Land Studies* 5:259-262.

Aubréville, A. (1949) *Climats, Forêts et Désertification de l'Afrique Tropicale*. Société d'Éditions Géographiques, Maritimes et Coloniales, Paris.

Ben-Asher, J., and Berliner, P.R. (1994) Runoff irrigation. *In* K.K. Tanji and B. Yaron (eds.) *Management of Water Use in Agriculture*. Springer Verlag, Berlin, Chapter 6, pp. 126-154.

Behnke Jr., R.H., I. Scoones and C. Kerven (eds.) (1993) *Range Ecology at Disequilibrium: New Models of Natural Variability and Pastoral Adaptation in African Savannas*. Overseas Development Institute (ODI), London.

Bruins, H.J. (1990) The impact of man and climate on the central Negev and northeastern Sinai deserts during the Late Holocene. *In* S. Bottema, G. Entjes-Nieborg and W. van Zeist (eds.) *Man's Role in the Shaping of the Eastern Mediterranean Landscape*. Balkema, Rotterdam, pp. 87-99.

Bruins, H.J (1993) Drought risk and water management in Israel: planning for the future. *In* D.A. Wilhite (ed.) *Drought Assessment, Management and Planning: Theory and Case Studies*. Kluwer Academic Publishers, Dordrecht, Chapter 8, pp. 133-155.

Bruins, H.J. (1994) Comparative Chronology of Climate and Human History in the Southern Levant from the Late Chalcolithic to the Early Arab Period. *In* O. Bar-Yosef & R. Kra (eds.) *Late Quaternary Chronology and Paleoclimates of the Eastern Mediterranean*. Radiocarbon, University of Arizona, Tucson; and Peabody Museum, Harvard University, Cambridge, MA, pp. 301-314.

Bruins, H.J. (1996) A rationale for drought contingency planning in Israel. *In* Y. Gradus and G. Lipshitz (eds.) *The Mosaic of Israeli Geography*. Ben-Gurion University of the Negev Press, The Negev Center for Regional Development, Beer-Sheva, pp. 345-353.

Bruins, H.J. (1997) Desertification. *In* G.A. Good (ed.) *Sciences of the Earth: An Encyclopedia of Events, People and Phenomena*. Garland Publishing, New York, Two Volumes (in press).

Bruins, H.J., M. Evenari and U. Nessler (1986) Rainwater-harvesting agriculture for food production in arid zones: The challenge of the African famine. *Applied Geography* 6:13-32.

Bruins, H.J., M. Evenari and A. Rogel (1987) Run-off Farming Management and Climate. *In* L. Berkofsky and M.G. Wurtele (eds.) *Progress in Desert Research*. Rowman & Littlefield, Totowa, N.J., pp. 3-14.

Bruins, H.J. and H. Lithwick (1997) Proactive planning and interactive management in arid-frontier development. *In* H.J. Bruins and H. Lithwick (eds.), *The Arid Frontier: Interactive Management of Environment and Development*. Kluwer Academic Publishers, Dordrecht, pp. 3-29.

Brutsaert ,W. (1982) *Evaporation into the Atmosphere. Theory, History and Applications*. D. Reidel Publishing Company, Dordrecht.

Budyko, M.I. (1956) *The Heat Balance of the Earth's Surface.* Office of Technical Services. U.S. Department of Commerce, Washington (English translation of: *Teplovi Balans Zennoi Poverkhnosti*).

Budyko, M.I., G.S. Golitsyn and Y.A. Izrael (1988) *Global Climatic Catastrophes.* Springer Verlag, Berlin.

Dahl, G. and A. Hjort (1979) Pastoral Change and the Role of Drought. Swedish Agency for Research Cooperation with Developing Countries, SAREC Report, Stockholm.

Dietz, A.J. (1987) *Pastoralists in Dire Straits: Survival Strategies and External Interventions in a Semi-arid Region at the Kenya/Uganda Border, Western Pokot, 1900-1986.* Netherlands Geographical Studies No. 49, Amsterdam.

Dietz, A.J., M. Put and S. Subbiah (1997) Drought risk assessment for dryland agriculture in semi-arid Telengana, Andhra Pradesh, India. *In* H.J. Bruins and H. Lithwick (eds.) *The Arid Frontier: Interactive Management of Environment and Development*, Kluwer Academic Publishers, Dordrecht, pp. 143-161.

Doorenbos, J & W.O. Pruitt (1984) Crop Water Requirements. FAO Irrigation and Drainage Paper 24, FAO, Rome.

Dregne, H.A. (1983) *Desertification of Arid Lands.* Harwood Academic Publishers, Churchill/New York.

El-Baz, F. and M.H.A. Hassan (eds.) (1986) *Physics of Desertification.* Martinus Nijhoff Publishers, Dordrecht.

El-Bihbety, H. and H. Lithwick (1997) Cost-Benefit Analysis of Water Management Mega-Projects in India and China. *In* H.J. Bruins and H. Lithwick (eds.) *The Arid Frontier: Interactive Management of Environment and Development*, Kluwer Academic Publishers, Dordrecht, pp. 295-317.

Evenari, M., L. Shanan and N.Tadmor (1982) *The Negev: The Challenge of a Desert*, Harvard University Press, Cambridge, Massachusetts (2nd edition).

Glantz, M.H. (ed.) (1977) *Desertification: Environmental Degradation in and around Arid Lands.* Westview Press, Boulder.

Glantz, M.H. and N. Orlovsky (1983) Desertification: A review of the concept. *Desertification Control Bulletin* 9:15-22.

Goudie, A.S. (1994) Deserts in a warmer world. *In* A.C. Millington and K. Pye (eds.) *Environmental Change in Drylands: Biogeographical and Geomorphological Perspectives.* John Wiley & Sons, New York, pp. 1-24.

Heathcote, R.L. (1983) *The Arid Lands: Their Use and Abuse.* Longman, London and New York.

Huntington, E. (1924) *Civilization and Climate.* Yale University Press, New Haven.

Issar, A.S., with Z. Peiyuan, H.J. Bruins, M. Wolf and Z. Ofer (1995) Impacts of Climate Variations on Water Management and Related Socio-Economic Systems. UNESCO, International Hydrological Programme, IHP-IV Project H-2.1, Paris, 97 pp.

Klintworth, H. Desert encroachment over the Karoo. *Farming in South Africa* 23(1948):723-728.

Köppen, W. (1931) *Die Klimate der Erde.* Walter de Gruyter, Berlin.

Lamb, H.H. (1982) *Climate, History and the Modern World.* London, Methuen.

Le Houérou, H.N. (1977) Man and desertization in the Mediterranean region. *Ambio* 6:363-365.

Lövenstein, H., P.R. Berliner and H. Van Keulen (1991). Runoff agroforestry in arid lands. *Forest Ecology and Management* 45:59-70.

Lowdermilk, W.C. (1935) Man made deserts. *Pacific Affairs* 8:409-419.

Lowdermilk, W.C. (1944) *Palestine: Land of Promise.* Harper & Brothers Publishers, New York.

Mainguet, M. (1991) *Desertification: Natural Background and Human Mismanagement.* Springer Series in Physical Environments, Vol. 9, Springer Verlag, Heidelberg.

Mainguet, M. and R. Létolle (1997) Human-made desertification in the Aral Sea Basin: Planning and management failures. *In* H.J. Bruins and H. Lithwick (eds.) *The Arid Frontier: Interactive Man-*

agement of Environment and Development. Kluwer Academic Publishers, Dordrecht, pp. 129-142.

Meigs, P. (1953) World distribution of arid and semi-arid homoclimates. *Arid Zone Hydrology*. UNESCO Arid Zone Research Series 1:203-209.

Meir, A. and H. Tsoar (1996) International borders and range ecology: the case of Bedouin transborder grazing. *Human Ecology* 24:39-63.

Meir, I.A. (1989) Autarkic Settlements: The Use of Pattern Language in the Design of Highly Self-sufficient Settlements. M.Sc. Thesis, Technion, Haifa.

Nicholson, S.E., and H. Flohn (1980) African environmental and climatic changes and the general atmospheric circulation in Late Pleistocene and Holocene. *Climatic Change* 2:313-348.

Orev, Y. (1986) A Practical Handbook on Desert Range Improvement Techniques. Interagency Group on Agricultural Biometeorology, FAO, UNEP, UNESCO, WMO, Geneva.

Penman, H.L. (1948) Natural evaporation from open water, bare soil and grass. *Proceedings of the Royal Society, Section-A* 193:120-145.

Perevolotsky, A. (1995) Conservation, Reclamation and Grazing in the Northern Negev: Contradictory or Complementary Concepts? Overseas Development Institute (ODI) Pastoral Development Network Paper 38a.

Reifenberg, A. (1955) *The Struggle Between the Desert and the Sown*. Jewish Agency Publishing Department, Jerusalem.

Reij, C., P. Mulder and L. Begemann (1988) Water Harvesting for Plant Production. World Bank Technical Paper, No. 91, The World Bank, Washington, D.C.

Roos, M. (1997) Drought and water management in California. *In* H.J. Bruins and H. Lithwick (eds.) *The Arid Frontier: Interactive Management of Environment and Development.* Kluwer Academic Publishers, Dordrecht, pp. 277-293.

Sandford, S. (1983) *Management of Pastoral Development in the Third World*. John Wiley & Sons, New York.

Sauerhaft, B., P.R. Berliner and T.L. Thurow (1997) The fuelwood crisis in arid zones: Runoff agriculture for renewable energy production. *In* H.J. Bruins and H. Lithwick (eds.) *The Arid Frontier: Interactive Management of Environment and Development,* Kluwer Academic Publishers, Dordrecht, pp. 351-363.

Schneider, S.H. (1976) *The Genesis Strategy: Climate and Global Survival*. Plenum Press, New York.

Sellers, W.D. (1965) *Physical Climatology.* The University of Chicago Press, Chicago.

Shanan, L. (1997) Environmental hazards of irrigation development: The need for proactive planning and interactive management. *In* H.J. Bruins and H. Lithwick (eds.) *The Arid Frontier: Interactive Management of Environment and Development,* Kluwer Academic Publishers, Dordrecht, pp. 251-276.

Stebbing, E.P. (1938) The man-made desert in Africa. *Journal of the Royal African Society* (Suppl.) 37(146):13.

Stern, P.H. (1979) *Small Scale Irrigation: A Manual of Low-cost Water Technology.* Intermediate Technology Publications Ltd, London & International Irrigation Information Center, Volcani Center, Bet Dagan, Israel.

Summers, K., B. Hutchinson, R. Varady, C. Casler, and D. Campbell (1991) *World Desertification Bibliography*. United Nations Environment Programme (UNEP), Nairobi; The Office of Arid Land Studies (OALS), University of Arizona, Tucson.

Thomas, D.S.G. and N.J. Middleton (1994) *Desertification: Exploding the Myth*. John Wiley & Sons, New York.

Thorthwaite, C.W. (1948) An approach towards a rational classification of climate. *Geographical Review* 38:55-94.

Tsoar, H. (1995) Desertification in Northern Sinai in the Eighteenth Century. *Climatic Change* 29: 429-438.

UNCCD (1995) United Nations Convention to Combat Desertification in Those Countries Experiencing Serious Drought and/or Desertification, Particularly in Africa UNEP, Interim Secretariat for the Convention to Combat Desertification, Geneva.

UNCED (1992) United Nations Conference on Environment and Development. Rio de Janeiro, June 1-12.

UNCOD (1977) Secretariat United Nations Conference on Desertification. *Desertification, Its Causes and Consequences*. Pergamon Press, New York.

UNEP (1992) *World Atlas of Desertification*, Edward Arnold, London.

UNESCO (1979) Map of the World Distribution of Arid Regions. UNESCO, Man and the Biosphere (MAB) Technical Notes 7, Paris.

Verstraete, M.M. (1986) The United Nations organisation and the issue of desertification. *In* F. El-Baz and M.H.A. Hassan (eds.) *Physics of Desertification*. Martinus Nijhoff Publishers, Dordrecht, pp. 42-51.

Warren, A. (1997) Environmental science and desertification at the arid frontier. *In* H.J. Bruins and H. Lithwick (eds.) *The Arid Frontier: Interactive Management of Environment and Development*. Kluwer Academic Publishers, Dordrecht, pp. 117-127.

Warren, A. and C. Agnew (1987) An assessment of desertification and land degradation in arid and semi-arid areas. Drylands paper 2, International Institute for Environment and Development, London.

White, K.D. (1970) *Roman Farming*. Thames and Hudson, London.

Whyte, R.O. (1961) Evolution of land use in south-western Asia. *In* L. Dudley Stamp (ed.) *A History of Land Use in Arid Regions*. UNESCO, Paris, pp. 57-118.

Wilhite, D.A. (1996) A methodology for drought preparedness. *Natural Hazards* 13:229-252.

Wilhite, D.A. and M.H. Glantz (1986) Understanding the drought phenomenon: The role of definitions. *Water International* 10:111-120.

Wilhite, D.A. and Hayes, M.J. (1997) Drought planning in the United States: Status and future directions. *In* H.J. Bruins and H. Lithwick (eds.) *The Arid Frontier: Interactive Management of Environment and Development*. Kluwer Academic Publishers, Dordrecht, pp. 33-54.

Williams, M.A.J. and R.C. Balling Jr (1996) *Interactions of Desertification and Climate*. Arnold, London.

ABOUT THE AUTHORS

Hendrik J. Bruins is Senior Researcher at the Jacob Blaustein Institute for Desert Research, in the Social Studies Center; he lectures at the Department of Geography and Environmental Development, and is Research Fellow at the Negev Center for Regional Development (Ben-Gurion University of the Negev). He obtained his M.Sc. *cum laude* at The Hebrew University of Jerusalem at the Institute of Earth Sciences, receiving the Dicker-Shraga Award. He obtained his Ph.D. at the Agricultural University of Wageningen (The Netherlands). His multi-disciplinary experience in the sphere of *Man and Desert* is directed at both historical and modern issues. New insights concerning the unresolved Exodus enigma were recently published in *Nature* (1996, Vol. 382, No. 6588). He is active in multilateral cooperation to combat desertification. Dr. Bruins heads a tripartite project on Drought Planning for Arid-Zone Pastoralists, funded by the Netherlands-Israel Development Research Programme (NIRP). The relations between environmental and social aspects in disaster planning, policy and management have his keen interest.

Dr. Hendrik J. Bruins
Ben-Gurion University of the Negev
Jacob Blaustein Institute for Desert Research
Social Studies Center
Sede Boker Campus, 84990
Israel (fax: 972-7-6596863; e-mail: hjbruins@bgumail.bgu.ac.il)

Pedro R. Berliner holds a Ph. D. in agricultural meteorology and soil physics from The Hebrew University of Jerusalem. His main interest is the movement of water in the soil-plant-atmosphere continuum. Dr. Berliner is the head of the Runoff Agriculture Unit at the Jacob Blaustein Institute for Desert Research (Ben-Gurion University of the Negev). The Unit maintains two experimental runoff farms: The Evenari Farm near Avdat and the Mashash Farm south of Beer-Sheva. In addition, fields at the Sede Boker Campus are used for simulated rainwater-harvesting investigations. Basic research related to runoff agroforestry systems has been carried out. An additional sphere of his present research is the quantification of radiation, water and carbon dioxide fluxes in runoff agroforestry systems.

Dr. Pedro R. Berliner
Wyler Laboratory for Arid Lands
A. Katz Center for Desert Agrobiology
Jacob Blaustein Institute for Desert Research
Ben Gurion University of the Negev
Sede Boker Campus, 84990
Israel

Chapter 6

Environmental Science and Desertification at the Arid Frontier

Andrew Warren

Introduction

Science, like other aspects of culture, enters new territory at the frontier. This paper is an attempt to uncover general patterns in this process. It compares approaches to environmental science in three situations: in a humid area, for reasons of perspective, close to a political 'core' (Britain); on a historical dry frontier (U.S.A.); and in the African Sahel.

The main focus is on the arid Sahel, because it is here that environmental science is currently being most severely challenged to deliver answers to problems of massive migration and extreme poverty. Ten million people were displaced within these areas in the mid-1980s, largely because of short- and long-term drought, but with an unknown component of land degradation (United Nations Office for Emergency Operations in Africa, UNOEOA, 1985). No other environmental problem has caused so much migration (though cyclonal flooding in the Bay of Bengal may come close). Aridity does constitute a frontier in development: of the 20 most disadvantaged countries in the world, as classified by the UNDP index, 16 have large portions of territory in dry zones.

The imperfection of the science that has been applied to these issues is linked to their marginality. A negating argument almost prevailed at UNCED (the UN Conference on Environment and Development) at Rio de Janiero in 1992: because 'desertification' had not been measured, the problems of the dry lands could be disregarded (Middleton, 1993). This highly environmentally disadvantaged part of the world has no direct access, as a result, to the large funds in the Global Environment Facility (which, though established before UNCED, was redirected to global issues after the conference). Global warming, ozone depletion and loss of biodiversity, which do have access to the fund, have uncertain and prospective, rather than actual, impacts compared to the known impact but uncertain cause of problems in the dry lands (Warren, 1993).

H. J. Bruins and H. Lithwick (eds.), The Arid Frontier, 117–127.

It is true that few environmental programs have a history of success. Campaigns to control carbon dioxide or ozone-destroying emissions, or to slow the rate at which species are lost can boast few achievements. Nonetheless the history of the desertification issue is one of even more conspicuous failure, and has little prospect of success. Since emerging on the world scene at the United Nations Conference on Desertification in 1977, there has been a widely acknowledged failure to control dry-land problems (Swift, 1996). There are many symptoms of inadequacy. Tolba (1987) estimated that $4.5 billion per year was required in 1987 to control desertification, yet only $600 million per year had been provided. Goldsmidt (1981) talked of 'unrelieved failure' in pastoral projects in Africa. The documents accompanying the negotiations for the new Desertification Convention admit candidly to the failure of the campaign since 1977 (International Institute for Sustainable Development, 1993).

This depressing record raises large questions. This chapter explores only one of these, namely the way in which science has been applied to the problem.

Neither the fact of nor the causes for the marginalization of dry-land science need much debate. Countries with large proportions of their territory in the arid zones, such as Australia, China, the United States, and Israel have invested a much smaller proportion of the national scientific effort in the dry than in the humid sectors. The reasons are very clear: low economic return from dry lands; relative strangeness of the environment; difficulty of access. Whether scientific effort is proportional to the economic value of dry lands is a more interesting question, though still unresearched. It is likely that the scientific investment in dry lands has a larger ratio to output than investment in humid zones. If so, this may indicate the perceived potential or strategic importance of the dry sector. However, this, too, is not at issue here. The focus in this chapter is the impact of marginalization on the ways in which science has been conducted in the dry lands.

The Development of Rural Environmental Science Close to a Political 'Core' (Britain)

The nature of environmental science, particularly soil science, in the dry-lands is put into relief by a short examination of its history nearer a political 'core'. The terms 'core' and 'periphery' are, of course, not unproblematical, but they are used here in the very restricted sense that the control and manipulation of science to practical ends is apparently more easily achieved near a core than on a frontier. It is undeniable that in recent centuries the centers of political power have been in the humid world and that most of the dry lands have been on the periphery. Britain, used here in the first case study, was for some of this time closer to a political 'core' than the other areas discussed in this chapter. Washington, D.C. and the United Nations system are termed 'cores' in the restricted sense that they were and are the centers in which important decisions have been made about the deployment of science on the arid periphery in the U.S.A. and in the Sahel, respectively.

British rural science at the end of World War II attempted to begin on a clean slate. Rural environmental management was to be planned on a 'rational' basis. Science had gained immense status in the war effort, and was now to be focused on recalcitrant civilian problems (Lowe, 1983). One of the elements of the new approach was the mapping of soils, but quite soon the approach was being critically appraised. Such a re-appraisal has been less easy to achieve at the dry frontier, where environmental science policy repeatedly made similar mistakes.

In Britain, the re-appraisal of soil survey as a means of achieving greater productivity introduced a fundamental issue in environmental science. Webster and Beckett (1968), as part of the re-appraisal, noted that because each unit on a soil map inevitably contains variability, the users have to specify the level of variation they require and the price they are willing to pay for detail. The nature of the users' response, of course, is related to the economic and social climate. This reveals a fundamental point: effective environmental science, right at its heart, must be a negotiation between the scientist and the user. Negotiation can and did occur where, as in Britain, the user is in some control; it has been more difficult on the frontier.

Radical changes did indeed occur in the role of science in the planning of rural land use in Britain. The main factors behind the changes included:

1. A shift from a belief in state planning towards belief that the market should be the main arbiter;
2. Decreasing importance of soil as a factor in agricultural production;
3. Increasing availability of field-by-field and year-by-year advice (largely divorced from soil surveys); and
4. Above all, the increasing responsiveness of farm economies to rapidly changing economic signals (which quickly outran the assumptions on which the soil maps had been made).

In the face of these changes, to which science policy responded relatively quickly, soil survey as a tool for agricultural planning had been virtually abandoned by the late 1980s. There is at present very little centralized and standardized environmental appraisal in Britain, though other rural land appraisal systems are still used as planning tools (though rarely for agricultural land use itself).

Dry-land Science in the United States

A longer look at land science in the dry parts of the United States takes the argument to the arid lands and to a zone further from scientific and political centers of power. There are two advantages of examining the American experience in the present context. First the dry parts of the U.S., in many analyses, have been seen as the quintessential arid frontier. Second, there has been extensive analysis of the scientific approach to that environment (White, 1986; Riebsame, 1987).

The nemesis of one phase of land science in dry America was the Dust Bowl of the Great Plains in the 1930s, after which the Roosevelt administration began to invest large sums of federal money to resurrect science, believing it to be the only sound basis for the reconquest of these zones (Lyles, 1985). Early post-Dust Bowl science had some distinctive characteristics, many of which were premonitions, both in style and methodology, of the approaches to the African dry lands, discussed below. For example, being directed from the core in Washington, D.C., rather than from the periphery on the Plains, it began by adopting a very large-scale approach and by relying on the opinions of surveyors rather than on detailed measurement (Figure 1). These assessments placed huge parts of the country into categories like 'severe sheet and gully erosion', judgments that have not been vindicated by subsequent history. Some of these very areas are now among the most productive on Earth.

Dubious survey techniques were mirrored in the sweeping nature of some of the solutions proposed from the center. The most amazing of these was Roosevelt's dream of a 260 km x 3,000 km shelterbelt along the 100th meridian (Zon, 1935), an uncanny precursor of a similar scheme in Africa. Although Roosevelt's green belt never materialized, there were, at its maximum in the 1960s, some 65,000 km^2 of shelterbelt on the Great Plains, financed by federal programs, many of which are now being grubbed up, for few proved their worth (Riebsame, 1987). The building of major dams and irrigation schemes with federal 'core' money, also gained momentum in the pro-science, pro-technology climate of opinion (El-Ashry & Gibbons, 1986).

Accompanying these central-government approaches to the dry lands were a number of fringe developments, which find echoes in the later history of the Sahel. This included pseudo-science, at its worst, flourishing in a generally pro-science environment. One of these, *'dry-farming'*, gained considerable credibility despite the hazard of wind erosion of soils. Another was the *'rain follows the plough'* concept, which was, if anything, even more misguided (Riebsame, 1987).

Many reasons have been given for this distinctiveness of dry-land science and technology in the United States. First there was the unfamiliarity (to those in Washington, D.C.) of the dry lands at the frontier, a facet explored in depth in relation to settlers (Hudson, 1986), and, at least in an exploratory way, in relation to scientists (Malin, 1952; Bogue, 1980). A particularly unfamiliar feature was the variability of the environment, a characteristic brought sharply into focus by the Malin/Clements debate. Clements (1937) suggested that the equilibrium 'climax' concept was an appropriate model for understanding grassland ecology. The climax was a concept that had some validity in humid environments (although even here it has since been radically questioned, Pickett *et al.*, 1987), but in retrospect was even less ap-

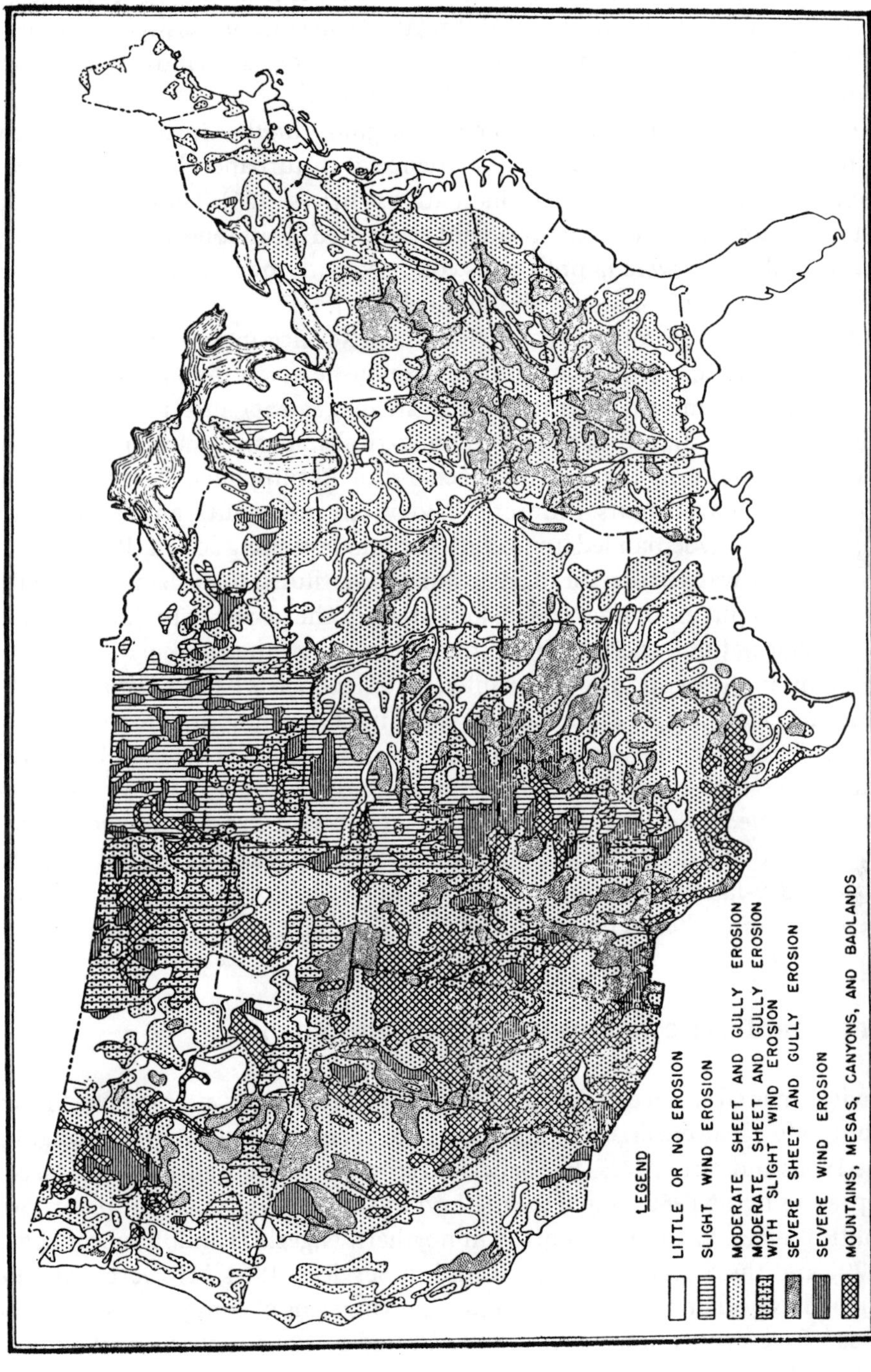

Figure 1. United States Department of Agriculture map of land damaged by erosion (after Bennett, 1939:65).

plicable to the variable environment of the drier parts of the continent. Nonetheless, Clements' followers made the climax model central in range management in these dry areas (Sampson, 1917). Malin (1952), in contrast, insisted that the inherent instability and long history of occupation of the dry lands were factors that could not be encompassed in Clements' framework.

A similar debate surrounded many of the gullies on the Great Plains, which were seen by early geomorphologists as evidence of disturbance, but which have since been shown to relate to climatic fluctuations (Brice, 1966). These early inspections of this unfamiliar environment had not accounted for its long-term variability. The effects of drought were in both cases confused with those of mismanagement, another pattern that is shared with the Sahel.

A second reason for the distinctiveness of science in dry America, partly as a consequence of its unfamiliarity and capriciousness, was the strong symbolism acquired by dry-land management (or mismanagement) after the Dust Bowl. The initial failure of the world's most technologically advanced society to control the dry environment, put high emotional charge on its subsequent better management. The unfamiliar threats inherent in dry environments and their apparent unpredictability, demanded immediate and unquestioning action (Riebsame, 1986). This attitude was a precursor to environmental attitudes to the Sahel in recent years, which again put emotional charge and urgency ahead of careful judgment.

Though natural science still dominates much of the thinking about land use problems in the dry parts of the United States, there has been a distinct move away from the early, centralized system towards methods of appraisal and control that are much more in the hands of individual farmers (Lyles & Tatarko, 1988). Erosion rates are now estimated by scientific measurement, rather than by rule-of-thumb (for example, Gillette *et al.*, 1992). The new 'Wind Erosion Prediction System' (WEPS) is deliberately designed for field-by-field appraisal (Hagen, 1991). In these respects dry-land management in the U.S.A. is seen to be converging to land management approaches in the humid parts of the country.

Dry-land Environmental Science in the Sahel

Environmental science in the Sahel, as in the United States after the Dust Bowl, was reborn after a severe crisis. In the case of the Sahel the crisis was the drought of the late 1960s and early 1970s. The Sahel had already experienced a phase of large-scale appraisals in the 1960s, when the method was still in vogue in some places closer to political cores. Examples occurred in northern Nigeria (Land Resources Division, 1970) and the Sudan (Hunting Technical Services, 1964). However, the method reached its peak in the 1980s and its apogee (in terms of comprehensiveness) in the Atlas of Desertification issued by the United Nations Environment Programme (UNEP, 1992).

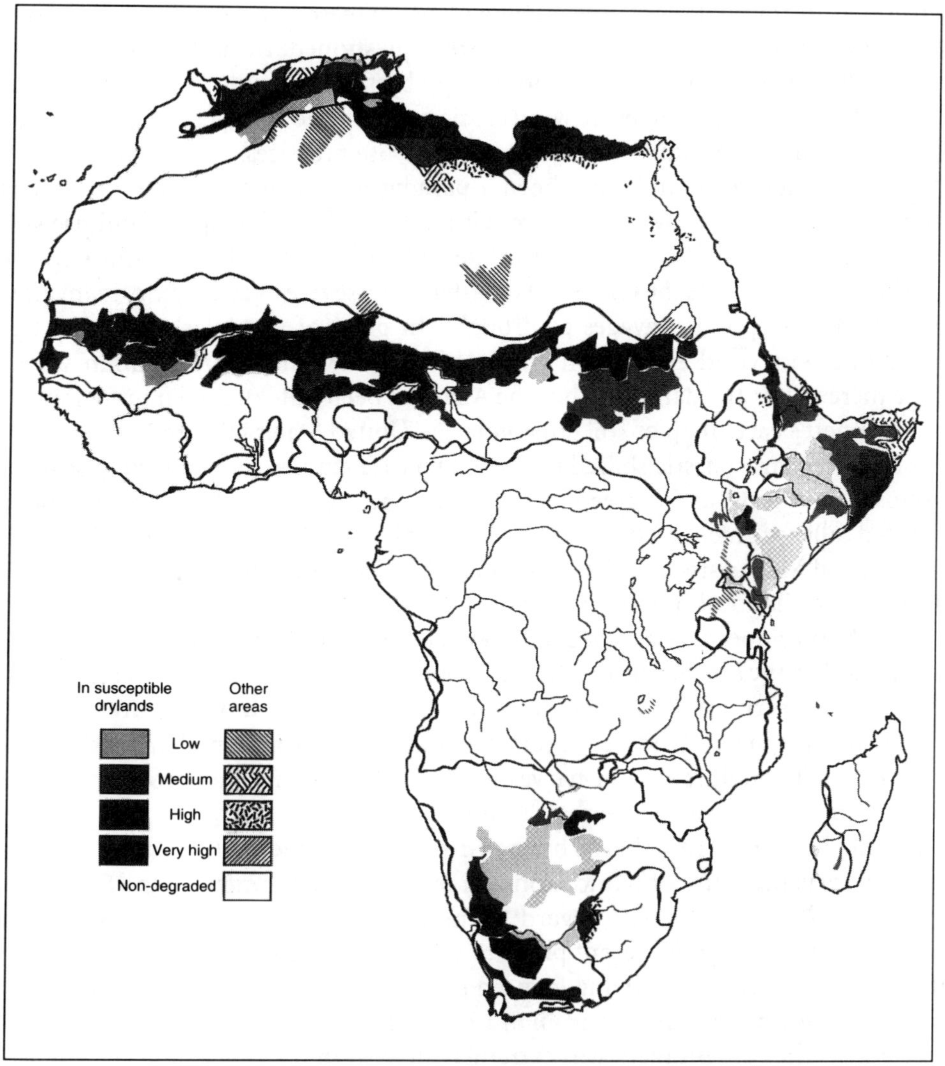

Figure 2. Map of wind erosion problems in Africa from the *World Desertification Atlas* (UNEP, 1992).

Like the appraisals of soil erosion in the U.S.A. of the 1930s, the Atlas purported to be an objective evaluation of the state of degradation of the dry lands. The Atlas's main basis was GLASOD, the Global Assessment of Soil Degradation, for which evaluation was based on the opinions of a panel of experts.

Figure 2 shows estimates of the damage of wind erosion in the Sahelian countries, taken from the GLASOD data base. The fragility of this evaluation can be gauged from the observation that there is virtually no scientific measurement of wind erosion in Africa. The opinions of the expert panel were based upon their perception and evaluation of indigenous land use in these areas, an evaluation which, only a few years of its making, is now severely questioned. Until the late 1980s and early 1990s African pastoralists, who occupy by far the greater part of the areas classified as suffering from severe wind erosion, were almost universally condemned by the scientific community for damaging land-use practices.

In the last few years there has been a paradigm shift in opinion among most range scientists, who now credit them with potentially well adapted land use systems (Warren, 1995). Where before the late 1990s range scientists and those whose opinions went to create the GLASOD data-base saw severe over-grazing, many now see only the temporary ravages of drought. In place of the Atlas's 'expert' judgments of severe wind erosion on pastoral lands, current 'expert' opinion would judge there to be little damage. Thus the Atlas has been outdated even more rapidly than Bennett's early map of soil erosion in the United States (Figure 1).

A second characteristic shared by these two case studies is the amazing scale of proposed solutions, as for greenbelts. In the Sahel, for example, there have been repeated calls for greenbelts to keep the desert at bay, the latest one being a greenbelt from central Mali eastwards to the Sudanese border (Rognon, 1991), a plan that is uncannily similar to those for the Great Plains. Third, in the Sahel, as on the Great Plains, a host of poorly researched, pseudo-scientific panaceas have been proposed, including climatic amelioration schemes (Glantz & Parton, 1976; Adams & Hughes, 1990) and indiscriminate tree-planting. However, trees demand moisture that may not be available in arid lands, even causing doubtful environmental impact (Cline-Cole, 1996). Though rainwater harvesting (runoff) systems may provide a viable solution in suitable areas (Sauerhaft *et al.*, this volume).

The final characteristic shared by the Great Plains of the 1930s and the Sahel in the 1970s was the sense of panic. Concern over the environment in the Sahel was first voiced in the 1930s with regard to the possibility of a Dust Bowl in Africa (Anderson, 1984). A more recent phase of concern began after the droughts of the late 1960s and early 1970s, being focused particularly in the United Nations Conference on Desertification in Nairobi in 1977. In each case the authorities were told by scientists that the problem was extremely urgent. In the case of the United Nations effort, the results were widely seen to have been hastily prepared and poorly considered (Reij, 1988; Warren & Khogali, 1992; Tiffen *et al.*, 1993; Swift, 1996).

Conclusion

This chapter has concentrated on two dry-land regions, and a very small scientific field. It would be valuable to extend the research into other subjects and geographical areas, but this is beyond the present scope. Nonetheless, a general pattern of dry-land environmental science appears to emerge:

1. The problems of arid lands, which are mostly peripheral to centers of power and decision-making, become only apparent to those at the core after bad years, usually drought. Solutions are then felt to be so urgently required that careful preliminary research is dispensed with. Peripherality, and the capriciousness of the environment and its frequently distressing consequences introduce a strong and seldom helpful element of urgency.
2. Peripherality encourages appraisals that are sweeping, and broad in scale, perhaps because, from the core, the arid frontier appears to have great uniformity.
3. The apparent prominence of the environment in dry-land problems encourages a deterministic view that natural science has a less restricted role in their solution than in the humid 'core'.
4. Distance also allows consideration of almost anything that claims to be based on science (like 'dry-farming' and indiscriminate tree-planting).
5. Peripherality, for all these reasons, also encourages development plans for the dry lands that are as sweeping as the appraisals.
6. Finally, peripherality discourages participatory negotiation with those who live on, understand and use the arid land.

In view of experience elsewhere in the world, gained in earlier periods, it is disturbing that this pattern recurred in the Sahel in the 1970s and 1980s. Methods that may have been excusable in the dry parts of the United States, Australia, the Soviet Union and even in the Sahel, when science and government knew little about these areas, can no longer be excused so easily. Land surveying, using arbitrary evaluations, was abandoned in the developed world not so much because it had served its purpose, but because it was irrelevant or even misleading. A similar history surrounds the other aspects of the history of science in dry lands discussed above. Their persistence in the dry parts of Africa is most readily explained by the extremity of their peripherality, at least from the perspective of certain political and scientific 'cores'.

References

Adams, W.M. and F.M.R. Hughes (1990) Irrigation development in desert environments. *In* A.S. Goudie (ed.) *Techniques for Desert Reclamation.* John Wiley and Sons, Chichester, pp. 135–159.

Anderson, D.M. (1984) Depression, dust bowl, demography and drought: the colonial state and soil conservation in East Africa. *African Affairs* 83:321–343.

Bennett, H.H. (1939) *Soil Conservation.* McGraw-Hill, New York.

Bogue, A.G. (1980) The heirs of James C. Malin: Grassland historiography. *Great Plains Quarterly* 1:105–131.

Brice, J.C. (1966) Erosion and deposition in the loess-mantled Great Plains, Medicine Creek drainage basin, Nebraska. United States Geological Survey, Professional Paper, 352-H:255–339.

Clements, F.E. (1937) *Environment and Life in the Great Plains.* Carnegie Institute, Washington, D.C.

Cline-Cole, R. (1996) Dryland forestry. *In* M. Leach and R. Mearns, R. (eds.) *The Lie of the Land: Challenging Received Wisdom on the African Environment.* Heinemann/James Currey, London, pp. 122–139.

El-Ashry, M., and D.C. Gibbons (eds.) (1988) *Water and the Arid Lands of the United States.* Cambridge University Press, Cambridge.

Gillette, D.A., G.J. Stensland, A.L. Williams, W. Barnard, D. Gatz, P.C. Sinclair, and T.C. Johnson (1992) Emissions of alkaline elements Calcium, Magnesium, Potassium and Sodium from open sources in the contiguous United States. *Global Biogeochemical Cycles* 6:437–457.

Glantz, M.H. and W. Parton (1976) Weather and climate modification and the future of the Sahara. *In* M.H. Glantz (ed.) *Politics of Natural Disaster: The Case of the Sahel Drought.* Praeger, New York, pp. 303–324.

Goldsmidt, W. (1981) The failure of pastoral development projects in Africa. *In* J.G. Galaty, D. Aronson, P.C. Saltzman, A. and Chouinard. *The Future of Nomadic Peoples.* International Development Research Centre, Ottawa, pp. 101–118.

Hagen, L.J. (1991) A wind erosion prediction system to meet user needs. *Journal of Soil and Water Conservation* 46:106–111.

Hudson, J.C. (1986) Who was forest man? *Great Plains Quarterly* 6:69–83.

Hunting Technical Services, Ltd. (1964). Report of the survey of geology, geomorphology and soils, vegetation and present land use: land and water use survey in Kordofan Province, Republic of the Sudan. DOX-SUD-A26, Doxiades Associates, Athens.

International Institute for Sustainable Development (IISD) (1993) Summary of the second session of the INC for the elaboration of an International Convention to Combat Desertification 13–24 September 1993. *Earth Negotiations Bulletin* 4(22), IISD, Winnipeg.

Land Resources Division (1970) *Land Resource Bibliography: Nigeria.* LRD, Overseas Development Administration (ODA), London, 3 volumes.

Lowe, P.D. (1983) Values and institutions in the history of British nature conservation. In A. Warren F.B. and Goldsmith (eds.) *Conservation in Perspective.* John Wiley and Sons, Chichester, pp. 329–352.

Lyles, L. (1985) Predicting and controlling wind erosion. *Agricultural History* 59:205–214.

Lyles, L. and J. Tatarko (1988) Soil wind erodibility index in 7 north-central states. *Transactions of the American Society of Agricultural Engineers* 31:1396–1397.

Malin, J.C. 1952. Man, the state of nature and climax, as illustrated by some problems of the North American grassland. *Scientific Monthly* 74:29–37.

Middleton, N. 1993. *Tears of the Crocodile: From Rio to Reality in the Developing World.* Pluto, London.

Pickett, S.T.A., S.L. Collins and J.J. Armesto (1987) Models, mechanisms and pathways of succession. *Botanical Review* 53:335–371.

Reij, C. (1988) L'état actuel de la conservation des eaux et du sol au Sahel, Comité permanente inter-états pour la Lutte contre la Sécheresse au Sahel, (CILSS)/Organization for Economic Co-operation and Development (OECD)/Club du Sahel, Paris, D(89)329.

Riebsame, W.E. (1986) The Dust Bowl: Historical image, psychological anchor and ecological taboo. *Great Plains Quarterly* 6:127–136.

Riebsame, W.E. (1987) Human transformation of the United States Great Plains: Patterns and causes. *In* B.L. Turner, II, W. Clark, R.W. Kates, J.F. Richards, J.T. Mathews, and W.B. Meyer (eds.) *The Earth as Transformed by Human Action.* Cambridge University Press, Cambridge, pp. 561–575.

Rognon, P. (1991) Un projet japonais de lutte contre la sécheresse au Sahel. *Sécheresse* 2:135–138.

Sampson, A.W. (1917) Plant succession in relation to range management. *Journal of Forestry* 15:593–596.

Swift, J. (1996) Desertification. *In* M. Leach and R. Mearns, R. (eds.) *The Lie of the Land: Challenging Received Wisdom on the African Environment.* Heinemann James Currey, London, pp. 73–90.

Tiffen, M., M. Mortimore, F. and Gichuki, F. (1993) *More People, Less Erosion: Environmental Recovery in Kenya.* John Wiley and Sons, Chichester.

Tolba, M.K. (1987) Preface. *In* Y.J. Ahmad and M. Kassas (eds.) *Desertification: Financial Support for the Biosphere.* Hodder and Stoughton, London, pp. vii–x.

United Nations Environment Programme (UNEP) (1992) *World Atlas of Desertification.* Edward Arnold, London.

United Nations Office for Emergency Operations in Africa (UNOEOA), (1985) Status report on the emergency situation in Africa as of 1st September, 1985, (3rd monthly report).

Warren, A. (1993) Desertification as a global environmental issue. *GeoJournal* 31:11–14.

Warren, A. (1995) in press. Changing understandings of African pastoralism and environmental paradigms. *Transactions of the Institute of British Geographers.*

Warren, A. and M. Khogali (1992) *Desertification and Drought in the Sudano-Sahelian Region 1985–1991.* United Nations Sudano-Sahelian Office (UNSO), New York.

Webster, R. and P.H.T. Beckett (1968) Quality and usefulness of soil maps. *Nature* 219:680–682.

White, G.F. (1986) 'The Future of the Great Plains' revisited. *Great Plains Quarterly* 6:84–93.

Zon, R. (1935) Shelterbelts: Futile dream or workable plan. *Science* 81:392.

ABOUT THE AUTHOR

Andrew Warren is Professor of Geography at University College London. His main research has been into sand dunes, on which he has worked in Israel, Sudan, Egypt, Libya, Niger, Algeria, Tunisia, Oman, Saudi Arabia and Nebraska. He has also worked on desertification, writing two reports for the UN and one for Greenpeace.

Prof. A. Warren
Department of Geography
University College London
26, Bedford Way
London WC1H 0AP
England.

Chapter 7

Human-made Desertification in the Aral Sea Basin: Planning and Management Failures

Monique Mainguet and René Létolle

Introduction

The ecological crisis in the Aral Sea basin, caused by bad planning and management of land and water resources, has attracted the attention of scientists during the last decade. This ecological disaster has resulted in changing the natural environment, causing severe damage to the economy of the five states in the Aral Sea basin and a deterioration in the living conditions of the population. The complexity of the disaster and the difficulty to reverse the land degradation, in the short or medium term, allow us in this case to speak of human-made desertification.

The term desertification has been defined in various ways. The *ad hoc* consultation meeting on Assessment of Global Desertification: Status and Methodologies, held at the United Nations Environment Programme (UNEP) in Nairobi (Feb. 1990) defined the concept as follows: "*Desertification is land degradation in arid, semi-arid and dry sub-humid areas resulting from adverse human impact.*" However, this definition was amended by the United Nations Conference on Environment and Development (UNCED) in Rio de Janeiro (July 1992, in Agenda 21): "*Desertification is land degradation in arid, semi-arid and dry sub-humid areas resulting from various factors including climatic variations and human activities.*" Understanding, assessing, and combating desertification require our ability to differentiate causes, processes, manifestations and consequences, as well as the degree of land degradation and the reduction of the resource potential.

The various arid regions on our planet, as mentioned in the above definition, have a considerable extent, comprising one third (37%) of the continental areas. In this chapter we focus on the causes and consequences of human-made desertification in the arid Aral Sea basin, situated in the drylands of Central Asia (Figure 1). The Aral Sea is located at a temperate latitude in between 43° and 47° N. The aridity of this intra-continental region is caused by its distance from the ocean and the resulting

H. J. Bruins and H. Lithwick (eds.), The Arid Frontier, 129–142.

diminution of rainfall, as well as its location leeward from formidable mountain chains which form a continuous range from the Kaukasus in the southwest, the mountains of Iran and Afghanistan in the south and the Tien Shan and Himalaya mountains to the east. The region has particularly cold winters, being unprotected in the north to the climatic influence of the Siberian plains.

The Aral Sea basin is the largest area on our planet affected by human-made desertification, having a size of 1.8 million km^2. The inland basin without any exoreic outlet is shared by five republics: Kazakhstan, Uzbekistan, Kyrgyzstan, Turkmenistan, and Tajikistan. The population affected by desertification totals about 40 million, which is more than twice the number of people that lived in the area in the early 1960s (Kharin, 1994).

Desertification may basically be considered as a process of environmental change, either naturally induced or human-made, through which an arid, semi-arid or subhumid landscape suffers degradation and may become more desert-like. The mechanisms of desertification have physical, chemical and biogenic aspects, while the consequences affect both the environment and human society. Desertification, therefore, whatever its cause, has not only physical dimensions but also socio-economic ramifications, as will be shown in this chapter about the Aral Sea Basin.

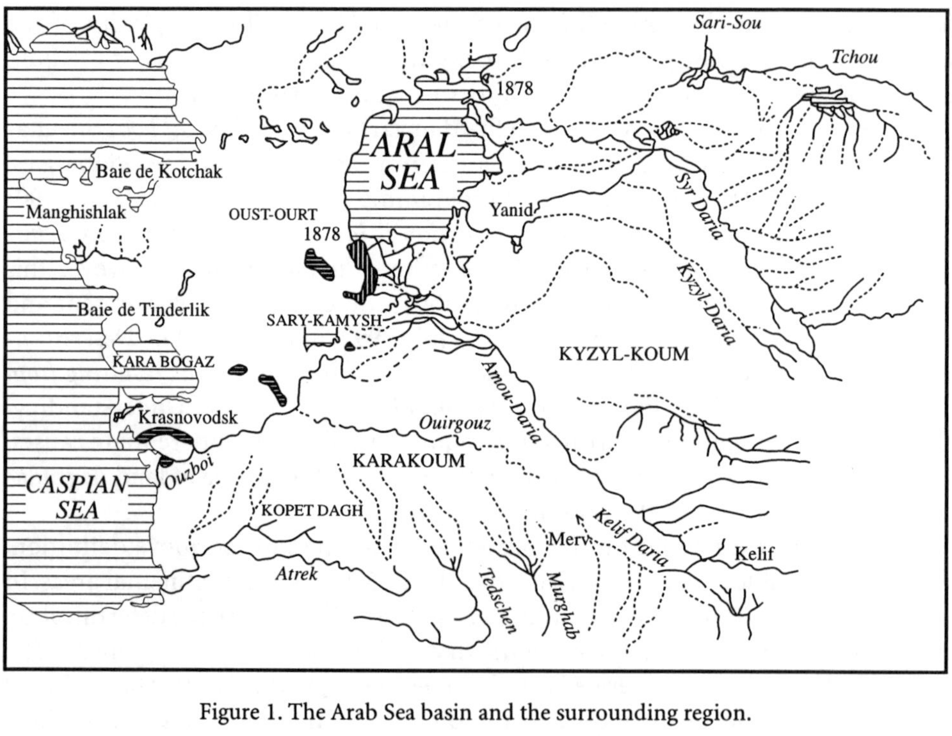

Figure 1. The Arab Sea basin and the surrounding region.

The term desertification should perhaps only be used when degradation reaches an irreversible degree on a human time scale, meaning that rehabilitation of the environment is technically or economically not possible within one generation (Mainguet, 1991). If we do not limit degradation and rehabilitation to one generation, then we believe that irreversibility is almost non-existent on a larger time scale and we are perhaps allowed to be more optimistic. If we define irreversibility as the incapacity of the generation responsible for land degradation to rehabilitate its land, due to financial or technical inability, then irreversibility is certainly the case in the Aral Sea basin.

The potential of ecological degradation as a result of development has been underestimated in the economic policy of the former USSR. This led to environmental problems and repercussions at the scale of a whole region. It is very important to understand the reasons for this crisis, as well as its consequences.

Causes and Extent of Human-made Desertification in the Aral Sea Basin

A distinction is made between two areas according to the severity of desertification: the Aral Sea basin in general, where large scale desertification takes place, and the area adjacent to the Aral Sea (also called pri-Aral), where the consequences of the ecological disaster are most strongly felt. Scientists from the NIS (Newly Independent States) sharing the Aral Sea Basin have distinguished at least six large sectors differently affected by anthropogenic influences.

The Aral Sea Basin administratively includes all the territory of Uzbekistan and Tajikistan, parts of the territory of Kazakhstan (Ksyl-Orda and Chimkent districts, south of Akmola district), Kyrgyzstan (Osh and Naryn districts), Turkmenistan (Dashkovuz, Chardzhou, Mary and Ashgabat districts) as well as parts of Northern Afghanistan and North-Eastern Iran. The total area of the former Soviet part of the region is about 1.4 to 1.6 million km^2 while that of the whole Basin is about 2 million km^2.

The pri-Aral Region includes the Karakalpakstan and Khoresm districts of Uzbekistan, the Dashkhovuz district of Turkmenistan, Ksyl-Orda and the southern part of Aktubinsk districts of Kazakhstan. The total area of the Aral Sea Region within these limits is about 480,000 km^2.

The socio-political errors responsible for desertification in the area originated in the early 1960s. An ambitious development plan was approved by the Supreme Presidium of the USSR, headed by Nikita Krushtchov, to conquer the drylands (Létolle & Mainguet, 1993). The strategy chosen was unlimited expansion of irrigation in an attempt to transform the vast, flat and dry endoreic ecosystem into the largest cotton belt of the planet. Thus the most gigantic irrigation structures on Earth were built, including the 1,300 km long Karakum canal in the south of Turk-

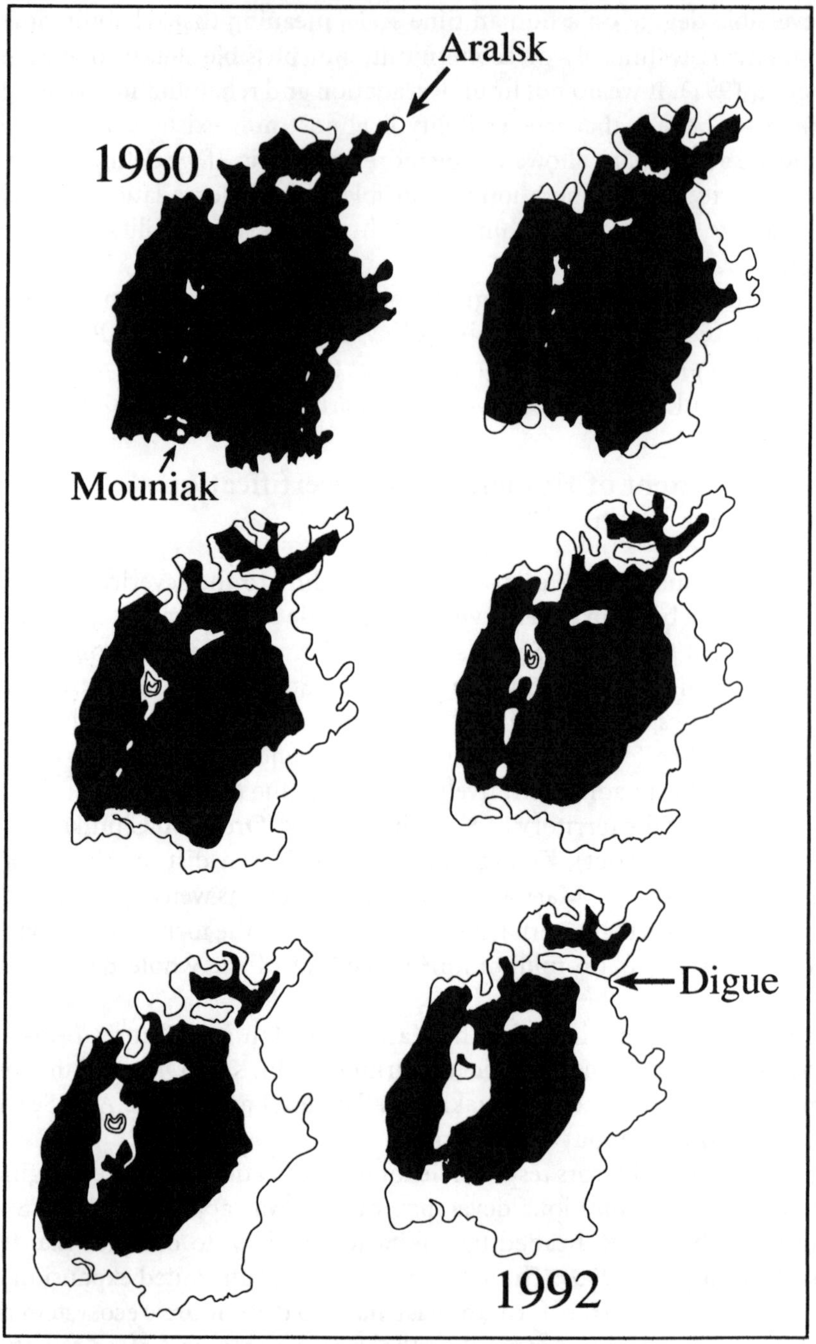

Figure 2. The shrinking Aral Sea, between 1960 and 1992.

menistan. About 4.5 million ha were already irrigated in 1965 in Soviet Central Asia, according to Kotlyakov (1990), consuming 50 to 55 km^3 of water per year. An additional 2.6 million ha were developed for cultivation during the following 25 years, requiring another 50 km^3 water. Therefore, virtually all the water resources of the only two permanent rivers which flow into the Aral Sea, the Amu Daria and Syr Daria, were used for irrigation. The devastating results of this gigantic development project, which caused human-made desertification, clearly shows the kind of development that should not be carried out in a dry ecosystem, where the surface water is allogenous and the hydrologic regime endoreic, forming a closed basin.

The water of the two main allogenous rivers, the Amu Daria and Syr Daria, beginning in the western mountain ranges of the Himalayas, is of glacial origin, due to the high topography. Hence the annual runoff from ice and snow melt is seasonal. The Amu Daria has an average annual flow of 85 million cubic meters (MCM) and the Syr Daria of 48 MCM at the foot of the mountains. However, almost none of this water has reached the Aral Sea since 1987.

The resulting desertification occurred in a very short time span, in one human generation, following the beginning of the colossal development project in the 1960s. An unusual complexity characterizes this desertification, as each type of human activity has engendered its own type of land degradation.

1. Overgrazing in the sand seas of the Karakum and Kyzylkum deserts has led to a degradation of the vegetation cover, resulting in dune reactivation and exacerbation of wind erosion.
2. Mega-irrigation in the valleys, plains and the two main deltas of the rivers Amu Daria and Syr Daria has, in this closed basin, inescapably led to salinization and waterlogging.
3. Industrial agrosystems, which replaced traditional agriculture, resulted in an overuse of all kinds of chemicals, causing toxicity of the soils, the water and the air in this dry ecosystem.

Classification of Desertification Processes in the Aral Sea Basin

The processes of desertification in the Aral Sea basin can be classified in physical, chemical and biogenic terms:

PHYSICAL ASPECTS OF DESERTIFICATION

Physical desertification in the Aral Sea basin mainly involves wind erosion and increased dust and sand storms (Mainguet, 1991), affecting more than 5 million ha, where aggravated aeolian processes and scattering of salts are the dominant processes. When, at the beginning of 1960s, the Aral Sea level began to drop, a strip of newly dry land appeared. At present the width of the dried strip in the north-eastern, eastern and south-eastern parts of the sea reaches up to 60–80 km, and in the area of the former bays 100–200 km. The total area of the dried bottom of the sea is more than 30,000 km^2.

The most specific feature of this landscape is the formation of barchanic sandy edifices and areas of erodible material. These are sources of deflation and massive dust transportation covering a total area of about 4800 km^2.

In the Aral Sea Region up to 10 strong dust storms are observed annually. Wind blown dust can travel a distance of 200–300 km (Razakov, 1990) and in some cases up to 500 km. The main direction of their movement is in the range N–S to NNE–SSW, towards the oases of the Amu Daria Delta and Khoresm (60%) and according to some scientists towards the west until the rangelands of the Ust Urt plateau (25%). We do not consider this last suggestion likely, for it is our understanding that the dust in the Ust Urt has a local origin. Therefore, in 85% of the cases, dust is transported above the sea, which has a significant reducing effect—the main part of the sandy aerosol is deposited on the water surface.

There is no single opinion about the quantity of dust material transported from the dried bottom of the sea. It is estimated to range from 8 to 30 million tons per year. Since the calculations have an estimative character, they probably are overstated, because of idealization of the effect of deflation on the newly dried sandy surfaces, the perception of long continuity of the strong dust storms (one day) and the suggested stability of the processes of particle transport.

According to these observations and assumptions, about 90–100 thousand tons (90–100 kg/ha) of particles (fine sand, silt, dust, salt) fall on an area of 10,000 km^2 annually in the Amu Daria delta area; and about 40–50 thousand tons (31–39 kg/ha) in the Ust Urt plateau with an area of 13,000 km^2. The total volume of particles exported from the area of intensive dust-salt storms (4,800 km^2) is 5–8 million tons per year. With such quantities of transported material, the rate of denudation of the dried sea bottom is about 2 mm/year. This is close to reality, because during the last three decades a top layer of 6–9 cm was removed by the wind from the newly-dried surface.

CHEMICAL ASPECTS OF DESERTIFICATION

Toxification and salinization of the soil are the two main processes of desertification in the Aral Sea basin. Salinization means abnormal concentration of salts at the surface of the soil, inside the soil, and in the groundwater. Salinization in the region is a consequence of irrigation, which is one of the most dangerous types of land use in this semi-arid ecosystem with internal drainage and a very high rate of evaporation. There is hardly an irrigated area without the potential hazard of salinization (Shanan, 1997, this volume).

Three main processes should be distinguished:

Increase in the salinization of the water of the lake as a result of reduction in the flow rate of the rivers Amu and Syr Daria towards the Aral Sea, as well as evaporation from the sea surface. The salinity of the water in the Aral Sea has multiplied by four in the last three decades, increasing from 10 to 12 g/l at the beginning of the 1960s to 40 g/l in the 1990s. At the northern and eastern part of the shrunken lake, 30 to 60 tons of salt have accumulated on each hectare of former lake bottom that has become dry land.

Salinization of the soils in the irrigated lands: 10 to 20 tons of salt per hectare accumulates annually on the soil surface and 15 to 25 t/ha/yr in the sub-soil, due to excessive irrigation, resulting in waterlogging and deposition of salts brought upwards by evaporation. Salinization affects 87% of the irrigated land in Turkmenistan, 60% in Uzbekistan (84% of Ferghana, 377,000 of the 485,000 ha of irrigated land are affected by salinization in Karakalpakstan), 35% in Tajikistan, 40% in Kyrghyzstan, and 60 to 70% in Kazakhstan (Khakimov, 1989 in Glazovsky, 1995).

The annual volume of drainage water collected by the Amu Daria river, having a salt content of 3.4–7.7 g/l, is more than 5.5 billion m^3. The salt content of water in the lower Amu Daria reaches up to 2.5–3.0 g/l as a result. Such a high mineral content of irrigation water led to the salinization of lands under irrigation. Taking into consideration the salt content of the water and the common irrigation quota for cotton, 10–33 tons of salt are brought annually with the irrigation water for each hectare of land.

Good drainage could eliminate 30 to 40 tons of salt per hectare a year, but this is difficult to achieve in the closed, flat Aral Sea basin. Drainage may in fact increase the risk of waterlogging and salinization downstream, due to the flat topography and endoreism.

The decline in the level of the Aral Sea caused a change of direction in the halochemical process. The Aral Sea used to be the salt receiver, which is the reason for the heavy salinization of the soil below the former sea surface. Today, the dried sea coast is the source of salt that is blown into the rangelands and irrigated lands. Some 0.2–5 tons of salts are deposited annually per hectare in the Amu Daria delta, as a result of sand storms.

Chemical pollution. The use of fertilizers, herbicides, pesticides and defoliants in the Central Asian lands under irrigation exceed the former Soviet Union standards by 10–15 times. Up to 54 kg per hectare of different herbicides, defoliants, *etc.* were used in Uzbekistan; and in the Dashkovuz district of Turkmenistan, each hectare of ploughed land received 8.6 kg of pesticides, 7.8 kg of defoliants, and 0.4 kg of herbicides annually. This has resulted in the pollution of soils, rivers and underground water. Toxic substances were not only discovered in water but also in food.

BIOGENIC ASPECTS OF DESERTIFICATION

Biogenic degradation concerns the vegetative cover as well as the macro- and micro-fauna. Desertification caused a diminution or destruction of the biological potential of the land (soils, plants, and animals) in the Aral Sea basin, just when the increased population requires enhanced productivity.

The degradation, transformation and markedly reduced biodiversity of the biological chain is one of the consequences of salinization. Some 20 species of edible fishes were living in the Aral Sea in the 1940s, but in the early 1990s, only 5 species remained, and at present only survives. The number of nesting bird species in the delta of the Amu Daria river has dropped from 319 to 168, while the amount of mammals species has diminished from 70 to 30. The riparian *tugai* forests, specific of the central Asian valleys, were reduced in size from 550,000 to 20,000 ha (Novikova, 1990; Zaletaev & Novikova, 1990).

Environmental Consequences of Desertification in the Aral Sea Basin

The massive diversion of water from the Amu Daria and Syr Daria rivers for irrigation purposes has caused numerous impacts on the environment. The manifestations of desertification in the Aral basin can be classified in three categories:

Those related to the water body of the lake:

1. The decline of the water level in the Aral Sea, which is a terminal lake, has progressed to such an extent that the lake has lost 60% of its volume, resulting in a regression of the coast line by 65 km in certain areas. The water volume of the Aral Sea has diminished from 1064 km^3 in 1960 to 310 km^3 in 1994 (Zonn, 1995), while its water surface has decreased from 66,000 km^2 in 1960 to 34,800 km^2 in 1990 (Belyaev, 1990; Kotlyakov, 1991).
2. The salinity of the water has risen from 10 g/l in 1942 to 40 g/l in the 1990s, resulting in the disappearance of the original water fauna and flora. Only one of the formerly 20 edible fish species survives at present.
3. A great part of the lake bottom has been exposed to aeolian deflation. Salts were blown several hundred kilometers away, mainly towards the south, polluting the air, surface water, soils and, indirectly, the underground water.

Those related to the Aral Region:

4. The desiccation of natural water courses in the deltas of the Amu Daria and Syr Daria rivers.
5. The number of dust storms in Karakalpakstan increased 60-fold between the 1950s and the 1970s, caused by the desiccation of the Aral Sea and the degradation of the natural vegetation in the region (UNEP, 1992).
6. The desiccation of the Aral Sea has resulted in climatic changes. The climate has become more continental, with an increased temperature range of 1.5 to 2.5°C between summer and winter, and an increase of 0.5 to 3.3°C between day and night. The lower winter temperatures, accompanied by earlier and later winter frosts, characterize a changed climate which is less favorable for the cultivation of the two main crops that were selected for the development of the basin: cotton and rice.

Those related to the river valleys and the Aral sea basin as a whole:

7. Natural lakes in the plains of the basin have disappeared, while irrigation wastewater lakes have formed in the surroundings of the irrigated areas, flooding 300,000 ha.
8. A rise of the groundwater table around irrigated areas and canals has resulted in salinization and waterlogging of the soils, spoiling vast areas of grazing land.

TABLE 1. Changing hydrologic parameters of the Aral Sea from 1960 to 1990 (after Zonn, 1995, the values for 1994 are modified)

Data	*1960*	*1994*
World ranking of Aral Sea (size)	fourth	seventh
Area (km^2)	66,900	28,000
Volume (km^3)	1,064	310
Water level (m)	53	34
Shoreline length (km)	4,430	3,950
Lake length (km)	428	240
Lake Width (km)	234–292	100
Depth max (m)	68–69	68–69
Depth average (m)	16.1	10.2
Salinity (g/l)	14	34
No. of islands in the Sea	12	2
Total area of islands (km^2)	2,230	4,000
Fishery (ton/yr)	30–40	0
Flow of Amu and Syr Daria rivers to Aral Sea (km^3)	52.9–56	20

Moreover, cotton plantations, occupying 60% of the irrigated land, are severely affected by an almost irreversible salinization: 15 to 90 tons/year of salt can accumulate in one hectare of irrigated soil (Kharin, 1994). The destruction of vast areas of grazing land is mainly due to the diversion of 30 km^3 per year of drainage water from the irrigated lands to natural desert rangelands. It causes waterlogging and salinization of 700,000 ha of desert grazing land.

9. The defective irrigation is not only responsible for the salinization of the soils, but has also caused pollution by excessive use of fertilizers, herbicides, pesticides and defoliants. The polluted drainage water cannot be evacuated out of the basin. Salts like gypsum, magnesium sulfite, sodium chlorides, which are toxic for human beings and animals above a certain threshold, will remain in the basin, polluting the underground water through infiltration.
10. The biodiversity has declined. Impoverishment of the terrestrial flora has taken place, as valuable plants were overtaken by unwanted species. The *tugai* forest, specific for the Amu Daria river banks, contained 576 species of which 29 are endemic to central Asia. Today 54 species are on the verge of extinction (UNEP, 1992). The terrestrial animal population has suffered similar damages.

Human Consequences of Desertification in the Aral Sea Basin

The impact of desertification in the area on human society can be summarized as follows:

1. Desertification in the Aral basin has resulted in the largest extent of transformed landscapes among all the drylands in the world.
2. Land degradation and loss of cultivable land occurred on a massive scale: over 7,200,000 ha of natural ecosystems were replaced by polluted agrosystems, 277,000 ha of plains were flooded by reservoirs, 40,000 ha were disturbed by canals. Unpopulated virgin lands were used for agricultural development. Human-made oases have been created as a result. The problems are related to the low quality of construction, over-exploitation of irrigation and drainage systems, excessive irrigation quotas and the introduction of monocultures of cotton and rice.
3. The loss of fishing potential, which used to be the main human activity in the coastal areas of the Aral Sea and the second activity after agriculture along the branches of the Amu Daria delta.
4. The population in the region is generally not serviced with centralized water-supply systems. In Turkmenistan only 85% of urban and 24% of rural population are provided with piped water. More than 50% of the population in the Aral Sea Region use rain water, polluted water from rivers, irrigation

canals and drainage collectors in order to meet their needs. Existing techniques of purification and distribution of water in the rural areas are not adequate to provide water of acceptable quality, meeting WHO health standards in terms of chemical content and bacteriological and toxicological indicators. The bad quality of drinking water due to pollution affects first and foremost the local population. In the collective farms *(kolkhoz)* the water contains high level of chemicals, mainly pesticides, as well as various salts.

5. The pollution of the air by dust and salts, combined with the consumption of bad quality water, adversely affects the health of the people. It has led to a reduction in human life expectancy, resulting in high rates of morbidity and mortality (Tables 2a, 2b).

Medical observations in the Aral Sea Region during last decade (1980–1990) show that mortality increased 15 times, cardio-vascular morbidity 1.6 times, tuberculosis 6 times, gastric diseases by a factor 5 and cancer of the esophagus by 7–10. The increase in infant mortality, according to existing data, is less related to bad medical care, than to the deterioration of the environment.

TABLE 2a. Morbidity in Karakalpakstan (per 100,000 inhabitants)

	1980	*1985*	*1989*
Cholelithiasis	8.5	50	58
Chronic gastricis	120	279	367
Kidney diseases	18	338	154
Arthrosis, Arthritis	7	12	26

	Karakalpakstan				*USSR*	
	1980	*1987*	*1988*	*1989*	*1988*	*1989*
Typhus, Paratyphus	26	17	13.5	13	4	3.3
Intestinal disorders	373	527	772	607	639	510
Viral hepatitis	584	1,503	543	771	251	316

TABLE 2b. Epidemiological Statistics (Nukus, 1990)

	USSR	*Turkmenistan*	*Bezirk Taschaus*
Average life expectancy	70	64.7	64.1
Puerperal mortality *	47.7	77.1	93.0
Infant mortality **	4.7	56.4	75.2
Hepatitis morbidity*	305.4	264.3	547.0
Malignant tumors*			295 (1985)
			334 (1988)
Newborn malformations			301 (1985)
			437 (1988)

* (per 100,000 inhabitants) ** (per 1,000 inhabitants)

It is difficult to ascertain whether the desertification status of the Aral Sea basin has reached the level of irreversibility from the scope of available economic and technical solutions. On the positive side, new laws on nature conservation and land tenure have been adopted. In Ashkabad (Turkmenistan) 500,000 bottles of mineral water are produced and delivered to the areas affected by water pollution, so that more people are enabled to drink safe water. Various efforts are being made to combat the effects of desertification, mainly by providing better drinking water to the population and a beginning to manage water scarcity through price adjustments and water-saving techniques. However, the scale and complexity of the desertification phenomena are so high that the resulting problems are difficult to solve. Positive corrective management involve reflooding of the deltas with drainage water and the building of windbreaks against sand encroachment.

The arid to semi-arid climate of the region increases the complexity, as it is difficult and costly to balance water supply and demand. The buffering effect of the modest biomass is insufficient to protect the landscape against severe wind erosion. This last aspect is often neglected.

Conclusions

Ambitious development planning and decision-making by a super-centralized state to create mega-irrigation in the arid ecosystem of the Aral Sea basin for the production of two wet tropical plants, cotton and rice, through an industrial agrosystem, has led to rapid ecological disequilibrium and human-made desertification in just one generation. Such mega development proved unsuited for the region, characterized by unfavorable climatic conditions, *i.e.* aridity, cold winters, frequent strong winds, as well as a restrictive flat topography without external drainage, in which two big rivers constitute the only supply of surface water, derived from precipitation falling outside the region. The human-made desertification seems irreversible within the time span of one generation, but may be reversed in the long run, although there are no cheap and simple solutions.

Major errors in design and planning of this mega-project were caused by the negligence to take into account the natural conditions, as well as the needs, wishes and suggestions of local populations. This bad example of human-made desertification in the Aral Sea basin demonstrates the difficulty and the risk of mega projects decided by politicians from outside the region. Moreover, it underlines the fact that sustainable development in an arid ecosystem is not possible without careful management (Bruins & Lithwick, 1997, this volume) of the water resources, the soils and the vegetation.

The economy of the Newly Independent States of the Aral Basin at the present time cannot support the allocation of large sums of money to combat desertfication.

Regular meetings of the Heads of the Central Asian States have been used to discuss a wide-range of problems affecting the Aral Basin. The formation of an *International Fund for the Aral Sea* (IFAS) to rescue the Aral were approved. Major tasks were identified to ameliorate the ecological situation of the Aral Sea Region within the next 4–5 years, taking into consideration the relevant social and economic development processes. The Statute on the *Interstate Council on the Aral Sea* (ICAS) was ratified and a leader for its executive body was appointed.

Last year, some socially-oriented projects were excuted to mitigate part of the negative consequences of the Aral Sea Region disaster. A 380 km long water-pipe operates in Uzbekistan (Tuijemoin-Nukus-Chimbal-Takhtakupir) with a capacity of 170,000 m^3 a day. Construction of gas-pipes was accomplished in the areas of Gazli-Nukus, Konghirat-Muynak, Kegheili-Bozatan . Hence 58% of the people could be supplied with natural gas and 52% with potable water. The government of Uzbekistan formed in February 1997 a *Coordinative Research Council for Study, Conservation and Rational Use of Central Asian Desert Resources* through its State Committee for Science and Technology.

References

Belyaev, A.V. (1990) Water balance of the Aral Sea and its man-induced changes. Prepared for International Conference 'The Aral Crisis: Causes, Consequences and Ways of Solution'. Nukus, Karakalpak, USSR, Oct. 2–5, 1990.

Bruins, H.J. and H. Lithwick (1997) Proactive planning and interactive management in arid-frontier development. *In* H.J. Bruins and H. Lithwick (eds.), *The Arid Frontier: Interactive Management of Environment and Development.* Kluwer Academic Publishers, Dordrecht, pp. 3-29.

Glantz, M.H., A.Z. Rubinstein and I. Zonn (1994) Tragedy in the Aral Sea Basin: Looking Back to Plan Ahead? *In* H. Malik (ed.) *Central Asia. Its Strategic Importance and Future Prospects.* St. Martin's Press, New York.

Glazovsky, N.F. (1995) The Aral Sea basin. Regions at risk: Comparison of threatened environments. *In* J.X. Kasperson, R.E. Kasperson and B.L. Turner II (ed.) *The Aral Sea Basin. Regions at Risk: Comparison of Threatened Environments.* United Nations University Press, pp. 92–139.

Kharin, N. (1994) Ecological catastrophe in Central Asia. *Our Planet* (UNEP) 6(5): 27–28.

Kotlyakov, V.M. (1991) The Aral sea basin. *Environment,* 33(1).

Létolle, R. and M. Mainguet (1993) *Aral.* Springer Verlag, Paris and (1996) *Der Aralsee. Eine ökologische Katastrophe.* Springer, Heidelberg (German).

Mainguet, M. (1991) *Desertification—Natural Background and Human Mismanagement.* Springer Verlag, Heidelberg, RFA, Springer Series in Physical Environments, Vol. 9.

Novikova, N.M. (1990) Problems of conservation of the ecosystems of river deltas in the Central Asia. Prepared for the International Conference 'The Aral Crisis: Causes, Consequences and Ways of Solution'. Nukus, Karakalpak, USSR, Oct. 2–5, 1990.

Razakov, R. (1990) The Aral sea and Aral zone: Ways and means of stabilizing the situation. Prepared for International Conference 'The Aral Crisis: Causes, Consequences and Ways of Solution'. Nukus, Karakalpak, USSR, Oct. 2–5, 1990.

Shanan, L. (1997) Environmental hazards of irrigation development: The need for proactive planning and interactive management. *In* H.J. Bruins and H. Lithwick (eds.) *The Arid Frontier: Interactive Management of Environment and Development*, Kluwer Academic Publishers, Dordrecht, pp. 251-276.

United Nations Conference on Desertification (1977) *Round-up, Plan of Action and Resolutions*. New York (1978).

UNEP (1992) Diagnostic study for the development of an action plan for the Aral Sea. Fourth meeting, Geneva, 3 August–4 Sept. 1992.

UNU (1992) Environmental management of the Aral Sea Region. Finding solutions to one of the world's major environmental disasters. Report of the International Symposium held at UNU Headquarters Building, Tokyo, Japan, Sept. 29, 1992.

Zaletaev, V.S. and N.M. Novikova (1990) Changes in biota of the Aral region as result of anthropogenic impacts in the period between 1950 and 1990. Prepared for International Conference 'The Aral Crisis: Causes, Consequences and Ways of Solution'. Nukus, Karakalpak, USSR, Oct. 2–5, 1990.

Zonn, I.S. (1995) Aral—From salvation of the sea to land restoration. *Proceedings* of the Forum on Caspian, Aral and Dead Sea Symposium on Aral Sea and the surrounding region. Tokyo.

ABOUT THE AUTHORS

Monique Mainguet is Professor of Geography at the University of Reims, Champagne-Ardenne. She is Director of the Laboratory for 'Géographie Zonale pour le Développement', created in 1973. She has worked since 1955 in the field of drylands, first investigating surface dynamics as a geomorphologist,which was published in the Global Wind Action System (GWAS). Next she studied land degradation and participated in the program to combat desertification, studying strategies for sustainable development of dry ecosystems. She has served on many international committees dealing with the human dimension of environmental changes.

Prof. Monique Mainguet
Professeur, University of Reims
Laboratory for "Géographie Zonale pour le Développement"
57 Rue Pierre Taittinger
51100 Reims
France

René Létolle is Professor Emeritus at the Pierre and Marie Curie University in Paris. He is a specialist in isotope geochemistry and water pollution. He supervised numerous students and published approximately 300 papers. He has been interested in the problem of the Aral Sea Basin since 1987, and has published two books and several papers on the subject with Prof. Mainguet.

Prof. René Létolle
Professeur, University Pierre and Marie Curie
75 252 Paris Cedex 05
France

Chapter 8

Drought Risk Assessment for Dryland Agriculture in Semi-arid Telangana, Andhra Pradesh, India

Ton J. Dietz, Marcel Put and S. Subbiah

Introduction[1]

The risk of drought in the semi-arid zone imposes on the farmers a meteorological-agricultural frontier situation, in which they struggle to obtain a reasonable yield, as the arid frontier changes in makeup and position from year to year. Semi-arid zones comprise risky environments and drought risks are prominent among a complex of environmental and socioeconomic risks in which farmers have to operate. However, the opposite of drought may also threaten farmers' livelihoods: floods or heavy rainfall can destroy crops when they are ripening. Pests can be a major risk. Labor shortage due to human disease problems or competing activities can also jeopardize the agricultural situation at the farm level. Price fluctuations, or insecure payments for crops which are marketed also form a risk, as well as price fluctuations and scarcity of necessary farm inputs. Political factors can also add to risk in some areas: from civil servants misusing their power to squeeze money out of farmers, to bandits and guerrilla fighters (*e.g.* Naxalites in the Indian context) stealing or claiming contributions from farmers. Drought risks are certainly not the only worries of dryland farmers. However, drought risks do matter and should not easily be dismissed as relatively minor, as political economists dealing with drylands sometimes do.

A systematic analysis of drought risks and a historical treatment of the rainfall variability can throw much light on the environmental context of farming in studies evaluating the performance of development projects, or in more general regional development studies dealing with dryland areas.

The relatively low rainfall is not the major problem for farmers in managing dryland agriculture in the semi-arid tropics. It does restrict cropping patterns to a few crops only: sorghum, millet, beans and cassava as food crops, and groundnuts,

H. J. Bruins and H. Lithwick (eds.), The Arid Frontier, 143–161.

castor or cotton as cash crops. The most worrying aspect for farmers is the unreliability of seasonal rainfall and the extreme variability of rainfall in time and place. One would expect major scientific attention focused on drought risks in areas where dryland agriculture has become the major lifeline of the rural population, and where relatively high population densities require optimal use of local resources. However, a systematic methodology to assess drought risks is lacking.

This chapter presents some of the results of a study in the Indian state of Andhra Pradesh, to assess the impact of dryland improvement projects. We will focus here on the southeastern part of Ranga Reddy district, in southern Telangana, where the Indian government, supported by the World Bank, attempted to improve dryland agricultural practices in the Maheswaram area from the mid-1980s onwards.

Figure 1. The location of Andhra Pradesh in India.

The southern part of Telangana is situated on the eastern side of the Deccan Plateau of Andhra Pradesh. It is not the driest region of India. The state of Rajasthan, in the west of India, as well as Rayalaseema in the southern part of Andhra Pradesh, receive less rainfall. However, southern Telangana has one of the highest rural population densities in the world in comparison to similar dryland areas. It has the additional burden of a mega city in its midst, the twin city Hyderabad-Secunderabad, with close to five million inhabitants. Our detailed analysis focuses on the eastern part of Ranga Reddy district (Figures 1 and 2). The study area receives between 600 and 800 mm of annual rainfall on average, with 50% of it during the months August-September (National Commission on Agriculture, 1976:15).

The geographical area of Ranga Reddy district amounts to 7,500 km^2, having a population of 2.6 million people (1991), of which 53% lives in rural areas. The population density is over 340/km^2. Literacy levels are rather high, 49% in 1991, due to the influence of Hyderabad (Government of AP, 1993:3–4, 22).

The geomorphology of the study area is rather hilly with isolated granite hills and two ranges. Elevations usually vary between 200 and 700 m above sea-level.

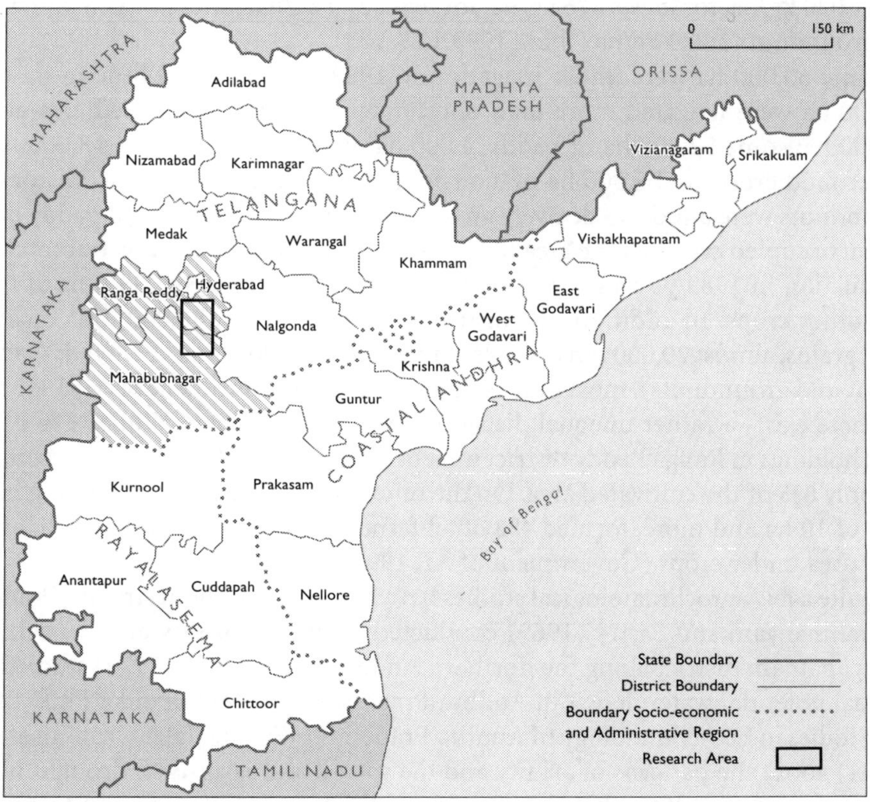

Figure 2. The location of Ranga Reddy in Andhra Pradesh.

Soils are mainly red sandy Alfisols, though mixed red and black soils (Vertisols) occur as well.

The total area of Ranga Reddy is 753,247 ha. The cultivated area in 1990/91 amounted to 40% of the geographical area; a 'net sown area' of 300,034 ha of which 10% was 'double cropped', giving a 'gross sown area' of 325,196 ha. Another 23% of the total area were 'current fallows' and 'other fallow land', while 4% was considered as 'culturable waste'. Other types of land use were: 'forests' (10%), 'permanent pastures and other grazing lands' (7%), 'land put to non-agricultural uses'(10%), 'barren and unculturable land' (5%) and 'land under miscellaneous tree crops and groves not included in the net area sown' (1%) (Government of AP, 1993, pp. 83–84). The irrigated area was 21% of the net sown area (Government of AP, 1993:150). Therefore, around 80% of the cultivated land can be regarded as 'drylands', which are in fact 'risky drylands' in the southeastern part of the district.

Food crop production dominates, occupying 82% of the 'gross cropped area'. In 1989/90, the main crops in the district were sorghum or *jowar* (32%), paddy (17%), pulses (16%), castor (8%), finger millet or *ragi* (3%), groundnuts (2%) and pearl millet or *bajra* (1%). Average crop yields in 1989/90 were: for rice 2,259 kg/ha, for *ragi* 1,020 kg/ha, for *jowar* 876 kg/ha, for *bajra* 340 kg/ha, for castor 468 kg/ha and for groundnuts 1,024 kg/ha (Ibid., 1993:122–123).

Some 60,000 ha were under irrigation in 1989/90 of which a little more than 20,000 ha were irrigated more than once, making the gross irrigated area about 81,000 ha: some 52,000 ha of paddy, 4,000 ha of 'other food grains', 18,000 ha of 'other food crops' and 8,000 ha of 'non-food crops'. Sorghum, millets, castor and groundnuts were almost exclusively grown as dryland crops. Irrigated paddy cultivation occupied only half the area of (non-irrigated) *jowar* and *bajra*, but total rice production in 1989/90 was some 125,000 tons against some 90,000 tons of these two other crops. In addition, the district produced almost 40,000 tons of other food grains, almost 20,000 tons of pulses and almost 40,000 tons of oilseeds (mostly castor and groundnuts), most of it as dryland crops (Government of AP, 1993:92).

There exists a rather unequal distribution of land holdings. While 39% of the land holdings in Ranga Reddy district were below one ha in 1986/87, they accounted for only 8% of the cultivated area. On the other hand, large farmers, owning holdings of 10 ha and more, formed 4% of all farmers, but they cultivated 29% of the total area under crops (Government of AP, 1993:142–143).

Quite a few agroclimatological studies have been written about Andhra Pradesh. Subrahmanyam and Sastry (1969) conducted a study of aridity and droughts in Vizagapatnam district along the northern Andhra Coast. Singh (1979) studied the annual potential water deficit in Andhra Pradesh. Hema Malini did a Ph.D thesis on 'Studies in Eco-climatology of Andhra Pradesh' (1979) and later wrote an article (1981) about the patterns of aridity and the spread and severity of drought in the state in the years 1966–1969. Application of the cumulative deviation technique from which severity indices were analyzed gave the duration and severity aspects of droughts.

Agrometeorological studies for the optional and planned use of surface irrigation and rainfall sources were carried out by Venkataraman (1981). Moisture adequacy indices were computed for 26 stations in the state by Subramanaian & Vidyanath (1984). The spatial patterns of moisture adequacy and water deficit were compared with the spatial patterns of intensity of cropping and of irrigation and crop diversification. The relationship between climate and crops in Rayalaseema has been analyzed by Venkateswara Rao (1984). An analysis of the rainfall patterns, droughts and water resources of the Cuddapah district, part of Rayalaseema, was studied by Krishna Reddy (1990). Kumar (1991) studied the correlation between agricultural potential and development in the drought-prone area of Anantapur district, also in Rayalaseema. He included the relationship between soil type and utilization potential with regard to surface and subsurface water resources. In their important book about 'Village and Household Economies in India's Semi-Arid Tropics' (1990:33–39), Walker & Ryan included an analysis of rainfall patterns and the consequences of variability for Aurepalle village in Mahabubnagar district. Only the last study places the consequences of the seasonality and variability of rainfall for agriculture at the center of its approach.

Rainfall Variability in Ranga Reddy District

Rain-dependent crop farmers in relatively dry areas have to cope with a variety of climate-related risks. The average low rainfall situation allows only a narrow range of crops. Lack of crop diversity diminishes the ability to diversify the agricultural economy. Average low yields of these crops, especially if combined with land or labor constraints, result in low food security margins. However, for rainfall-dependent farmers in dry areas the most problematic element of farming is not the average low rainfall and resulting average low yields, but the large fluctuations in rainfall, from place to place, from year to year and within the rainy season.

The rainfall stations situated in the vicinity of the driest part of Ranga Reddy are Ibrahimpatnam (data for 1957–1986), nearby Maheswaram (data for 1981–83 and 1985–93) and Kandukur (data for 1985–92). However, there are no temperature data, evaporation measurements or evapotranspiration assessments for these stations. Long-term data on rainfall, evapotranspiration and temperature are available for the district headquarter, Hyderabad Airport (1901–1973) and ICRISAT station (1974–92). For Ibrahimpatnam/Maheswaram Figure 3 presents a graphic illustration of the rainfall variability.

Ranga Reddy District lies in the semi-arid zone, according to rainfall (P) and potential evapotranspiration (ETP) data of the district headquarters (in Hyderabad)

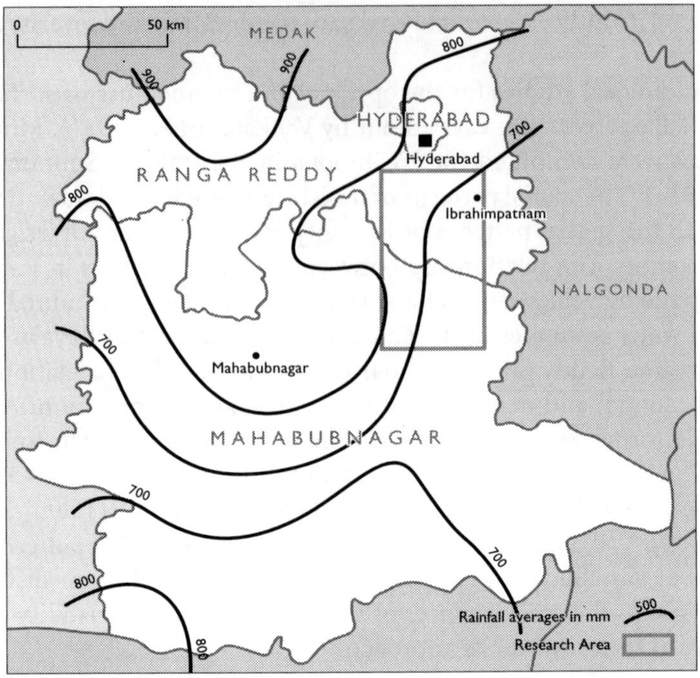

Figure 3. Ibrahimpatnam/Maheswaram: The variability of monthly rainfall, 1957–1993.

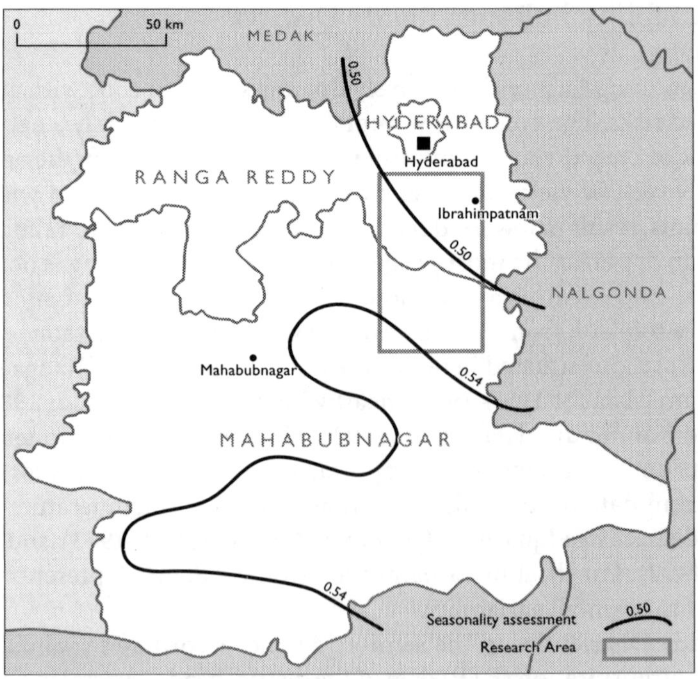

Figure 4. Average rainfall isohyets in Ranga Reddy District.[2]

and the UNESCO (1979) terminology, based on the P/ETP classification[3].The aridity index for Hyderabad Town is 0.45 (P 796 mm; ETP 1758 mm). The aridity index is about 0.40 in the driest parts of the district, near Ibrahimpatnam/Maheswaram (P 675 mm; ETP estimate: 1,700 mm). Average rainfall isohyets are presented in Figure 4.

In Hyderabad the period between 1920 and 1950 was relatively dry, particularly the 1920s (Table 1). After 1950 the decadal average gradually increased, peaking during the 1980s. The most extreme years in Hyderabad were 1915 and 1916, with a rainfall maximum of 1431 and 1425 mm, respectively, and 1972 with a minimum of 416 mm. The 'wet' and the 'dry' periods stand out clearly, if we take all the annual rainfall figures and make five groups of 20% each. For Hyderabad the categories are based on the annual rainfall data for 92 years (1901–1992):

++	wet	>925 mm
+	moist	824–925 mm
o	average	708–823 mm
–	dry	639–707 mm
–	very dry	<639 mm

The first three decades show a negative situation with 11 wet and moist years, against 16 very dry and dry years. The period 1930–1960 shows a balanced picture: 10 moist/wet and also 10 dry/very dry years. The period 1960–1990 shows a positive picture with 15 moist/wet and 10 dry/very dry years.

TABLE 1. Decadal averages of monthly rainfall in Hyderabad*, 1901–1990

Year	Jan	Feb	Mar	Apr	May	Jun	Jul	Aug	Sep	Oct	Nov	Dec	TOTAL	*KHARIF* SEASON **
1901–10	11	6	5	32	30	111	144	159	177	74	14	8	771	591
1911–20	7	11	22	27	35	122	193	136	185	81	39	6	864	636
1921–30	18	10	1	20	18	85	147	127	164	79	36	6	711	523
1931–40	3	22	12	31	38	99	166	113	152	70	34	10	750	530
1941–50	3	9	13	18	22	119	157	140	176	54	32	4	747	592
1951–60	0	1	15	20	31	101	179	190	160	97	17	2	813	630
1961–70	3	4	12	27	33	90	197	147	205	71	12	14	815	639
1971–80	5	7	5	22	36	111	139	215	165	94	21	1	821	630
1981–90	9	5	21	25	32	123	195	177	168	83	31	4	873	663
Average	7	8	12	25	31	107	169	156	172	78	26	6	796	604

Notes: * Hyderabad Airport (Begumpet) until 1972; Hyderabad ICRISAT from 1972 onwards (there are indications .that the ICRISAT station receives slightly more rain than the Begumpet Station).

** *Kharif* season: June–September .

Source: ICRISAT files.

For the drier southeastern part of Ranga Reddy the improving rainfall trend can also be confirmed. Figure 5 presents the evidence, for the 1957–1993 period.

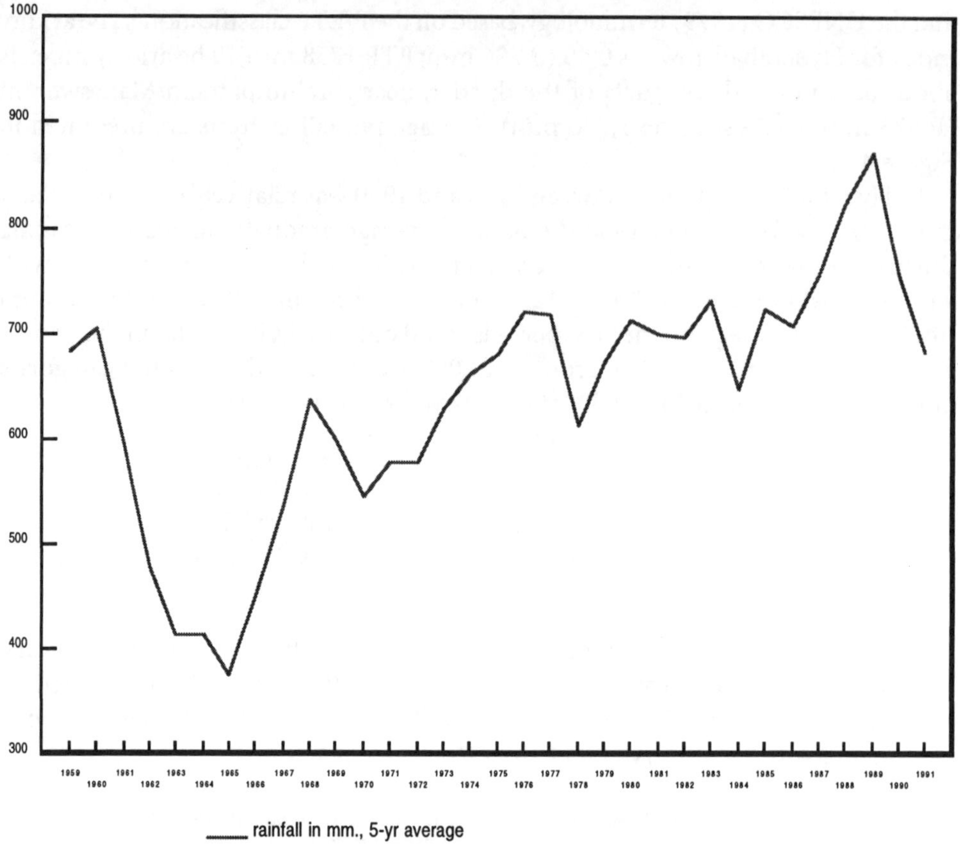

Figure 5. Ibrahimpatnam/Maheswaram: Long-term rainfall trends, 1957–1991.

Rainfall Seasonality in Ranga Reddy District

The rainfall pattern of Ranga Reddy is clearly seasonal. The greater part of the rainfall is received during the southwest monsoon, from mid-June to mid-October. Average monthly rainfall and evapotranspiration figures for Hyderabad are given in Table 2. They show that there are only three months in which ETPm[4] exceeds Pm: July, August and September. The rainy season in Ranga Reddy usually starts in June and ends in September or October. This period is called '*kharif*'. The period July-October has a higher rainfall than the period June-September in 30% of the years. Pre-monsoon showers often occur in May, prior to the real start of the rainy season, 'which helps in seeding the sorghum crop in the light textured soils of the region' (AP-DoA, 1987:3).

The period between October and December, called '*rabi*', has much less rainfall, although some farmers try to get a second harvest on dryland. The months January and February constitute the 'winter', when it is relatively cold and very dry. It is the slack season in agriculture, although some harvesting takes place of castor beans. The hot and dry period from March to May, called 'summer', is largely devoid of agricultural activity, but at the end of this season all farmers start field preparations.

TABLE 2. Average monthly rainfall and potential evapotranspiration data for Hyderabad

	Jan	Feb	Mar	Apr	May	Jun	Jul	Aug	Sep	Oct	Nov	Dec	Annual Total	Rainy Season*
P	7	8	12	25	31	109	168	157	170	77	26	6	796	625
ETP	110	130	182	198	220	198	140	136	120	124	104	100	1758	592
0,5ETP **	55	65	91	99	110	99	70	68	60	62	52	50		
Dry = P below ***	74	85	98	112	119	106	96	93	93	90	79	73		
Wet =P above ***	114	130	150	171	182	163	146	143	143	137	122	111		
P/ETP	.06	.06	.07	.13	.14	.55	1.20	1.15	1.42	.62	.25	.06	.45	1.06

Notes: * P based on annual figures for June-September or July-October, whichever is highest; ETP based on June–September.
** Needed for FAO approach to drought risk assessment, see later.
*** Needed for Bailey's approach to drought risk assessment, see later.
Source: ICRISAT files. Rainfall based on 1901–1992; potential evapotranspiration (most probably) on 1930–1960.

A number of 'seasonality indexes' appear in the literature. The simplest indexing method for tropical areas (Walsh, 1981), defines a 'dry month' as one with less than 100 mm average rainfall. Thus, the area of Ranga Reddy district has eight to nine dry months per year. The southern part of Andhra Pradesh is more extreme, with Anantapur district having ten dry months. Coastal Andhra Pradesh towards the east is more humid, with six to seven dry months.

Foeken (1994:29–30) formulates three other seasonality assessments, two of which are based on Walsh (1981). The first one is called the (relative) *seasonality ratio*, defined as the difference between mean rainfall of the wettest month and the mean rainfall of the driest month, divided by mean annual rainfall. Values range from 0, for equal intra-annual rainfall distribution, to 1, for complete concentration of rainfall in one month only. For Ranga Reddy the seasonality ratio is 0.21–0.23. This suggests a rather weak seasonality. The second index, developed entirely by Foeken, involves the three months with most rainfall and the three months with least rainfall. The difference between M3max and M3min is then divided by the mean annual rainfall. Foeken calls it the *Degree of Rainfall Seasonality*, DRS, with values from 0% (equal distribution) to 100% (all rainfall in three months). For Hyderabad the DRS would be 60%, which suggests a rather clear seasonality. Foeken also presents data based on Walsh's *'seasonality index'*, defined as the sum of the absolute deviations of mean monthly rainfall from the overall mean, divided by mean annual rainfall. Its value ranges between 0 (for rainfall equality) and 1.83 (a theoretical maximum indicating all rain to be in one month only). For Hyderabad the seasonality index is 0.88, somewhat in between.

In this contribution we go one step further. We base our own *'seasonality coefficient"* also on the statistical 'index of diversity', but adjusted it to arrive at values between 0 (maximum distribution) and 1 (maximum concentration).

$$D = \frac{\sum_{1}^{n} |x_i - x_a|}{2(n-1)x_a} \qquad \text{Using } n = 12: \quad S = \frac{\sum_{1}^{12} |x_i - x_a|}{22x_a}$$

D = Diversity Index; S = Seasonality Coefficient
where x_i = value for month i (1 for January, 12 for December)
where x_a = average monthly value

A seasonality coefficient that would even better describe the seasonal variability would be based on weekly figures (then n = 52).

TABLE 3. Hyderabad 1900–1990, seasonality coefficient

Decade	Average annual rainfall (in mm)	Average rainfall in June-Sept (in mm)	Average rainfall June-Sept or July-Oct (in mm)	R/Y in %*	Seasonality coefficient
1901–10	771	591	620	80	0.49
1911–20	864	636	662	77	0.45
1921–30	711	523	551	77	0.47
1931–40	750	530	547	73	0.42
1941–50	747	592	592	79	0.50
1951–60	813	630	663	82	0.52
1961–70	815	639	662	81	0.50
1971–80	821	630	658	80	0.51
1981–90	873	663	673	77	0.48
Average	796	604	625	79	0.48

R= Average rainfall in June-Sept. or July-Oct. (in mm), whichever is highest; Y=AverageAnnual Rainfall.

Our seasonality coefficients for Hyderabad are presented in Table 3. The percentage of rain falling in the 'rainy season', here defined as 'June-September or July-October, whichever is highest' is given as well.

The seasonality coefficient for Hyderabad on a decadal basis varies around 0.50. There was an exceptional period of lower seasonality during the 1930s, making the already dry decade even more problematic for agriculture. If we look at the decadal rainfall averages for the rainy season we can see that the 1950s and 1960s have been exceptionally wet and the 1920s very dry. Figure 6 presents a seasonality coefficient for the whole of Ranga Reddy, based on average figures.

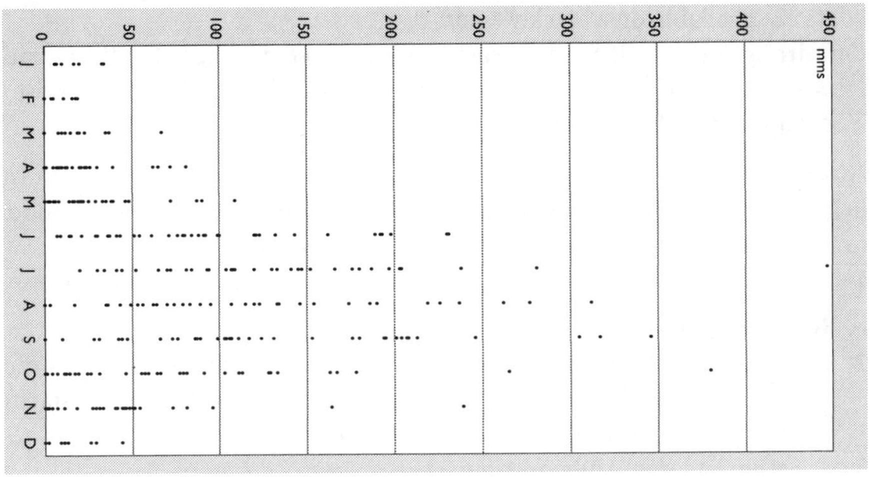

Figure 6. Rainfall seasonality in Ranga Reddy district.[5]

Drought Risk Assessment: A Methodology

Average climatic data for the area might suggest an environment that easily enables a sorghum and millet cropping pattern on dryland. However, the variability of the rainfall, its distribution over the months, and the insecure start of the rainy season present big risks for the farmers who cultivate dryland crops. This is a well-documented fact for semi-arid climates (*cf.* Reitsma *et al.*, 1992). The standard deviation of rainfall in the driest parts of the Deccan, with 600 mm average annual precipitation, is around 200 mm (Bakker, 1992:75). The Department of Agriculture of the State Government of Andhra Pradesh states: 'it has been observed that the potential of dryland crop(ping) is not significantly reduced even when only two-thirds of the normal rainfall is received provided it is well distributed' (AP-DoA:3). Any drought risk assessment, therefore, should look at the actual rainfall distribution during the months (or, if data would allow, during the weeks).

We have developed a drought risk index, based on two earlier indexes (Bailey, 1979 and FAO, 1980). Bailey's method used monthly rainfall figures and mean monthly temperature figures to arrive at a 'moisture index' in which he differentiates 'wet' months, 'neutral' months and 'dry' months. The FAO approach, developed for its Agro-Ecological Zones Project, used monthly rainfall figures and monthly potential evapotranspiration (ETP) figures and included soil moisture storage of a 'rain' month for a following 'dry' month.[6] So, for this drought risk index one needs: monthly rainfall data, (average) monthly evapotranspiration data, (average) monthly temperature data and a good calculator. It is preferable if rainfall data are available on a weekly basis.

Our drought risk index uses a scale from zero to ten, adapted to the situation of the study area. Zero indicates least drought risk, while ten indicates the most severe risk of crop failure due to drought.

No drought risk

0 At least four consecutive 'wet' months, according to Bailey's method; normal start of the rainy season (Ranga Reddy: June)

1 Same; late start of the rainy season: first 'wet' month only in July or August

Very light drought risk

2 At least four consecutive months which are, according to Bailey's method, at least 'neutral', but at least one of them is not 'wet'; normal start of the rainy season (June)

3 Same; late start (July or August)

Light drought risk

4 At least three consecutive months which are 'neutral' or 'wet' (Bailey); normal start (June)

5 Same; late start (July or August)

Moderate drought risk

6 According to Bailey's method there are no three consecutive 'neutral' or 'wet' months; but according to the less strict FAO method there are three consecutive 'rain' months; normal start (June)

7 Same; late start (July or August))

Severe drought risk

8 According to the FAO method there are no three consecutive 'rain' months; however a 'dry' month is sufficiently compensated by storage of rain of the previous month (storage: P-ETP; available rain = $P_m + P_{m-1} - ETP_{m-1}$); normal start of the rainy season (June)

9 Same; late start (July or August)

Extreme drought risk

10 Not enough rainfall or moisture storage for crop growth, according to the FAO method.

Besides drought farmers also face the risk of extremely high rainfall, which can damage the crop. 'Too much rainfall received at the wrong time can have a devastating effect on yield', as observed by Walker & Ryan (1990:49) in their Aurepalle case study in Mahabubnagar, to the southwest of our research area. 'Abundant and sustained rain in late September and early October precipitated a sharp rise in insect and disease infestation'. We will not include an analysis about 'harvest spoilage risks' in this chapter. Farmers generally try to avoid late harvests, knowing that rainfall in the last part of the rainy season can be damaging. Therefore all farmers try to sow sorghum in the last week of May or the first week of June, before the start of the rainy season in order to harvest in September and not in October. If the situation in September/October after the *kharif* period is still good, some try to get a second *rabi* harvest, somewhere in December. Most farmers are more relaxed concerning castor beans: they start in July, and harvest in the December-February period.

Analysis of Drought Risk in Southeastern Ranga Reddy District

Based on the evidence for Hyderabad,[7] Ranga Reddy farmers experienced during this century 5 years with virtually no drought risks (DR = 0), 23 years with a very light drought risk (DR = 1 or 2), 21 years with light drought risks (DR = 3 or 4), 28 years with moderate drought risks (DR = 6 or 7), 10 years with severe drought risks (DR = 8 and 9) and 3 years with extreme drought risks (DR = 10) (Table 4). However, as we have indicated, the situation in Hyderabad is relatively favorable compared to the area more towards the southeast of the district.

As an example of the drought risk assessment approach we will use data for the driest part of the district, based on monthly rainfall figures for Ibrahimpatnam/ Maheswaram for the years 1957–1993. We had to make use of proxy data for average monthly evapotranspiration and average monthly temperatures, based on the situation in Hyderabad.

It is obvious that the rainfall situation in the Maheswaram area is indeed drier than Hyderabad Town. The average annual rainfall is only 638 mm, with about 500 mm falling in the rainy season. The average drought risk is moderate, with an average drought risk assessment index of 7.2, which is considerably higher than around the district headquarters. Table 5 presents a comparison between the drought risk in Hyderabad (based on data for 1900–1992) and the drought risk in Ibrahim–patnam/ Maheswaram (based on data for 1957–1993).

For Hyderabad the risk of a moderate, severe or extreme drought risk during the periods 1901–1920, 1931–1950 and 1961–1980 was more than 50%. However, that risk was much higher in the southeastern, driest parts of Ranga Reddy district, around Ibrahimpatnam and Maheswaram: 80% in the 1960s and 70% in the 1970s

TABLE 4. Rainfall and drought risks in Ibrahimpatnam/Maheswaram, 1957–1993

Year	*P(an)*	*P(rs)*	*P(rs) typ.*	*DR*
1957	711	629	+	6
1958	754	458	o	9
1959	682	597	+	5
1960	602	451	o	8
1961	667	512	+	7
1962	822	688	++	3
1963	202	143	--	10
1964	95	88	--	10
1965	283	283	--	10
1966	671	496	o	3
1967	633	597	+	9
1968	560	452	o	10
1969	545	390	-	6
1970	782	611	+	6
1971	471	387	-	10
1972	363	239	--	10
1973	728	491	o	7*
1974	544	478	o	9**
1975	1044	891	++	3
1976	630	504	o	4
1977	449	384	-	9
1978	932	707	++	2
1979	530	379	-	10
1980	524	423	-	6
1981	932	645	+	6
1982	640	450	-	8
1983	872	764	++	1
1984	520	360	--	6
1985	690	445	-	9
1986	522	424	-	6
1987	1019	623	+	9
1988	778	731	++	5
1989	763	731	++	9
1990	1030	737	++	2
1991	771	655	+	6
1992	435	278	--	10
1993	422	373	--	9
Aver.	638	500	o	7.2

Notes: * start in August; ** start only in September.
P(an) = Annual rainfall; 1957–85 for Ibrahimpatnam; 1986–93 for Maheswaram.
P(rs) = Rainfall during the rainy season: June–September or July–October whichever is higher.
P(rs)typ = Same, typology: -- = < 379 mm; - = 379 to 450 mm; o = 451 to 511 mm;
+ = 512 to 688 mm; ++ = > 688 mm.
DR = Drought Risk Assessment, based on methodology explained in the text.

TABLE 5. Drought risk assessment, comparison between Hyderabad and Ibrahimpatnam/Maheswaram

Drought risk assessment	*Ranga Reddy/ Hyderabad (risk in %)*	*Ranga Reddy/ Ibrahimpatnam Maheswaram (risk in %)*
0 & 1 none	5	3
2 & 3 very light	26	14
4 & 5 light	24	9
6 & 7 moderate	30	30
8 & 9 severe	11	32
10 extreme	3	22

and 1980s. When two or more years with severe or extreme drought risks follow each other, dryland farmers are faced by very threatening circumstances. In Ibrahimpatnam/Maheswaram this happened in 1963–65, 1967/68, 1971/72 and 1992/93.

Reliable data on real harvests are difficult to obtain. The Ranga Reddy district Annual Handbooks of Statistics do give estimates of harvests for a variety of *kharif* crops. However, these figures include the more humid western parts of the district and do not differentiate between irrigated and dryland areas. They are notoriously unreliable, because 'area cropped' and 'yields' are based on civil servants' guesswork. Hence, it is not useful to compare these data with the drought risk assessment figures of Hyderabad and Ibrahimpatnam/Maheswaram. A more restricted scale of time and place can present some useful evidence, though. For three research villages near Ibrahimpatnam detailed comparable farm-level data are available for the years 1989, 1990 and 1991.[6] According to the drought risk assessment for Maheswaram, 1989 was a year with a severe drought risk (although the rainfall in the rainy season was relatively high). 1990 was a wet year, with only very light drought risk; 1991 had a moderate drought risk. Table 6 presents data about the percentage of farmers who had sorghum yields above certain thresholds in three villages combined (source: IDPAD team, supervised by M. Put).

The research data show that the yields were relatively low for all three years. The year with most adequate rainfall, 1990, indeed gives the highest percentages of sorghum yields obtained by farmers above all thresholds used. The drought risk assessment for 1989 would suggest a higher risk than for 1991. However, this is only

TABLE 6. Sorghum yields 1989–1991,Ghatpally, Mansanpally and Debbadaguda combined

Year	*% without jowar **	*n jowar*	*%>200 kg/ha*	*%>300 kg/ha*	*%>400 kg/ha*	*%>500 kg/ha*	*%>800 kg/ha*
1989	20	233	95	74	51	20	3
1990	23	226	95	79	68	36	8
1991	27	215	85	63	51	23	7

* jowar = sorghum

the case for the higher yields. For the lowest yield thresholds it is the other way around.

Concluding Remarks

Socioeconomic studies of dryland agriculture in semi-arid areas do regard drought as a threat (among other threats) to farmer's livelihood situations, but often in very general terms. Hardly ever do these studies present an analysis of the rainfall data to arrive at a more sophisticated assessment of the drought risk for dryland agriculture. Our study attempted to develop such a drought risk assessment scale for a densely-populated semi-arid region in the southeastern part of Ranga Reddy district in the Indian state of Andhra Pradesh, not far from Hyderabad Town. An attempt to validate the approach, as a 'predictor' of dryland crop yields could not be more than a first preliminary trial. It is hoped that the methodology presented will be applied by other scholars and tested by using reliable data on dryland crop harvests. If rainfall data permit, it would be good to use weekly figures, and of course, if evapotranspiration and temperature data can be based on the same weekly (or monthly) periods as the rainfall data that would further refine the assessment. For long-term reconstruction of rainfall contexts for dryland farmers such refinement is obviously impossible. The proxy approach which we suggested (based on monthly data) does make it possible, though, to reconstruct long-term trends and fluctuations of one of the most important environmental variables for dryland agriculture. The approach can also help to understand the effect of external development interventions to improve dryland agriculture, as the adoption or non-adoption of suggested improvements will at least partly be based on yields in relation to rainfall and droughts before, during and after project implementation.

Notes

1. Acknowledgements: this publication is based on research funded by the Indo-Dutch Programme on Alternatives in Development, IDPAD (see Put & Van Dijk, 1996). Figures were drawn by UvA Kaartenmakers.
2. Based on average monthly rainfall data for many rainfall stations in the district, provided by the Indian Meteorological Department, probably based on 1930–1960; own calculations of seasonality coefficient; see text.
3. UNESCO (1979, p. 11) suggests the following categories for Average Rainfall (P) divided by Average Evepotranspiration (ETP):

> 0.75	humid
0.50-0.75	subhumid
0.20-0.50	semi-arid
0.03-0.20	arid
< 0.03	hyper-arid

The figures for evapotranspiration for Hyderabad were provided by the Indian Meteorological Department (New Delhi), and are probably based on Penman's methodology.

4. Pm= average rainfall for the particular month; ETPm= average evapotranspiration for the particular month.
5. FAO: when Pm > 0.5 ETPm a month is sufficiently wet for a crop that is adapted to dry conditions; when P>ETP, the surplus rainfall can be stored for the next month as soil moisture (with a maximum storage period of one month).
   Bailey's moisture index: $Si = 0.18P/(1.045)^t$, where precipitation P is in cm and t is in °C. A month is 'wet' when Si > 0.81 and 'dry' when Si < 0.53. Months are called 'neutral' in between 0.53 and 0.81 (see Dietz, 1987 for earlier applications).
6. Detailed data can be found in Put & Van Dijk, 1996.
7. Results do not seem to be reliable. Further analysis is needed, not presented here.

References

AP-DoA (nd) Andhra Pradesh, Department of Agriculture: Pilot Project for Watershed Development in Rainfed Areas, Maheswaram Project; Soil and Land use Survey Report. Hyderabad.

AP-DoA (1987) Andhra Pradesh, Department of Agriculture: Pilot Project for Watershed Development in Rainfed Areas, Maheswaram Project; Action Plan for 1987–88. Maheswaram, Hyderabad.

Bakker, E. J. (1992) Rainfall and Risk in India's Agriculture. An Ex-Ante Evaluation of Rainfall Insurance. Ph.D. Thesis, Wolters-Noordhoff, Groningen.

Bailey, H.P. (1979) Semi-arid climates: Their definition and distribution. *In* A.E. Hall,, G.H. Cannell & H.W. Lawton (eds.) *Agriculture in Semi-Arid Environments.* Springer, Berlin, Ecological Studies 34, pp. 73–97.

Dietz, A.J. (1987) *Pastoralists in Dire Straits. Survival Strategies and External Interventions in a Semi-Arid Region at the Kenya/Uganda Border: Western Pokot, 1900–1986.* Netherlands Geographical Studies, Utrecht/Amsterdam.

FAO (1980) Report on the second FAO/UNFPA Expert Consultation on Land Resources for Populations of the Future. FAO, Rome.

Foeken, D. (1994) Climatic seasonality in Kenya with special reference to Coast Province. *Eastern & Southern Africa Geographical Journal* 5(1):28–43.

Government of Andhra Pradesh (1993) *Statistical Abstract 1990–91.* Hyderabad.

Government of India (1992) *Statistical Outline of India 1992/93.* New Delhi.

Hema Malini, B. (1979) Studies in Eco-climatology of Andhra Pradesh. Unpublished Ph.D. Thesis, Andhra University, Waltair.

Hema Malini, B. (1981) Patterns of aridity, spread and severity of drought in Andhra Pradesh. *Indian Geographical Journal* 56(1):30–36.

Krishna Reddy, P.R. (1990) *Some Aspects of Agroclimatology, Landuse and Cropping Pattern of the Cuddapah District, Andhra Pradesh, Using Remote Sensing Data.* Sri Krishnadevaraya University, Anantapur.

Kumar, Y.V. (1991) Agricultural Potential and Development in the Drought Prone Area of Anantapur District, Andhra Pradesh. Unpublished Ph.D. Thesis, Sri Krishnadevaraya University, Anantapur.

Penchaliah, T. (1989) An Analysis of Rainfall, Droughts and Water Resources of the Cuddapah District. Unpublished M.Phil. Thesis, Sri Krishnadevaraya University, Anantapur.

Put, M. and M.P. van Dijk (1996) Government and NGO Interventions in Dryland Agriculture. A study in Andhra Pradesh. IDPAD: Amsterdam; final report.

Radhakrishna, R. and S. Reddy (1990) *Incidence of Poverty in Andhra Pradesh.* CESS, Hyderabad.

Ramakrishna Reddy, V. (1987) *Economic History of Hyderabad State (Warangal Suba: 1911–1950).* Gian Publishing House, New Delhi.

Ranga Reddy District (1985, *etc.*) *Handbook of Statistics 1982–83 & 1983–84, 1985–86, 1986–87 & 1987–88, 1988–89, 1989–90, 1990–91 & 1991–92.* Office of the Chief Planning Officer, Ranga Reddy District.

Rao, G.R.S and P.K.S. Madhavan (1988) Voluntary Action in Andhra Pradesh. Summary on the study of the State Andhra Pradesh. Hyderabad (typescripts).

Reitsma, H.A., A.J. Dietz and L.J. de Haan (1992) *Coping with Semi-Aridity.* Netherlands Geographical Studies, Utrecht/Amsterdam.

Rupa Kumar (1980) Agroclimatic Conditions in Andhra Pradesh and the Effects of Weather Anomalies in Major Crops. Unpublished Ph.D. Thesis, Andhra University, Vizagapatnam.

Singh, N.K. (1979) Annual potential water deficit in Andhra Pradesh. *In* Reddy, N.B. K.(ed.), *Proceedings of All India Symposium on Drought Prone Areas in India.* Rayalaseema Geographical Society. S.V. University, Tirupati.

Sitapathi Rao, C., M.V.K. Sivamohan, S. Chandy and R. Sudhakar Rao (1990) Pilot Project for Watershed Development in Rainfed Areas Maheswaram. Special Studies on Soil and Moisture Conservation Activities, Crop Production Programme and Pasture and Fodder Development programme. Administrative Staff College of India. Hyderabad.

Subrahmanyam, V.P. and C.V. Sastry (1969) A study of aridity and droughts in Vizag. *Annals of Arid Zone* 8(1):18–20.

Subramanaian, M.J. and V. Vidyanath (1984) Moisture adequacy and cropping pattern in Andhra Pradesh, India. *Indian Geographical Journal* 59(2).

UNESCO (1979) Map of the world distribution of arid regions. Explanatory note and Map. MAB Technical Notes No. 7, Paris.

Venkataraman, I. (1981) Agrometeorological Studies for the Optional and Planned Use of Surface Irrigation and Rainfall Resources. Unpublished Ph.D. Thesis, Andhra University, Waltair.

Venkateswara Rao, M. (1984) Some Studies on Climate and Crops in Rayalaseema, Andhra Pradesh, India. Unpublished Ph.D. Thesis, Andhra University, Waltair.

Walker, T. S. and J. G. Ryan (1990) *Village and Household Economies in India's Semi-Arid Tropics.* The Johns Hopkins University Press, Baltimore and London.

Walsh, R.P.D. (1981) The nature of climatic seasonality. *In* R. Chambers *et al., Seasonal Dimensions to Rural Poverty.* Francis Pinter, London.

ABOUT THE AUTHORS

Ton Dietz is Professor of Political Environmental Geography, Faculty of Environmental Sciences, University of Amsterdam, the Netherlands. His Ph.D. thesis (*Pastoralists in Dire Straits,* 1987, Amsterdam) was about the breakdown of the pastoral economy of lowland Pokot living in the semi-arid northwest of Kenya and northeast part of Uganda (the result of drought, livestock diseases and violence) and the survival strategies developed by these (former) pastoralists. His research experience is mainly in Kenya, Zimbabwe and India. Currently, he is a member of the directorate of the Netherlands Research School for Resources and Development, CERES; a member of the Steering Committee of the Netherlands-Israel Development Research Programme (NIRP); and coordinator of a program to strengthen the School of Environmental Studies, Moi University, Kenya.

Marcel Put is a human geographer at the Faculty of Environmental Sciences, University of Amsterdam, the Netherlands. He obtained his M.A. from the same university in 1987, with a thesis on the food and labor situation in villages in India's semi-arid tropics, which was based on field research in interior Tamil Nadu. From 1986 until 1989, he worked at the Royal Tropical Institute, Amsterdam, on a study of soil and water conservation practices and crop management by dryland farmers in southern India. In addition, he participated in an evaluation study for the Operations Review Unit, Dutch Ministry of Development Cooperation, on the effects of food aid on agricultural development in Zambia. Currently he is finalizing his Ph.D. on the effects of a government and an NGO project, both aiming to induce farmers in the semi-arid tropics of Andhra Pradesh (India) to adopt innovations for dryland agriculture.

Prof. Ton J. Dietz
Mr. Marcel Put
Department of Human Geography
Faculty of Environmental Sciences
University of Amsterdam
Nieuwe Prinsengracht 130
1018 VZ Amsterdam
The Netherlands

S. Subbiah is Professor at the Department of Geography, University of Madras, India. He received his Ph.D. in 1982 from the Jawaharlal Nehru University, New Delhi, on the subject of regional variations in agricultural productivity and its determinants in Tamil Nadu. He has published on many geographical topics, such as oceanography, climatology, agricultural geography, and livelihood systems in Indian villages. He is the team leader of the GIS laboratory in his department and editor of the *Indian Geographical Journal* since 1980.

Prof. S. Subbiah
Department of Geography
University of Madras
Chepauk
Madras 600 005
India

Part Four

Drought, Food and Land Management in Africa

Chapter 9

Traditional Management of Drought and Famine in Kenya

Joshua J. Akong'a and Joram N. Kareithi

Introduction

There is general unanimity that drought and famine are recurring events in Sub-Saharan Africa. People's patterns of thinking, feeling and potential behavior are influenced by their environment. Hence recurring events are institutionalized and become an important aspect of culture (Hofstede, 1991). Fleuret noted: "history tells us that drought and consequent food shortage are regularly recurring events which have stimulated adaptive responses. Further, these responses do not remain static and unchanged, but affect and are affected by other events and changes occurring in society" (Fleuret, 1986:224; Akong'a, 1988).

In Kenya, severe droughts and famines are widely identified as major historical land marks used in various societies to remember other events or age as they are remembered in a chronological order (Akong'a, 1988:101–106). Among some of the Luyia sub-ethnic groups such as the Bukusu in the Western Province of Kenya, droughts and famines are reflected in people's names for example *Nasimiyu* (female) and *Simiyu* (male) born during the dry season, *Wanjala* (male) and *Nanjala* (female) born during the hunger season and so on. In fact, among the Luyia, there are two main seasons during the year: (1) *esubwe* corresponding with the post long rains season when food is in relative abundant supply and (2) *erotso* (severe food shortage) corresponding with the post short rains season when less food is available. This implies the existence of a recurring hunger season every year. In addition, the years are punctuated by periods of severe droughts and famines experienced on regional or national levels. The stress of drought and famine is experienced differentially by individuals, families and communities depending on their social, political, historical and physical circumstances *vis-à-vis* those of their neighbors.

> Societies in areas prone to recurrent shortages have developed diverse coping strategies, means of reducing the impact of food deficits on the population while these differ in detail from one group to another, in all cases such strate-

H. J. Bruins and H. Lithwick (eds.), The Arid Frontier, 165–184.

> gies are integral components of rural production systems. They are not independent behaviors resorted to only in times of crisis, rather they are activities which are available continuously but assume greater importance during periods of food deficit (Campbell, 1989:7).

In his analysis of the human responses to severe food shortages, Robert Dirks (1980) indicates that famine has both biological and social dimensions with interpersonal and psychological implications. From the social point of view, there are three phases that characterize response to famine conditions. When people first learn of the onset of famine, they seek and give assistance through the culturally patterned reciprocity networks where they exist. The persistence of the famine and corresponding depletion of food stocks leads to the erosion of the values and norms associated with reciprocity. The family rather than the community becomes the unit of survival. This is the time any food is normally acquired on behalf of children as adults try to persevere. Later, extreme forms of selfishness and individualism set in probably at the time famine related deaths begin to occur. The individual rather than the family becomes the unit of survival and it is probably the time when items not before defined as food are re-defined as food, if only to keep alive. The threshold for stress varies from one individual to another, but it is true to say that the majority of those in this stage of survival would be found living in a state of shock biologically, socially and psychologically. During this stage, those who have not died or become refugees in neighboring regions, will probably have started living as if they had no culture before (see Turnbull, 1972). This is therefore, a situation that should be avoided as much as possible as it is extremely dehumanizing and yet it is very common on the African continent as a consequence of drought and war.

In Sub-Saharan Africa there are three main modes of subsistence: agriculture, pastoralism and hunter-gathering. The latter activity is almost negligible as a full time source of subsistence. Hunter-gathering communities in many Sub-Saharan countries, Kenya included, have been under extreme pressure to abandon their mode of subsistence. This is partly due to the policy of national governments to evict them from forested areas, which are some of the few remaining protected areas. In Kenya the Okiek (Dorobo) of the Rift Valley province have at times been forcefully evicted from protected areas. Moreover, the areas occupied by hunting and gathering communities have not been marginal in all cases. Encroachment by agriculturists and pastoralists in need of more land to satisfy their larger populations, has also caused pressure on hunters and gatherers.

This review focuses on pastoral and agricultural modes of livelihood. It is not entirely possible to distinguish between the purely pastoral management strategies and those used by agriculturists because of the historical symbiotic relationships that have existed between pastoral and agricultural societies in an attempt to diversify their resource base. The chapter takes a more general view of drought and famine management strategies to include institutionalized behavior patterns that have a bearing on survival. These strategies include: (1) organization of production that has risk avoidance as one of its objectives; (2) processing of the harvest and storage management strategies and (3) the actual drought and famine management strategies, being the contingency measures resorted to during severe drought and famine. This approach has been taken in view of over-emphasis in the literature on drought and famine coping or management strategies. Our Study, therefore, basically deals with adaptive strategies in peasant communities (Barlett, 1980).

Organization and Management of Production in Cultivating Societies

Agriculture as a mode of production is widely practiced among Kenyan societies. However, the methods of cultivation and crops grown vary from region to region. The hoe and ox-plough are the most common tools used by small-scale subsistence farmers. Modern farming equipment is mainly used by the large-scale commercial cultivators.

The household provides the bulk of the labor required in subsistence cultivation. Labor availability, however, varies in accordance with the nature, size and life cycle of the particular family on the one hand and the season.

The practice of polygamous marriages enables the recruitment of more labor sources than the monogamous households. Boserup (1970:39) notes: "although households with large numbers of wives seem to have more or less disappeared in most of Africa, polygamy is still widespread and is considered an economic advantage in many rural areas." According to the Kenya Fertility Survey of 1977/78, 30% of all married women aged 15 to 50 lived in polygynous unions in Kenya. A figure of 25% by the Kenya Contraceptive Prevalence Survey, was recorded in 1984, while the Kenya Demographic and Health Survey of 1989 put it at 23%. Acquisition of additional wives is a traditional method of expanding production through provision of more labor, especially given the low technology and the cost of hired labor. Polygamy is currently practiced mainly by the older generation and varies by ethnic group, age and education.

A household with more children, especially sons can be assured of more labor hence the expansion of its production. Children in Africa are important not only to the immediate family, but also to the whole total kin group (Kayongo-Male, 1984:6). Large numbers of children are highly valued, while a childless marriage is thought of as troublesome and seems meaningless *(Ibid.)*. Children are a source of labor in the short-term and in the long-term become the security of parents during old age through parent child wealth flows in different forms. Child labor in the household can involve direct productive activities (farm work, herding) or indirect productive activities (running errands, childcare, *etc.*). These activities are important to the household's production, since it releases the adults to concentrate on production activities. However, having more wives and children in modern circumstances does not necessarily lead to higher levels of production and productivity, but increases maintenance costs to the household.

Cultivating households may experience labor shortages during peak labor demands even when they have a large number of productive members. Extra-household labor sources are therefore needed for specific tasks or for the regular activities in the production system of the household. Traditionally, mechanisms exist to allow for the recruitment of labor outside the household. These channels of labor recruitment usually take the form of cooperative or reciprocal labor. The category of persons involved largely consists of relatives, neighbors, or friends who work for each other in rotation or on request. Social networks of a household are significant for purposes of labor acquisitions outside the immediate household and in sharing of the produce, especially during times of food shortage. A significant feature of cooperative reciprocal labor is the absence of cash payment. However, expenses in form of feasting and gifts may be incurred by the host households as a show of hospitality and as a mechanism of enhancing relationships with participants in the provision of labor.

Despite the rapid population increase in recent times, recruitment of labor has been severely constrained by monetization of the economy. Most small scale cultivators cannot afford to hire extra labor. Rural to urban migration of able bodied men and women is aimed at obtaining alternative and supplementary sources of income. School enrollment has demobilized a significant proportion of child labor. The fragmentation of land into portions that have become economically and environmentally unviable has become a major problem. Over 70% of land holdings in Kenya are below 2 acres in size.

In cultivating societies, domestic animals are kept as part of the farming system. These animals serve a wide range of functions. Apart from the use of milk and milk products, cultivators derive a minimum of subsistence from domestic animals. The animals also serve ritual and symbolic functions, as in bridewealth exchange and in sacrifices. During periods of hardship, such as drought, famine and funerals, animals become an important source of food and currency in addition to being exchanged for grains.

AGRO-TECHNICAL STRATEGIES IN THE MANAGEMENT OF DROUGHT

The impact of drought on the households subsistence can be mitigated through production management decisions and practices. Studies have shown that rural communities have developed some elements of 'disaster preparedness' built into their production systems to minimize risks to food supply and livelihood (Rahmato, 1991; Eshete, 1995; Reitsma *et al.*, 1992). The strategies adopted by individual households are geared towards maximizing production and minimizing loss through good storage.

Households cultivate various plots of land located in different micro-climates as a strategy to spread risk and minimize loss due to drought. This is a strategy that has become minimal since independence due to the national land policy of land consolidation and registration. They also practice intercropping (Riley and Brokensha, 1988:209; Reitsma *et al.*, 1992:102). The subtle gradations in elevation, topography, vegetation and soils within a region may cause variations in drought vulnerability and suitability for certain crops (Ambler, 1988:50). The location of fields in different ecological zones, serves as partial insurance against total loss of crops due to drought. This strategy also ensures a household access to the cropping advantages of varied micro-environments, allowing a diversity of crops to be grown. Among the Kitui Kamba for example, women cultivate several scattered gardens (O'leary, 1984:69). The various strips of cultivated land are normally located in different ecological zones to take advantage of the micro-environmental variations. Among the Marakwet, every household tries to acquire a piece of land in the different ecological zones (Reitsma, 1992:102). Maize and dairy cattle are kept in the highlands, finger millet, sorghum, cassava and bananas are grown on the slopes, while irrigated agriculture of sorghum, millet, cassava and bananas is practiced in the valley zone. The strategy of staggered planting dates is also employed to counteract risks connected with uncertainties in precipitation patterns. The spreading of planting dates increases the chances of utilizing the precipitation during the growing cycle of the crops (Reitsma, 1992:32). Such considerations explain the fragmented nature of the household's land parcels in both the pre-colonial and colonial Kenya among cultivators. In modern times, cultivators try to own land parcels in different localities through purchase or leasing.

Mixed cropping or intercropping is practiced by cultivators as a strategy to minimize risks of crop failures. The reliance on a diversity of crops rather than a single species cushions the cultivators. If one crop fails due to inadequate rainfall or disease, another crop may still produce a yield. Mixed cropping also provides the household with a balanced diet by increasing the range of food crops accessible to the family, while reducing labor peaks (Nadar, 1980:2). The adverse effects of pests is reduced, through mixed cropping since various crop species are prone to different pests. Mixed crops may provide favorable micro-environments for each other (*e.g.* maize, millet, cowpeas and beans). Some fix nutrients in the soil useful for other plants.

Some communities developed indigenous irrigation systems to reduce vulnerability to drought and famine. Indigenous irrigation schemes in Kenya include the Elgeyo-Marakwet system in Keiyo and Marakwet districts, the Pokot system in West Pokot district, the Mwatate system in Taita-Taveta district, and the North Iveti Hills system in Machakos district (Arao, 1988:225). The operations of these indigenous irrigation systems is organized around the traditional social forms of each particular society. Among the Pokot, the council of elders deals with the issues of justice, especially regarding land and water distribution of the system (Reitsma *et al.*, 1992:98). The irrigation systems provide subsistence, even when rainfed cultivation is affected by drought. The system also allows the production of crops (such as fruits), which otherwise would not do well given the rainfall characteristics of the regions.

Traditional crops which tolerate or evade drought have been widely grown in the past in various parts of Kenya (Anyango *et al.*, 1989:175). These crops include cereals such as sorghum and millet, legumes such as beans and cowpeas, as well as root crops such as cassava, yams and sweet potatoes. Cassava and potatoes grow and yield better on low fertility soils, in comparison to other crops. They have an ability to withstand hailstorms, locust attacks, and drought, while only requiring low labor cost requirements. (Dei, 1992:65).

Cassava is a staple food crop in Kenya among the Mijikenda at the Kenyan coast. It is also widely used among the Luyia of Western province and the Luo of Nyanza province, besides other groups. The crop is usually harvested only when required for immediate consumption as food, but it can remain in the ground for up to five years without going bad! If the cassava is harvested for later use, it is peeled, sun-dried and stored to be ground into flour for the staple meal, *ugali,* which can also be prepared from maize, millet and sorghum flour, either separately or in combination.

The traditional cereal and root crops, are increasingly being neglected by the various communities in Kenya due to various factors (Anyango *et al.*, 1989:175). Taste preferences especially among young people favor maize meal, which they consider more palatable and civilized (Riley &Brokensha, 1988:228). Lack of a well developed market for these crops, lack of child labor for chasing birds from millet and sorghum gardens, land scarcity and overemphasis of cash crops in extension services are partly to blame. People transporting millet and sorghum could be arrested and prosecuted in recent years on suspicion that it is intended for preparation of the illicit brew—Busaa and Chang'aa.

Maize crop has a low level of tolerance for either drought or too much rainfall. Compounded by the fact that maize does not keep in storage as long as the other cereals, the incidence of famine is potentially as frequent as drought. However, the intervention of the government and international Relief Food Agencies have in recent years minimized the deleterious impact of drought, so that full fledged famine did not develop.

FAMINE MANAGEMENT

Cultivating communities have developed measures to reduce the risks of food shortages, although the possibility of famine always remains a threat. Food stocks are maintained At the household level, especially grains. These provisions are judiciously managed in order to off-set food shortages that may affect a household if the stores are exhausted before the new crop is harvested. Studies on drought impacts and responses in Central and Eastern Kenya (Kamau *et al.*, 1989) revealed that each of the wives in a polygamous household had her own granary, but heads of the households kept a reserve store of grains to last a season or more. Among the Mbeere, the grains produced on the land of the household heads is stored separately and serves as a food security measure for the household (Mwaruvie, 1991:149). The man's food store becomes a strategic food reserve in case of a drought, as it is separated from the wife's stores which supply the family its daily rations of food.

Proper storage facilities and preservation techniques of food reserves is part of food security management, in order to prevent damage from pests. The granaries are prepared before the new crop is stored, using plants which serve as repellants to insect pests to minimize infestation of the stored grains (Fernandez, 1994:10). The stored foods are further protected from pests and rats through specially designed storage houses built on raised supporting pillars. This also gives protection against rain-runoff on the ground. Storage near or above the kitchen hearth is used in some communities. This keeps the grains dry and pest free, due to the heat and smoke from the fire place. Grain containers designed to keep away moisture, such as well sealed guards and hand woven baskets, are also widely used to store grains. Cereal grains continue to give off carbon-dioxide after harvest which displaces oxygen. Therefore, sealed containers are an elementary form of fumigation chambers (Parrish, 1994:8). Among the Luo and Luyia, ash serves as a preservative for grains, especially beans and peas, following thorough sundrying.

There are two forms of storage to ensure economy of use of the harvested crop. Out-door storage granaries with thatched roofs were reserved for maize on the cob and millet or sorghum on fingers. Among the Luyia, these are the first to be used at the discretion of the wife/mother. Grains in bulk were stored indoors in pots or special granaries. These were alternatively located on the veranda of the house where each granary was sealed using a pot and cow dung serving as cement. This made the granary air tight and therefore too hot for pests to survive inside. Sorghum in particular could keep for more than 10 years in such a store. The sealing method also gave some protection against stealing, making it noticeable if someone had tampered with the store.

Other forms of food preservation techniques exist in various agricultural communities. Among the Luo and Luyia, surplus meat was cut into thin strips, salted, sun-dried and stored for future use. In some cases, such meat is hung in the roof above the fire place to further enhance the loss of moisture (Were &Soper, 1986:50). Vegetables, Mushrooms, and cowpeas leaves are also sun-dried and stored. Locusts were a delicacy in Western province among the Luyia. During the period of locusts, families went out, especially in cassava and potato gardens, to catch the locusts, which were boiled in salted (alkaline) water from ash and subsequently sundried and stored. Those who had lactating animals extracted butter that was boiled into a liquid to transform it into ghee, which would last for years without going bad. Among the Tharaka of Northern Kitui District, honey stored in wooden containers could last for up to five years. Among the riverine and lacustrian regions of Kenya and Uganda, there has been a long tradition of sun drying and smoking fish, both as a preservative to keep it as food for a longer period and also to transform its taste.

Some communities controlled the number of meals children and adults, respectively, could have during a day. Other communities, especially pastoralists focused on the principle of mutual exclusiveness: one is not permitted to consume both milk and eat meat during a meal or at the same seating. It is also not permitted in pastoral communities, to look after livestock and participate in fishing without performing a cleansing ritual, as the two are viewed as mutually exclusive forms of livelihood. In fact, a person is not expected to eat the meat of a wild animal and drink milk or eat the meat of a domestic animal at the same time (Were & Akong'a, 1986).

SOCIAL NETWORKS

The most widely used strategies for coping with famine are based on social networks built over time by individuals, households or even whole communities. The networks should not be perceived as *ad hoc* crisis management, but are proactive long range measures which people build and maintain during normal times. These social networks become crucial for survival during a crisis like famine.

Exchange systems existed among communities in Central Kenya, which allowed individuals suffering from famine to seek assistance from neighbors who were less affected. Various arrangements existed in which labor provided by people from famine stricken areas was exchanged for food in the less affected regions (Ambler, 1988:61).

In some cases men and women worked briefly in exchange for food from farmers who were not affected by the famine. The arrangement could also involve travelling to neighboring pastoral districts where they were contracted to work in exchange for livestock. Such contract workers became sometimes longstanding clients of their hosts and settled permanently in the societies which recruited them.

Young boys could be absorbed by households away from home as indentured servants in exchange for food, while adult men pawned dependent relatives (especially women) when food shortages occurred. The expectation was that such pawned relatives would be redeemed when the food situation at home improved.

Children were sent away as 'visitors' to relatives in places less affected by famine. The adults in the family could make 'hunger-trips' for a few days to relatives or acquaintances in areas not affected to work for food and return home. The strategies outlined above are still in operation in many communities today but in modified forms and have visibly been reported in the press in the course of the current (1996/97) drought and famine that has affected almost all the districts of Kenya.

The dispersal of household members in the above strategies sometimes did not yield food directly to the household, but it helped to relieve the pressure on the scarce food resources available by reducing the number of mouths to be fed.

In more extreme cases, wives were pawned in exchange for food and girls married out or betrothed so that their families could receive food credits or livestock as part of bridewealth payment. This is a common strategy among the Kamba, who usually obtained food credits from the Kikuyu during famine periods.

Therefore, the strategies aimed at dispersing dependants as well as acquiring food during a severe famine were facilitated by social networks involving extended family relationships and pseudo-kinship networks that a household established and maintained during normal times. The institution of *Giciaro* (blood brotherhood) was especially significant to the household in times of hardships among central Kenya communities,(Ambler, 1988:82).

Through *Giciaro* men became brothers by a ritual rebirth which united the families of the two men. Blood brotherhood relationships entail all the rights, obligations and prohibitions that would occur in a relationship between real siblings. A blood brother (*muciarwa*) expects practical assistance from his partner and his family during times of need especially in a crisis like a famine.

Among sedentary agricultural communities, temporary migration of single adults in search of work is usually an early response to the threat of famine. Remittances from migrant workers during periods of stress represent a crucial entitlement support for vulnerable households in Africa (Dreze & Sen 1989:78). Permanent migration of entire families is rare in agricultural communities and when it occurs it is usually a last resort option.

Temporary relocation of an entire family is however an established traditional response to severe and prolonged drought and famine (Ambler, 1988:126). Families who move out of their normal residence typically make use of personal networks and traditional land tenure systems to find places for their relatives.

Concerning the absorption of relocating families, the Kikuyu customary land tenure recognized the cultivation and settlement rights of *muhoi* (squatter), *muthami* (sojourner), *muthoni* (in-law) and *muciarwa* (blood brother) (Kenyatta, 1978). The families assisted under these arrangements were expected to behave well and to respect the rights of the members of the land owning group where they settled. The refuge seeking families typically settled down for a few months or years in the hosts property and raised their own food on the portion of land temporarily allocated to them. The host gained from the 'visitors' in terms of labor, because the accommodated families were obliged to help the host family in any emergency work. Members of relocated families often tried to keep together as a unit, but in some circumstances they got dispersed among a number of homesteads with disastrous consequences for the whole family. Due to the changes in land tenure system, the above settlement rights although recognized have become difficult to operationalize today.

Communities faced with the threat of famine also try to cope by adjusting their dietary habits. Food rations are scaled down, while the number of meals per day is reduced to the minimum in an attempt to stretch the available food resources over a longer period of time. Local substitutes for processed products, *e.g.* tree barks in place of tea leaves, are used to minimize cash expenditure. Additional methods of food processing are adopted so that parts of food crops, which are discarded under normal circumstances, become edible (Dei, 1992:73).

Reliance on wild plants and game as sources of food increase during a famine. Gathered wild fruits sometimes become an individual survival strategy, as individuals, including children, collect and eat wild fruits on their own. However, hunting, which is more demanding in terms of organization, may involve team work with the aim of obtaining meat for the whole family.

Changing dietary habits and food recipe experimentation during famine often leads to increased incidences of food poisoning and gastric problems, as individuals and groups deviate from the norm and re-define what constitutes food.

Pastoral Household Economics

Pastoralism in Africa forms the main land use in the Arid and Semi-arid lands (ASAL). In Kenya the ASAL area constitutes about 80% of the country, supporting some 25% of the entire livestock population (Kenya, 1979:48). Rainfall in this large area, occupied by pastoralists, is usually low and erratic in distribution. Pastoral societies in Kenya include the Maasai, Turkana, Samburu, Rendille, Orma, Pokot, Boran, Gabra, Somali and others. The main animals reared are cattle, sheep, goats, donkeys and camels.

The household is the basic unit of production (Dahl & Hjort, 1976), as it owns and manages the herd. Rights of livestock within the household, are complex, due to multiple and overlapping rights regarding specific animals in the general household herd (Toulmin, 1983:9). Customarily the animals are allocated to the wives and sometimes to dependent women who from time to time may join a household. The women have use rights of the livestock allocated to them. When dependent women join a household they bring along 1/4 of the livestock allocated to them in the past by husbands (Gulliver, 1951:89). They use this livestock for subsistence purposes.

The distribution of animals to the wives and dependent women is usually unequal: the first wife getting a far larger share of the herd than the others (Gulliver,

1951:91). This unequal allocation emanates from the fact that the first wife represents the husband and indeed the whole household in matters of hospitality to visitors, and she also is the custodian of the important animals in the herd such as serving males. However, it is expected that mutual assistance exists between the wives in sharing livestock products whenever a need arises. Children too may have specific rights of certain animals. Among the Turkana for example, young children are allocated animals, especially goats from which they can drink milk straight from the teats (Watson, 1988:48) while they are out in the fields herding the animals. Each son of a Turkana herd owner is given responsibility over a particular species of animals, and at the death of the father each son inherits a disproportionately bigger share of animals from the species assigned to him when he was young (Dyson-Hudson, 1982:17). Therefore, in a Turkana household, individual members have overlapping rights of the animals owned by the household as a unit. Members of a pastoral household in Kenya feel attached to the animals and participate in livestock management, due to these overlapping rights.

The de facto ownership of the household herd lies with the male household head, but the usufruct rights of specific animals overlaps in the domestic group. Thus the male household head cannot slaughter, give away, or sell particular animals without the consent of the wife and sometimes the children (Nakhabi, 1991:41), because they may have a stake in specific animals of the general herd. This shows that women in pastoral societies have a great deal of influence on the household assets. It is wrong to assume that women are a passive and disadvantaged group within the pastoral economy.

ADAPTIVE STRATEGIES TO ENVIRONMENTAL LIMITATIONS IN PRODUCTION

Climatic characteristics of the areas inhabited by pastoralists provide severe limitations on the mode of production. Inadequate rainfall, erratic distribution of rainfall or total rainfall failure are matters which even the most knowledgeable herdsmen cannot alter. However, the knowledge about the ecological constraints on the pastoral economy, accumulated over the years, is an asset. Herdsmen employ this indigenous knowledge to tune animal husbandry and herd management practices to the environmental limitations.

Pastoralists in Kenya generally rely on a diversified herd rather than on a single species. Each household endeavors to keep a herd composed of both small stocks (goats and sheep) and large stocks (Cattle and/or camels). The various animals have different productive characteristics regarding milk, blood, fat and offspring. A diversified herd, therefore, enhances the household's food security and extends the production of food stuffs. The species of animals mentioned above have different feeding habits. A household with a diversified herd is able to take advantage of the various micro-ecological characteristics and vegetation types in the rangelands in a better way than a single species herd. The small stocks, which browse on vegetation not eaten by cattle, are drought resistant (especially goats) and are less prone to tick-borne diseases (Nunow, 1994). Goats and sheep have a higher reproductive potential and are therefore best suited in post-drought recovery and herd reconstruction (Dahl & Hjort, 1976). Camels are slow in reproduction, but are most drought resistant, while lactating for longer periods. Socially, each of the species play a role in the social functions of the household and in its interaction with the wider community. The various animals enable the household to meet social, religious and ritual obligations and hence to function respectably in the society. Diversification of the herd has ecological, economic and social rationale.

Pastoralists often keep large herds as a safety measure against the threat of destitution from drought conditions, coupled with the subsistence requirements of the household. This practice has sometimes been labelled irrational and has been used to explain the degradation of the rangelands, especially by government officers. It should be noted, however, that large herds are a rational calculation to ensure household viability both in the short and long-term. In maintaining large herds, the pastoralist proactively plans to mitigate the impact of drought as at least some of the animals will survive to form a nucleus for rebuilding the herd. However, severe and extensive droughts have at times wiped out all the livestock of households regardless of herd size, thereby defeating the above logic of the herdsmen. Nevertheless, large herds may often give more security in the face of environmental uncertainties in arid and semi-arid regions.

Herd mobility is a practice employed by pastoralists to take advantage of the micro-environments which result from the uneven distribution of rainfall. Generally pastoralists use different geographical grazing locations for the dry season and for the wet season. The usual pattern is to move out to more remote and often drier areas during the wet season; during the dry season they move to areas where pastures and water are available, for example, along river valleys. However, the decision of pastoralists to move is not always in response to the environment, but is also influenced by social and political factors (McCabe, 1992:8).

The consequences of mobility of the pastoral economy is a greater demand on household labor resources. The required labor is not always available during certain phases of the family life cycle, meaning that a household has to look for mechanisms of recruiting this labor from other sources. Labor bottlenecks among the Maasai are often resolved through the use of mechanisms such as herding cooperative groups (Grandin, de Leeuw & Ole Pasha, 1991:62), in which several herds, from different households are merged together for purposes of grazing. The herding cooperative groups differ in their duration and composition. Some are short-term ad hoc arrangements constituted during the dry season when mobility is high, while others are long-term and last for years (Grandin, de Leeuw & Ole Pasha, 1991:77). The 'fora' System, among the Gabra and Boran of Marsabit district which involves the division of the herd into two sections, is used in herd management to resolve labor problems encountered during the dry season (Legesse, 1989:267). In this system, a small milk herd is kept and grazed in the vicinity of the camp, while the larger dry herd is grazed away from the camp sites for long durations.

Pastoralists often practice herd splitting and herd dispersal as an insurance system against localized risks such as predatory animals, raids or animal disease epidemics (Norconsult, 1990:6). The general household herd is split into smaller herds which are dispersed geographically to friends (stock associates) and relatives. This strategy enables the use of good pastures elsewhere, as well as labor resources and livestock management skills in other households. The practice of polygamy among pastoralists facilitates this strategy, as the various wives may be located in different regions within the general territory. Each wife becomes the nucleus of a small herd. The practice of herd splitting and dispersal serves to distribute grazing pressure more broadly and evenly in the pastoralist's territory (Turkana Drought Contigency Planning Unit, 1992:15).

Customarily, land among Kenyan pastoralists is communally utilized as a unit. However, this does not mean that it is a 'free-for-all' mechanism of land use. Galaty (1984:23) notes, that few pastures are truly 'common' or 'open' since specific groups exercise property rights through use or occupation. General rights of access to resources in a given territory are determined by membership of a household in a clan, territorial section or other specifiable social group occupying and utilizing a particular region. The right to pastures in a given territory is, however, not undisputed and the individual has to be ready to defend his claim to such rights (Storas, 1987:140). Territorial borders are constantly defined and redefined through confrontations between social units. Individual members of different units can negotiate usufruct rights either temporary or permanently in the territory of another unit *(ibid.)*. Access to extensive territory is necessary for the survival of herds and the viability of a pastoral economy in dry regions.

Limits are set by the elders about the degree and manner of the use of resources within the territory occupied and utilized by a particular group of pastoralists (Legesse, 1989:264). Certain areas are recognized and set aside as drought reserves (McCabe, 1992:16). The elders ensure that these areas are not encroached upon during normal times and are utilized only during the dry seasons. Social rebuke and avoidance of camps and households, which do not observe proper resource use, is applied among some communities. This form of ostracism ensures that individual herders do not compromise the sustainability of the common resources through selfish over-grazing or abuse. The Boran have taboos against the killing of trees which are ritually protected because of their economic value. *Acacia Tortilis,* which produces pods used as fodder for sheep and goats, is one such species of trees (Legesse, 1989:264). Tree resource management among the Turkana involves household ownership and control of specific pod bearing *Acacia Tortilis* trees, used for animal fodder, especially in the dry season (Gufu, 1987:15). Herdsmen use long hooked sticks to shake down the pods from these trees to feed their animals and avoid cutting the tree down.

RESPONSES TO DROUGHT AND FAMINE

Livestock loan systems in varying forms are one of the strategies pastoralists use to cope with drought and famine (Nunow, 1994; Turkana Drought Contingency Planning Unit, 1992). Livestock is transferred to relatives or stock friends (stock associates) even during normal times without hardships. Its logic is rooted in the need for self-insurance in an environment that is extremely vulnerable to climatic vagaries and an economy that is extremely susceptible to many risks. Livestock loan systems serve three main functions: (1) balancing economic risks, (2) establishing social bonds and (3) redistributing and reallotting labor (Schlee, 1989:56). Distributing animal gifts in different households amounts to a dispersal of one's total livestock possession and therefore spreads out the risks. The social bonds established through the transfer of loan animals are investments which the giver can turn to for help when confronted by animal loss. A herdsman who suffers losses can expect to receive assistance from the network of friends established through livestock gifts.

Livestock loan system enables a household to meet its subsistence requirements even when its actual herd size has fallen below viability levels. The recipient household is able to slowly rebuild its own herd by sharing in the offspring of the loaned animals. A household also benefits from cross breeding its own herd with loaned animals and thereby improving the genetic characteristics of its herd (Odegi-Awoundo, 1990:41). Pastoralists practice selective breeding in attempts to improve the adaptive characteristics of their animals, by using animals which have survived droughts: their offspring are said to be well-adapted to the environment (Nunow, 1994). Males with such characteristics may be in short supply among pastoralists who experience frequent droughts, as these are sold off first to procure food or they are castrated. The latter option is practiced by pastoralists in the dry season since castrated animals have a better capacity to withstand drought conditions (Turkana Drought Contingency Planning Unit, 1992:15). There are less flies during the dry season which is, therefore, an opportune period for animal castration, reducing the risk of wound infection. Selecting males of good progeny is facilitated by the traditional loan system of the pastoralists.

During drought and famine hardship, pastoralists reduce the pressure on the available livestock resources by reducing the number of persons directly dependent on the household herd. Members of the household are split up and some move to live temporarily with better off relatives or they engage in other activities such as farming (Norconsult, 1990:14). Cultivation has been used by pastoralists as a fall back activity, during which impoverished herders try to rebuild herds and make a return to pastoralism (Toulmin, 1982:45). The cultivation alternative is becoming an increasingly difficult option for pastoralists, because arable areas at the margin of rangelands are now occupied by traditionally agricultural societies or by wealthy herders who have diversified their investment activities to include farming (Little, 1987). Competition for agricultural land at the margin of rangelands makes it difficult for pastoralists not only to shift back to cultivation, even on a temporary basis but also to deploy the full range of coping strategies formerly available. When the need arises it is usually possible only in the least productive areas (Mace, 1993:371).

Spencer (1973) has detailed the way Samburu and Rendille symbiotic relationships operate, involving the movement of people to alleviate economic hardships. The Rendille camel economy grows slowly and supports fewer people than the Samburu cattle economy, which grows relatively more quickly. The surplus population of the Rendille society is ejected to the Samburu economy as a result producing an intermediate economy of the Ariaal (Odegi-Awoundo, 1990:18). The Ariaal physically occupy the territory between the Samburu and the Rendille and the population is composed of people of mixed Samburu and Rendille origin. The Ariaal keep both camels and cattle, which renders a more flexible economy than either that of the Samburu or Rendille proper.

The Gabra and Boran of Marsabit district also practice symbiotic interdependence (Legesse, 1989:269). The Boran inhabit the Moyale-Sololo escarpment and the Marsabit mountain regions, having a comparatively wet territory where cattle does well. The Gabra occupy the dry environment around the Chalbi desert and mainly raise camels. In their framework of mutual risk diversification the Gabra raise cattle among the Boran, while the Boran also keep camels among Gabra relatives and stock associates (*Ibid.*:261). The two societies help each other in this way to take advantage of both environments and to reduce the impact of drought

Cross-ethnic ties in Northern Kenya form a dense web among the various societies. Clans often occur in one or two societies (Schlee, 1989:2). These relationships are reactivated when the need arises, especially for economic relief in times of hardship. A household can move from an area severely affected by drought to the territory of a totally different society, gaining access to resources by invoking latent clan ties. Marriage relationships among pastoral groups can be used to gain access to pasture resources across international boundaries during times of drought and famine.

Pastoral communities make use of internal population controls to maintain a viable balance between people, livestock and land (Toulmin, 1983:9). The Gada system, which operates among the Boran and the Gabra is a demographic control mechanism which limits the rate of population growth (Legesse, 1989:262). Marriage is delayed until an advanced stage in the life cycle: with approximately 50% of the women remaining unmarried into their thirties. Once married, the women are culturally subjected to several years of *post-partum* sexual taboo after the birth of a child, while the nursing period is an extended one. Until recently this led to a slow natural population growth, which ensures a balance between livestock, human and natural resources.

During famine situations, pastoralists used to resort to hunting and collection of wild fruits, also as forms of individual survival strategies. These options are on the decline due to the decrease of wild fruits, resulting from over-exploitation and the expansion of agricultural lands, as well as the strict anti-game hunting laws instituted by the government as part of the wildlife conservation policy.

Pastoral groups have historically not been isolated. They have usually maintained symbiotic relations among themselves as already stated, and between neighboring agricultural groups based on the exchange of livestock for grains. The Kikuyu for example have had long standing symbiotic relations with the Maasai. Intermarriages, resulting from this relationship, led to a stronger basis for interdependence that persists today, especially in Kajiado, Narok and Transmara Districts. In many African communities, feuds were usually suspended during severe drought and famine to permit even enemies to get access to survival resources, such as food, grazing areas, water and salt licks (Saina, 1995). Hungry persons in Uganda were permitted by convention to enter a banana grove and eat ripe bananas, on the spot without carrying away any. In many Kenyan societies, a hungry sojourner was free to help himself to sugarcane before proceeding. Normally water to drink and fire are not denied anyone. A hungry sojourner asking for any of the two items was thought to be asking for food as well. As a result, some societies socialized their children into knowing survival strategies. Among the Kitui Kamba, for example, children in circumcision camps were taught things related to personal hygiene, folk history and also how to survive hunger, even if it meant stealing from another person's garden.

Conclusions

1. Drought and famine are recurrent features of the African landscape. Both agricultural and pastoral societies have historically evolved drought and famine coping strategies relevant to their environments and modes of livelihood.

2. These strategies portray useful indigenous knowledge systems that need to be conserved in the absence of guaranteed food security. However, the institutionalization of coping strategies is not in any way an indication that they always ensured food security, however, they provided avenues for survival that individuals and families could control themselves. In fact, the study by Akong'a (1988) in Kitui District showed that coping strategies differ from one drought, famine and physical environmental situation to another, depending on the resources available for survival.

3. Similar critical resources are used in both agricultural and pastoral societies for survival. These resources include labor, land, livestock, as well as networks of relatives and friends.

4. The drought and famine coping strategies discussed are mainly of the risk aversion type. They have, therefore, not contributed to long term food security, which is a pity. This is why a country like Kenya with a diversity of traditional modes of livelihood and thus, staple food stuffs requires to put in place a pragmatic food policy that emphasizes food security both at the family and national levels in order to avoid future incidents of massive food relief supplies by the government and NGOs as experienced in recent years such as 1980, 1985, 1993 and 1997.

References

Akong'a, J. (1988) Drought and Famine Management in Kitui District Kenya. *In* D.W. Brokensha and P. Little (eds.) *Anthropology of Development and Change in East Africa.* Westview Press: Boulder, pp. 99–120.

Ambler, C.H. (1988) *Kenya Communities in the Age of Imperialism.* Yale University Press: Haven.

Anyango, G. *et al.* (1989) Drought vulnerability in Central and Eastern Kenya. *In* T.E. Downing, K.W. Gitu and C.M. Kamau (eds.) *Coping with Drought in Kenya: Local and National Strategies.* Lynne, Rienner Publishers, Boulder, pp. 169–210.

Arao, A.L. (1990) Types of Irrigation Systems in Kenya. *In* G.M. Ruigu and M. Rukumi (eds.) *Irrigation Policy in Kenya and Zimbabwe.* IDS, University of Nairobi, pp. 224–228.

Barlett, P.F. (1980) Adaptive strategies in peasant agricultural production. *Annual Review of Anthropology* 9:545–73.

Boserup, E. (1970) *Women's Role in Economic Development.* Earthscan Publications Ltd.,London.

Campbell, D.J. (1989) *Community-based Strategies for Coping with Scarcity: A Role in African Famine Early Warning African Studies Centre.* Michigan State University, U.S.A.

Dahl, G. and A. Hjort (1976) *Having Herds: Pastoral Herd Growth and Household Economics.* Liber Tryck, Stockholm.

Dei, G.J.S. (1992) The indigenous responses of a Ghanian community to seasonal food supply cycles and socio-environmental stresses of the 1980's. *In* D.R.F. Taylor and F. Mackenzie (eds.) *Development from Within: Survival in Rural Africa.* Routledge, New York, pp. 58–81.

Dirks, R. (1980) Social responses during severe food shortages and famine. *Current Anthropology* 21(1):21-44.

Dre'ze, J. and A. Sen (1989) *Hunger and Public Action.* Clarendon Press, Oxford

Dyson-Hudson, R. (1982) South Turkana Herd Structures and Human/Livestock Ratios. Report on a Survey of 75 Ngisonyoka Awis June-August 1982, Lodwar.

Eshete, D. (1995) Differential Socio-economic Impact of Food Shortages and Household coping Strategies: A case of Wolaita District in Southern Ethiopia. *Africa Development* 20(1) 1995.

Fernandez, P. G. (1994) Indigenous Seed Practices for Sustainable Agriculture. *Indigenous Knowledge and Development Monitor.* September 1994 2:9-12

Fleuret, A. (1986) Indigenous Responses to Drought in Sub-Saharan Africa. *Disasters* 10(3)86: 224–229.

Galaty, J. G. (1984) Cultural Perspectives on Nomadic Societies. *Nomadic Peoples 16:15-29.*

Grandin, B.E, P.N. de Leeuw and I. Ole Pasha (1991) The Study Area: Socio-Spatial Organization and Land Use. *In* Bekure S. *et al.* (eds.) *Maasai Herding: An Analysis of the Livestock production System of the Maasai Pastoralists in Eastern Kajiado, Kenya.* ILCA Systems Study 4, ILCA Addis Ababa, Ethiopia pp. 57–70.

Government of Kenya (1979) Major Highlights of the Kenya Fertility Survey Vol. 4 No. 2 December 1979 Nairobi, Ministry of Economic Planning and Development.

Government of Kenya (1989) Kenya Demographic and Health Survey Nairobi. National Council For Population and Development, Ministry of Home Affairs and National Heritage and Institute of Resource Development and Macrosystems Inco. Columbia, MD.

Gulliver P. H. (1951) *A Preliminary Survey of the Turkana.* New Series No.26 Commonwealth School of African Studies, Capetown.

Gufu, O. (1987) Basis for Longterm Monitoring of Arid Ecosystem of Central Turkana: Some Practical Considerations. Report presented at KALRES/NORAD District Orientation Seminar 3–24 June 1987, Lodwar.

Hofstede, G. (1991) *Cultures and Organizations: Intercultural Co-operation and its importance for Survival.* Harper Collins Publishers, London.

Kayongo-Male, D. (1984) *The Sociology of the African Family.* Longman, London.

Kamau, C. M. *et al.* (1989) Case Studies of Drought Impacts and Responses in Central and Eastern Kenya. *In* T.E. Downing, K. W. Gitu and C. M. Kamau (eds.) *Coping with Drought in Kenya, National and Local Strategies.* Lynne Rienner Publishers, Boulder.

Kenyatta, J. (1978) *Facing Mount Kenya: The Traditional Life of the Gikuyu.* Heinemann Educational Books, Nairobi.

Legesse, A. (1989) Adaptation, Drought and Development: Boran and Gabbra Pastoralists of Northern, Kenya. *In* Huss-Ashmore Rebecca and S.H. Katz (eds.) *African Food Systems in Crisis.* Gordon and Beach Science Publishers, Philadelphia pp. 261–279.

Little, P.D. (1987) Land use conflict in the Agricultural/Pastoral Border Lands: The case of Kenya. *In* Little, P. D. and M. M. Horowitz (eds.) *Lands at Risk in the Third World: Local Level Perspectives.* Westview Press, Boulder, pp. 195–212.

Mace, R. (1993) Transitions Between Cultivation and Pastoralism in Sub-Saharan Africa. *Current Anthropology* 14(4):363-382.

Mwaruvie, J. M. (1991) *An Economic History of the Mbeere (Siakago Division) CAD 1500–1914.* Unpublished, M.Phil Thesis, Department of History, Moi University.

McCabe, T.J. (1992) Land use Among African Pastoralists: A case study of Turkana of Kenya. Paper presented at the Conference on tribal and peasant Pastoral people, Pavia, Italy June, 1992.

Nadar, H. M. (1980) Intercropping under Marginal Rainfall conditions in Kenya. Paper prepared for Second Symposium on Intercropping for Semi-Arid areas 4–7 August 1980 Morogoro Tanzania Dryland Cropping Systems Research project USAID/KARI.

Nakhabi, E.M.W. (1991) *Strengthening the Management of Pastoralism and the Livestock Sector in Turkana District, Kenya.* Unpublished M.Sc. Thesis, Silsoe College.

Norconsult (1990) *Environmental Study of Turkana District Volume II.* Norconsult Ltd. Nairobi.

O'leary, M. (1984) *The Kitui Kamba: Economic and Social Change in Semi-arid Kenya.* Heineman Educational Books, Nairobi.

Odegi-Awuondo, C. (1990) *Life on the Balance: Ecological Sociology of Turkana Nomads.* ACTS Press, Nairobi.

Parrish, A.M. (1994) Indigenous Post-Harvest Knowledge in an Egyptian Oasis. *Indigenous Knowledge and Development Monitor* 2:7-9.

Rahmato D. (1991) *Famine and Survival Strategies: A case study from North, East Ethiopia.* Nordiska Afrikainstitutet, Uppsala.

Nunow, A.A. (1994) *The structure and Functioning of the Somali Pastoral Economy, with specific Reference to Agro-Pastoralism: A case study from Central Division, Garissa District, Kenya.* Unpublished M.Phil Thesis School of Environmental Studies, Moi University.

Reitsma, H., T. Dietz and H. de Leo (eds.) (1992)*Coping with Semi-aridity: How the Poor Survive in Dry-Season Environments.* Netherlands Geographical Studies 146.

Riley, W.B. *The Mbeere in Kenya Vol.1:* Brokensha (1988) *Changing Rural Ecology.* University Press of America Inc.

Saina, C.K. (1995) *Drought and Famine Coping Strategies and the Environment among the Keiyo Living in Kerio Valley, Kenya.* M.Phil Thesis, Moi University School of Environmental Studies.

Schlee, G. (1989) *Identities on the move: Clanism and Pastoralism in Northern Kenya.* Gideon S. Were Press, Nairobi.

Storas, F. (1987) Intention of Implication—Effects of Turkana Social Organization on Ecological Balances. *In* Baxter P.T.W. and Richard Hogg (eds.) *Property, Poverty and People. Direct Design Bornemouth Ltd,* Dorset pp. 137–146.

Spencer, P. (1973) *Nomads in Alliance: Symbiosis and Growth Among the Rendille and Samburu of Kenya.* Oxford University Press, New York.

Toulmin, C. (1983) Economic Behavior Among Livestock Keeping Peoples: A Review of the Literature on the Economics of Pastoral Production in the Semi-Arid Zones of Africa. Development Studies Occasional Paper No.2 School of Development Studies, University of East Anglia.

Turnbull, C. (1972) *The Mountain People.* New York, Simon and Schuster.

Turkana Drought Contingency Planning Unit (1992) *Turkana District Drought Manual Version 2.* Turkana Drought Contingency Planning Unit, Lodwar Watson, C. (1988) *The Development Needs of Turkana Women.* A Report for Oxfam and Public Law Institute, Nairobi.

Were, G. and R. Soper (eds.) (1986) *Kenya Socio-Cultural profiles: Busia District.* Institute of African Studies and Government of Kenya.

Were, G. and J. Akong'a (eds.) (1986) *Baringo District Socio-Cultural profile.* Institute of African Studies and Government of Kenya.

ABOUT THE AUTHORS

Joshua J. Akong'a is Professor of Anthropology at Moi University, Eldoret, Kenya. As a pioneer anthropologist in Kenya, he has set up Departments of Anthropology at the Institute of African Studies, University of Nairobi, and at Moi University. He has a long standing interest in drought and famine management, socialization and socio-cultural concomitants of development. He has contributed to various books and journals including David Brokensha and Peter Little (eds.) *Anthropology and Development and Change in East Africa* (Boulder, 1988); Orieko and Roberta Mutiso (eds.) *Working with Rural Communities* (University of Nairobi Press, 1991); Michael Coy and Leonrd Plotricor (eds.) *African and African American Sensitivity* (ethnology Monograph No. 15, 1995); Jos van der Klei (ed.) *Popular Culture: Africa, Asia and Europe* (CERES, 1995); Jan-Olof Drangert *et al.* (eds.) *Safe Water and Environments*, Report No. 24 1996 and Tom Weisner *et al.* (eds.) *African Families and the Crisis of Social change* (Greenwood Press 1977); T.E. Downing *et al.* (eds.) *Impact of Climatic Variations on Agriculture, Vol. 2.* (Kluwer Academic Publishers, Dordrecht, 1988) among others.
He is currently Dean of the School of Social, Cultural and Development Studies (BA programs) and Head of the Department of Anthropology at Moi University.

Joram N. Kareithi received his B.A. Hons (Anthropology) at the University of Nairobi in 1989 and M.Phil. (Anthropology) at the university of Cambridge in 1991. Between 1991–93, he worked in the Ministry of Culture and Social Services in the government of Kenya. He is currently a lecturer in the department of Anthropology at Moi University, Kenya. His research is on socio-economic changes and food security in the arid lands of Kenya. He is currently undertaking Ph.D. research on the Impact of Socio-Economic Change on Household Food Security Among the Turkana Pastoralists of North-Western Kenya.

Prof. J.J. Akong'a, Head
Mr. J.N. Kareithi
Department of Anthropology
Moi University
Box 3900
Eldoret
Kenya

Chapter 10

Land Tenure Frontiers and Food Security among Maasai Pastoralists in Kenya

Marcel M.E.M. Rutten

Introduction

The number of undernourished people in Africa has increased by 46% since 1970, reaching 175 million in 1995 (IFPRI, 1995). Most of the food production gains in Africa during the last decades resulted from an extension of the agricultural frontier into drier regions (Platteau, 1988). However, at present this arid frontier can only be moved marginally, while it has already reached its sustainable limit in some countries. What can be done to cover the difference between food needs and production in Sub-saharan Africa aside from importing food? The IFPRI 2020 vision report mentions a whole range of actions including enhancing natural resource management by securing and reforming property rights (IFPRI, 1995:49).

A reform in property rights is considered by some as the first and foremost action needed: 'land tenure and land use policy issues will be of strategic importance in the 1980s and 1990s as the frontier phase is exhausted' (Eicher, 1984:455–56). Privatization and individualization of landholdings has also been propagated in the past by institutions like the World Bank.

This process of tenure change, which started in the humid areas of Africa, has reached the border zone between sub-humid and semi-arid areas. The frontier of private and individual tenure rights is increasingly creeping into the formerly held communal and public arid and semi-arid lands of Africa (Dufour, 1971; Rutten, 1992; Besteman, 1994; Hesseling, 1994). Kenya's Kajiado District, home of Maasai pastoralists, is illustrative of this process. The district comprises 22,000 km^2, inhabited in 1989 by some 265,000 people, slightly half of them Maasai pastoralists. Soils, rainfall regimes and vegetation cover differ significantly. Only 8% has reasonable potential for cultivation (Agro-climatic zone IV (50–65% r/Et). Most of this higher potential land is situated along the upland fringes of the district. Zone V (30–50% r/Et) covers foremost the northern and central regions and comprises a total of 56% of the area. The remaining 36% is in zone VI (20–30% r/Et) encom-

H. J. Bruins and H. Lithwick (eds.), The Arid Frontier, 185–208.

passing the southern rangelands. Today the importance of cultivation is increasing rapidly, both in terms of land occupation and, even more profoundly, by the number of producers involved. The estimated availability of Good Agricultural Land per Person, expressed in High-Potential Land Equivalents, declined from 0.45 to 0.18 ha from 1969 to 1989, particularly due to immigration. The land tenure and agricultural frontier in Kajiado District has almost enclosed the sub-humid/semi-arid zone IV and is now creeping into semi-arid zone V.

The Maasai were once called 'the most wealthy tribe in East Africa, both in land and the stock they were able to sustain' (KLC, 1934). Now their nomadic future and even their pastoral future has turned bleak. The processes of structural impoverishment and acute major crises of hunger and starvation for man and his animals seem to have become more frequent in recent times.

Rebuilding their herds and returning to a full pastoral way of life is a goal that is no longer fulfilled by growing numbers of Maasai. Systems of mutual assistance needed for restocking are falling apart. Livestock diseases result in forced slaughterings. Cash needs for food acquisition and other non-food needs, such as schooling, require the sale of animals. The main causes for this growing crisis in the pastoral population include: (1) relatively less livestock numbers in relation to the larger human population, and (2) stratification of livestock ownership.

The human population growth is viewed by some researchers as the main cause of the economic problems of today's pastoralists. Droughts, rangeland degradation, reduced access and control over land, as well as an unfavorable, even hostile, political and economic environment are other major factors. Opinions as to the relative importance of each of these causes differ between several schools of thought within and between disciplines (Behnke, 1985). However, they all share the view that traditional nomadic pastoralism is losing ground. This chapter makes an assessment of changes in the land tenure frontier and its impact on the food security of Maasai pastoralists. The effects of commoditizing land in Kajiado District on the food entitlements of the Maasai population is studied in particular.

Land Availability, Tenure and Food Security: A Conceptual Approach

LAND AVAILABILITY

Though landlessness is foremost a phenomenon of Asian countries, the increase in landlessness during the 1980s was highest in Africa. The emerging problem of landlessness in Africa is related to a combination of several factors. The large increase of the agricultural population and the growing tendency towards the privatization of land rights are two main factors (WCARRD, 1988:41).

Land availability to nomadic pastoralists was abundant in the past. Dry season grazing areas were considered to be laying idle throughout most of the year by agriculturalists and administrators. There has been continuous outside pressure to turn these essential links in the mobility pattern of livestock into cultivated lands. Land available to pastoral households is increasingly being recognized in the last decades as becoming a scarce resource. Non-livestock activities or changing land tenure arrangements are responsible for adding to the growing set of obstacles to maintain a viable pastoral way of life.

LAND TENURE

Three broad trends of policy governing access to land have prevailed during the early 1980s in Africa:

1. There have been shifts in some countries towards a socialization of land through cooperatives, collectives and state farms, sometimes coupled with villagization programs (as in Ethiopia, Mozambique, the Congo).
2. There have been shifts towards or a continuation of the privatization of land, as in Malawi, Kenya, Liberia, and Reunion.
3. There have been adaptations of existing customary tenures as in the Gambia and Lesotho (WCARRD, 1988:38).

Within the first group of countries, the need has been recognized for greater allocation of resources towards the peasant sector and cooperative sector, as state farms are not achieving their intended results. In the second group of countries, programs of accelerated adjudication and registration of individual titles continued. These countries are reported to face an increasing number of landless people, rural-urban migration and settlement in marginal areas (Kenya, Malawi).

In the third group of countries, the adaptation of customary tenure has taken various forms such as the establishment of land allocation committees chaired by the chiefs with regulations and leaseholds (Lesotho). These latter arrangement represents a steady expansion of state power in the regulation of customary tenure.

A switch to the adoption of the second type of land policy can be recorded among a growing number of African countries. This will probably sharpen the inequality in the land distribution which, though less pronounced as in Latin America or Asia, does exist in Africa. Some of the factors contributing to this imbalance seem to have originated from the privatization of lands held under customary tenure. Indeed the highest inequality of land distribution in Africa is found among the group of countries propagating the privatization and individualization of land, *e.g.* Liberia, Kenya, Reunion (WCARRD, 1988:38).

It can be concluded that the trend in most countries of the developing world is one of further deterioration in access to land for a majority of farmers. Groups such as pastoralists, hunter-gatherers, the illiterate and female agriculturalists are particularly victims of the capitalization of land. These groups cannot compete with the vested interests of powerful groups of rich farmers, ranchers or multinationals. Sometimes the dispossessed oppose each other over the same lands, for example marginalized farmers competing with pastoralists. This struggle for land can also be recorded among members belonging to the same group between age groups or clans. Increasing competition evolves into open conflict.

FOOD SECURITY

The Food and Agriculture Organization of the UN defines household food security as access to food at all times as necessary for a healthy life for all members of the household (FAO/WHO, 1992). The achievement of household food security has three dimensions: adequacy, stability and sustainability (Frankenberger *et al.*, 1993). Adequacy refers to foods which are adequate in quality and quantity, free from toxic and harmful substances, and culturally acceptable. The supply of food should have a reasonable degree of stability throughout the year, as well as from one year to another. A household should be able to bounce back and regain an adequate food supply in case of a shortfall, often referred to as resilience. Finally, the supply of food should be sustainable; the ability to generate access to sufficient food while maintaining the endowment of resources over an extended period of time (Brouwer, 1994; Maxwell, 1995).

A model developed by IFAD distinguishes immediate, underlying and basic conditioning factors resulting in adequate, stable and sustainable food supplies at the household level (Brouwer, 1994:13 and Figure 1). Immediate factors are access to an adequate food base (all the foods available from different sources) and effectiveness in food handling (*i.e.* all processes necessary for increasing the life of food products and their safety such as storage, conservation and processing). Underlying factors are the stable access to essential resources, the management of resources and the pattern of social support. Finally, basic conditioning factors are the structures and mechanisms for control and management of resources, local agro-ecological conditions and the existing marketing system.

The linkage between tenure rights and food security at the household level is first and foremost one of agricultural production; access to suitable land enables the production of food, and hence its consumption. Tenure institutions indirectly affect food security at the regional or national level through overall food availability and hence food prices. Maxwell (1995:1) postulates that:

> enhancing the security of rights to land should, presumably, enhance food security. There is strong evidence of a direct negative link; that is, how reduction in tenure security or outright loss of access in an agrarian society leads directly to loss of food security, or alternatively, how severe food entitlement failure or famine results in the distress sales of land and other productive assets.

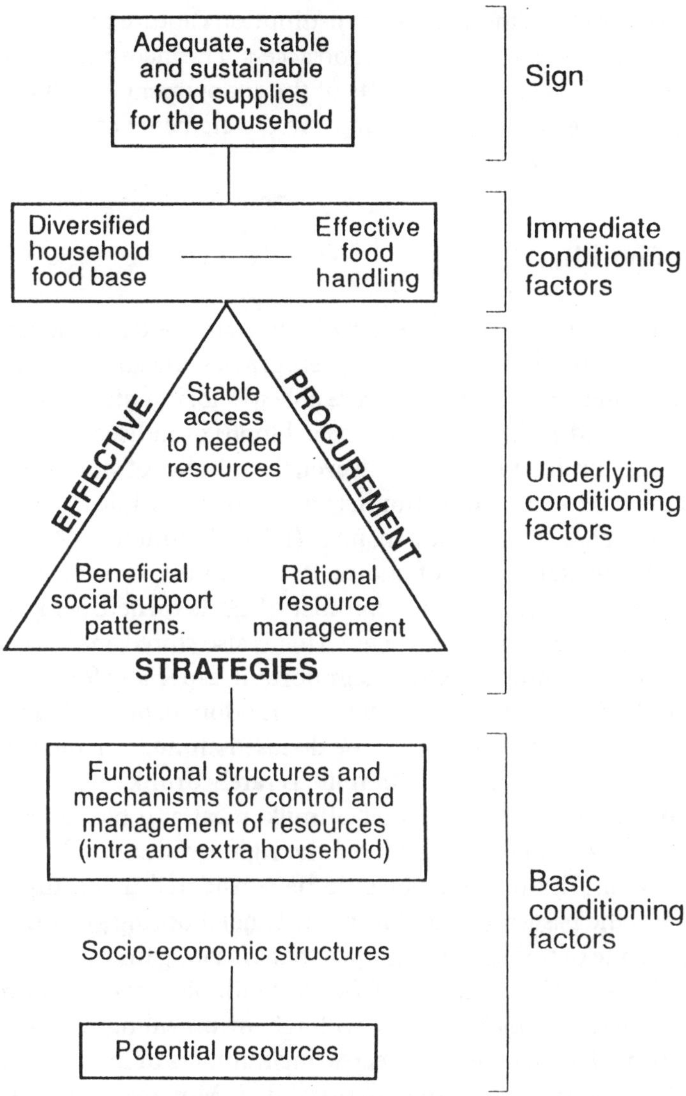

Figure 1. Factors affecting food security (see Frankenberger *et al.*, 1993).

The evidence of a positive linkage, on the other hand, is subject to numerous empirical questions; how is food access obtained (direct production, income and market access, food transfers through barter, begging and the availability of food stocks); in which way link livelihood strategies land use and food consumption; is this linkage employed in a sustainable way; and how do changes in land security affect livelihood strategies and the sustainability of resource usage (Maxwell, 1995).

Methodologically, 'household food security' is very difficult to operationalize. Consumption is usually taken as a proxy for the much more complicated definition offered above. At best, measures of consumption capture only the elements of food sufficiency and, to a partial degree, food access. The incorporation of dynamic elements such as of rights to land as parts of the endowment on which household food entitlements relies would help to capture the elements of sustainability and uncertainty outlined above.

The Maasai Food Base

The main aim for every Maasai household is to secure food in sufficient amounts. Livestock is used as an intermediate to convert grasses and shrubs into milk, meat and blood. For a long time the Maasai were able to subsist entirely on the produce of their large herds, either directly (milk, meat, blood) or indirectly through (lucrative) barter for agricultural products. However, the number of stock units per capita has dwindled over the years. More mouths have to be fed, while livestock numbers increase at a lower pace in an erratic manner (Dietz & Rutten, 1989).

Milk is the Maasai staple food of choice, its availability varies seasonally and is heavily dependent on rainfall and the number of cattle owned. Milk is consumed fresh as well as skimmed or mixed with tea. Milk is also stored and turned into sour milk. Fat (*i.e.* butter or animal fat) is also important, especially for young children.

The consumption of meat is low and taken at random depending on the number of (forced) slaughtering because of old or diseased animals or the occurrence of ceremonies. Blood, fresh or mixed with milk, is rarely drunk.

According to Århem (1989:77) a diet of milk, meat and blood from domestic livestock, cattle in particular, is considered the right or ideal kind of food among the pastoral Maasai. All other foods would be considered unworthy. Such a diet also distinguishes the pastoral Maasai from other neighboring groups who eat cultivated crops. In spite of this ideal it is known that for a long time Maasai pastoralists have consumed cereals (Lawren, 1968; Gower, 1948). Moreover, this strong feeling for a traditional livestock-based diet is, willingly or unwillingly, eroding due to a number of reasons. The Maasai have decreasing numbers of livestock available per capita to feed themselves on a purely pastoral diet. Moreover, Maasai youngsters increasingly join schools where other types of food are served. Food habits are also changing due to assimilation of many outsiders with agricultural backgrounds into the Maasai households. All these factors seem to change the negative attitude to cultivation and its products.

The change in food habits is exemplified by the gradual acceptation of chicken, rice and maize in the Maasai diet, mainly as porridge or cooked maize meal. Taboos on fish and pork are still strong. Food restrictions also apply to certain periods in the life of the Maasai. For example, pregnant women reduce food intake from the seventh month onwards to prevent babies becoming too big. Maasai warriors are not allowed to eat food containing fat in the presence of women. Finally, the change in diet is also influenced by the increased availability of maize and maize meal, through a better infrastructure, improved transportation and the expanding number of shops in the rural areas. Rainfed and irrigated cultivation is also growing in importance in the Maasai area. Nowadays some 50% of dietary energy intake comes from non-livestock foods (Homewood, 1992).

The Maasai diet results in a very high protein intake but an insufficient energy intake (65–80% Required Daily Intake), according to certain official standards. However, some authors raise questions as to whether FAO's recommended daily intake for energy is set too high to be applicable to Maasai pastoralists (Bekure *et al.*, 1991:111).

Intra-household food distribution should be taken into account, when analyzing household food security, because adult men and adolescents are favored at the expense of women and children. The former have better access to food (especially meat) than the latter. The rules of food distribution dictate that the elders are served first. They demand respect for having gone through all ritual stages from boyhood, via warriorship, through wars, raids and suffering, to the current status of junior or senior elder, awaiting their final journey to the ancestral world (Rutten & Tonkei, 1995). Moreover, they control finances enabling them to buy meat from local butcheries and drink beer in the bars, especially on market days. Men regularly gather to discuss issues and might decide to slaughter an animal and share the meat. Animals that died during times of drought are consumed by the adolescents accompanying them. Warriors to keep their strength as they join in so called meat-camps *(olpul)* at the end of the dry season to consume enormous amounts of meat from animals provided by the community.

Mature women, by contrast, are the chief participants in the production and maintenance of food security in their homes. If they fail in this respect, they might face severe punishments. Their efforts are fully directed to obtain food for all household members (except the warriors), through milking, buying, borrowing and, if need be, by begging. They are the last persons to be served. The wife will stay hungry, if the amount of food is not sufficient.

The position of children in intra household food distribution is less clear. Nestel (1989:28) reports that undernutrition is mainly manifest in early childhood as a result of an inappropriate diet (milk predominately) for 2–5 year olds. Severe physical stresses associated with herding practices of the 9–14 age-group may relate to insufficient food intake. They walk up to 20 km a day. The main meal is in the evening, while they eat wild roots, berries and tubers during the day time.

Maasai Livelihood Strategies in Securing Access to Food

Food security in a pastoral setting is to a large extent related to the ability of the herder to keep a large and varied enough herd for direct milk production, and to a lesser extent meat and blood. In addition, the sale of animal products such as milk, meat, manure, bones, hides and skins enables the purchase of food, *i.e.* agricultural products, sugar, tea, *etc.*

Productivity levels in the arid and semi-arid rangelands of Africa are more difficult to predict and/or to control, as compared to temperate zone pastures. The former areas are marked by intra and inter-annual fluctuations between wet and dry seasons. Fodder and water availability fluctuate in time and space, forcing pastoralists to trek with their animals in search for new grazing grounds and water. Mobility of herds and people in search of fodder and water is but one of these strategies. Other factors affecting mobility are the need to avoid disease affected and insecure areas. Actual migration patterns are also influenced by the location of trading centers, agricultural areas and quarantine regulations. Pastoralists are used to seasonal variations in food availability accompanied by a dwindling milk production in the dry season. Besides a sufficiently large herd for direct subsistence needs, pastoralists need to maintain enough breeding animals for reproductive reasons, as an insurance against losses due to diseases and droughts, as well as for social obligations (bridewealth, assisting poorer kinsmen, *etc.*). Thus management of African rangelands cannot solely be directed towards achieving maximum commercial output.

In this non-equilibrium environment, animals mediate between man and the natural environment, converting the scarce resources of water and pasture into milk, meat, blood and other animal products for direct consumption or sale. The risks involved necessitate the herd owners to engage in a whole range of stress coping strategies against droughts, diseases, and predators as part of their overall livelihood strategy geared at accumulation of wealth, betterment, sustenance and, in exceptional circumstances of crisis, survival (Dietz, 1991).[1]

The strategies employed ultimately aim to secure the physical well being of the household by defending the accumulated wealth or by restoring it. Thus both, preventive and curative elements can be distinguished. Preventive elements are preservatives, applied (long) before the actual problem arises. Curative elements are applied as a 'cure action' when problems are there to be solved. Preventive strategies among Maasai pastoralists (Rutten, 1995c) are mainly aimed at securing the underlying and basic factors affecting food security (*i.e.* herd management and resources management). The relative importance of each of these strategies varies in time and place. Some of the elements involved are listed below:

HERD MANAGEMENT

- Keeping sufficiently large herds to accommodate anticipated losses resulting from droughts and diseases.
- Herd mobility enhances the optimal use of a heterogeneous environment.

- Destocking through sales before and during a crisis (*e.g.* a severe drought) and restocking when pasture is available again after a drought.
- Managing a variety of livestock species (cattle, camels, sheep and goats) with specific ecological niches (grasses, shrubs, trees) and production characteristics (*i.e.* different periods of milk recovery, disease susceptibility).
- Allocation of animals with friends and relatives on special loan arrangements allowing the reclaiming of these animals in times of need;
- Breeding animals with low basal metabolic rates and high milk and meat output.
- Dipping the animals in acaracide baths to prevent tick-borne diseases; installing tsetse fly traps.
- Castrating animals which helps to fatten them in order to make them more drought resistant.

RESOURCE MANAGEMENT

- Certain dry season grazing areas (*e.g.* hilly areas) are set aside in cooperative agreement with other households.
- Individual households refrain from using certain patches of land nearby their permanent homesteads. These fenced areas *(olkeri)* are used for calves and sick animals.
- Making improvements in water availability through shallow wells and boreholes.
- Food preservation through drying of meat *(sureeni)* or storing of cheese *(encholati)*. These traditional practices are on the decline, foremost because cereals are easily available nowadays.
- Diversification of the household economy through engagement in rainfed and irrigated cultivation, wage employment and commercial business.
- Bank accounts are increasingly used to store money for less fortunate days.

Emergency strategies in managing the Maasai food base are implemented during and following times of stress or severe crises. The most important of these strategies are differentiated as follows:

CHANGE IN FOOD HABITS

- Reduction in food intake by the pastoralists. Less home produced milk is available. Less food is eaten especially by herdsmen.
- Change in food intake. Milk is mixed with herbs. More goats and sheep are milked. Less favored parts of the animal such as hub, liver, tongue, pancreas

and heart are eaten. Terribly emaciated meat from dead animals is boiled, roasted or cooked. Blood consumption increases (cooked or mixed with milk). Sales of animals and skins increase to buy food from the shops mainly maize meal and sugar or barter from agriculturalists.
- Hunting of antelopes and elands, as well as gathering of wild fruits, berries and roots.

CHANGE IN HERD AND RESOURCE MANAGEMENT

- Elderly people take over livestock herding, while younger Maasai search for distant pastures.
- Use of dry season areas, renting of outside pastures or long distance trekking to areas controlled by other groups.
- Collection of grasses, tree pods and leaves as fodder, especially for young calves and rams.
- Animal intake of drinking water is changed from every day to once in two days.
- Milking practices are changed as more milk is left for calves and less milk is consumed by people.

SOCIAL SUPPORT NETWORK

- Assistance is sought from friends and relatives (money, labor, animals, food).
- Praying and sacrificing a ram of specific colors (said to be on the decrease).
- Relief programs organized by the Government and Church either in the form of free hand outs or as Food for Work Programs as introduced in 1994. Such relief also alleviates the need to sell livestock. FFW might undermine the natural recovery cycle of the pastures.

During 1994, the men contributed less in FFW activities, took less responsibility for household food security and instead lived a luxurious life in the small rural centers. Food relief programs have become a new safety net within the Maasai society to an extent that young Maasai no longer are aware of the term *emperi* expressing a disastrous situation whereby in addition to animals human beings die as well (Rutten, 1995c).

Access to and control over land is an essential precondition for a successful implementation of many of the food security strategies mentioned above. Let me, therefore, present an analysis of the Maasai land resource and its changes this century.

Maasai Land Availability and Tenure: A Historical Review

COLONIAL PERIOD

Toward the beginning of the 20th century plans were made by administrators in unofficial alliance with influential settlers to remove the Maasai from their best grazing areas. In 1902, the First Crown Lands Ordinance allowed the local administration to issue settlers with 99 year leases and the possibility of selling land. The decision as to whether land was vacant or not was in the hands of the British administration. In practice this meant a denial of traditionally established African rights over land. The land policy was successful in attracting white settlers. Large concessions were granted to South African and British entrepreneurs to an extent that some sub-ordinates questioned Commissioner Eliot's policy of integration by inducing the Maasai to settle in small villages among Europeans. They preferred to divide the pastures and set apart a reserve for the Maasai where settlers could not interfere. Treaties were signed in 1904 and 1911 which made it possible to remove the Maasai from their best pastures in the north to a southern reserve. This area of less quality, about 36,000 km^2, is only half the size of their former territory. In fact over half of the areas settled by Europeans used to be former Maasai lands.

Agricultural groups like the Kikuyu, Kamba and Chagga were very much interested in the areas suitable for cultivation within Maasailand. Individual Maasai households allowed these groups to cultivate in these high potential zones on a share cropping basis. However, a special permit was needed to enter the Maasai Reserve. The official policy was to prosecute and return all Kikuyu who could not claim, through long residence or circumcision rights, to have become Maasai.

The Kenya Land Commission, installed in 1932 to review African land grievances, had established the boundaries of the Reserves for every group in Kenya. The Maasai unsuccessfully repeated their protest against the loss of land. The Kenya Land Commission reaffirmed the policy of the Administration towards pastoralists, totally opposing any extension of their land. The Commission even blamed the Maasai for having so much land to the detriment of Europeans and other African groups.

The British Administration fostered that traditional grazing control among the Maasai had been inadequate. The British aim was to implement a grazing control plan for each Maasai section, restricting the movement of stock from one section to another in order to enforce an economic limit within the boundaries of each section. The Maasai reserve was subdivided in 14 sections *(iloshon)*, each having a specific territory with well-defined boundaries. Livestock movements between sections needed approval of the section's authorities. In June 1946 the Maasai decided the abolition of section boundaries which allowed free grazing for the Maasai pastoralists throughout the whole of Maasailand in order to overcome a drought situation. By the early 1950s this decision was withdrawn. Large grazing schemes were established during the 1950s. Grazing principles were based on the traditional use of the area appended with strict grazing controls enforced on a clan basis by Grazing Committees. However, the Administration fully realized that 'owing to the vagaries of the rains it cannot be expected that an even pattern for grazing control will ever be achieved' (KDAR, 1957:16). This was proved in 1959 when any controls had to be abandoned as cattle based on the scheme had to go far afield in search of grazing even trespassing into neighboring districts and into Tanganyika.

Since 1954, local Maasai politicians in Kajiado District had started acquiring large individual ranches without a legal basis. Support for these individuals was, nevertheless, provided by the Local County Council, often because of private interests.

In addition to the grabbing of land from within, the land resource base available for Maasai livestock herding dwindled further as more agriculturalists started to infiltrate and occupy the high-potential dry season grazing pastures. The establishment of the Amboseli Game Reserve and Nairobi National Park, partly in response to an international conservationist lobby, put further strain on the availability of land for livestock.

POST-COLONIAL PERIOD

The Maasai feared that independence would end the Maasai Treaties of 1904 and 1911 which gave them exclusive rights to occupy Kajiado and Narok Districts. They went as far as to request the United Nations to recognize a Maasailand which comprised of Tanzanian and Kenyan Maasai. This proposal was as much condemned to fail, as was the Maasai request to the British Government to hand back the former 'White Highlands'.

Discussions concerning the status of Maasailand intensified in the early 1960s. The land tenure debate was particularly significant among the Kaputiei Maasai. A wide variety of propositions were discussed, such as the registration of all of the Kaputiei section area under one title deed or the creation of a fringe of individual ranches near the northern boundary to stop illegal infiltration. The Colonial Authorities had disagreed with the one title deed option. The idea of individual ranches suggested by young and educated Maasai was supported by the Kajiado County Council. Elderly Maasai opposed the move to individualization. Support for the latter was given by the Lawrance Mission of 1965 that criticized the haphazard approach of the government to the Maasai land question and the illegal approval of the creation of individual ranches. However, in 1969, all of the Maasai sections accepted the group ranch concept as introduced by the Kenya Livestock Development Project sponsored by the World Bank (see Figure 2).

Figure 2. Kajiado District group ranches.

A group ranch may be defined as a certain piece of land communally owned by a group of people who are registered as the legal owners through membership of the particular ranch. An attempt was made, through the provision of loans for infrastructural development and steer fattening, to transform the semi-nomadic, subsistence-oriented production of the Maasai pastoralists into a sedentary, more commercial system. This market-oriented production was intended to bring about a destocking of the Maasai pastures while at the same time providing meat for the national and international market. Phase one of the Kenya Livestock Development Project (KLDP I) was officially started in 1969. By 1970, 14 group ranches, covering over 10% of the district's area were recorded in the Kaputiei area. Two more phases followed in 1974 and 1980 which brought the total number of group ranches to 51 covering some 15,200 km^2 or some 75% of the Kajiado District area.

The performance of the Kajiado group ranches has been eagerly followed by several scholars and review missions. The introduction of the group ranch had various organizational, juridical and economical consequences: (1) The Kajiado District group ranches were effective to stop the allocation of individual ranches, to be taken from communal lands, by an educated elite of Maasai. The feared massive influx by non-Maasai was also hampered. Outsiders could buy only land from individual ranchers, as in the Ngong and Loitokitok areas. (2) The group ranches generated the construction of boreholes, dams, troughs, tanks, pipelines and cattle dips. (3) Wildlife was enabled to continue roaming freely over large parts of Kajiado District.

Besides these accomplishments, several problems and failures of the group ranches can be mentioned:

(a) Delays and problems in project implementation;
(b) Disappointing rates of investment and difficulties in loan repayment;
(c) Continuing trespassing of group ranch boundaries;
(d) Refusal to de-stock ranches;
(e) No real transformation to a market-oriented livestock production; and
(f) Corruption among several group ranch committees.

The group ranch concept was an artificial creation which lacked a firm traditional, sociological and ecological basis. Moreover, the implementation of this change in land tenure was over-ambitious as a method to cause the destocking of the pastures and the commercialization of production. Hardly any account was taken of the pastoralists own strategies and household needs. These problems and frustrations increased the wish among many Maasai for the subdivision of the group ranch into individually owned shares.

By 1990 a total of 40 group ranches had made the decision to dissolve their ranches. Only 4 group ranches opposed the idea of subdivision, while another 7 ranches had not yet decided. In other words, 78% of the ranches had ceased to exist or were intending to do so.

An overall positive attitude towards the process of subdivision and plot allocation was recorded among the Olkinos and Embolioi ranchers. Approximately 1 out of 5 households, however, had mixed to negative feelings concerning the size of the plots allocated. The group of non-registered group ranch members, mainly Maasai up to 30 years of age, also protested as they did not receive their own plot of land.

Figure 3. Olkinos subdivided group ranch 1986, 1990.

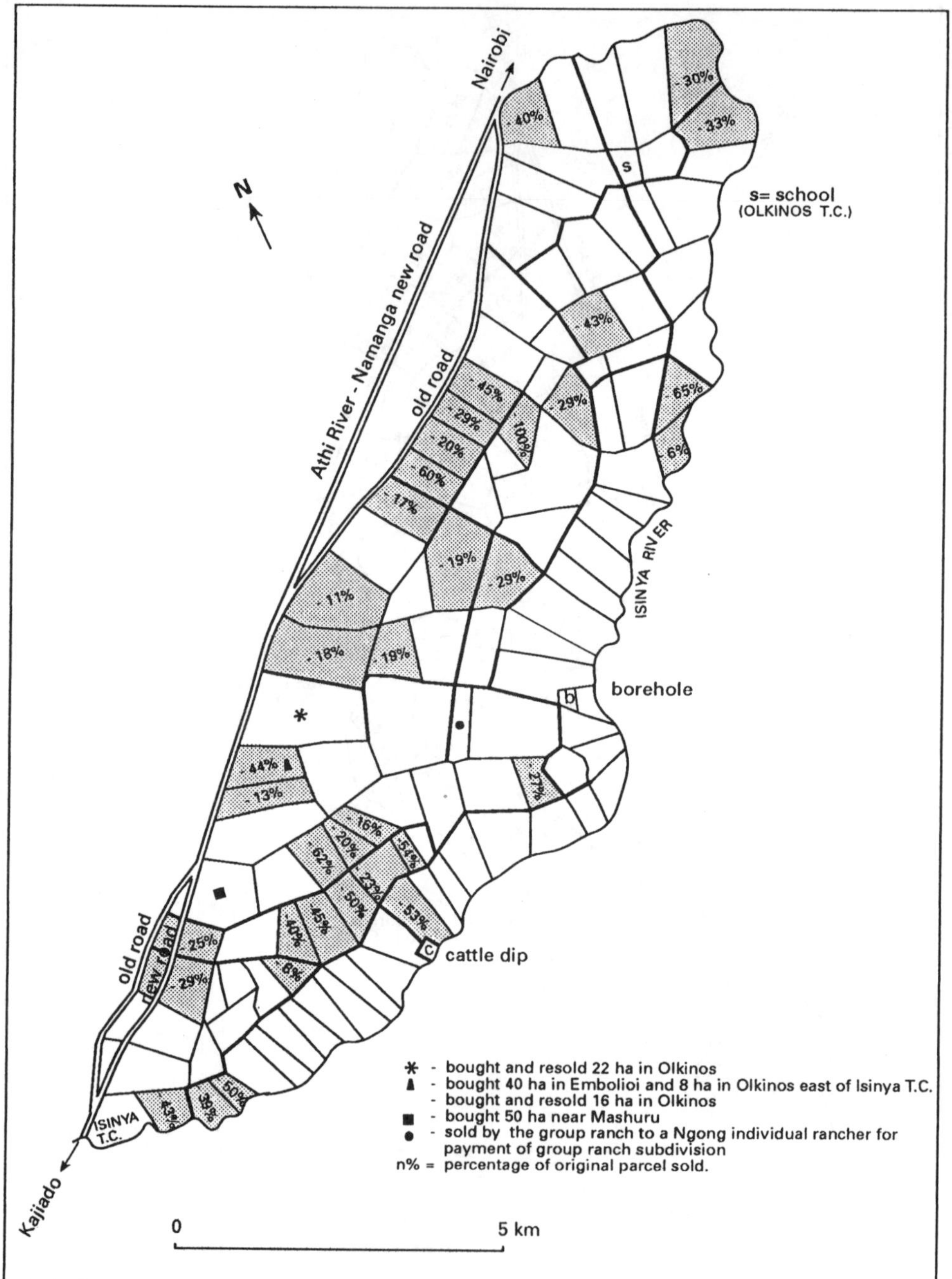

Figure 4. Olkinos selling of land, up to February 1990.

The former Olkinos group ranch members possessed after subdivision ranches with a size ranging from 11 to 132 ha per parcel, 46.7 ha on average (see Figure 3). The process of subdivision had been most favorable for the most powerful within the Maasai group ranch constellation. Plot sizes obtained by the committee-members averaged 81.8 ha.

Landownership per person decreased over the 1986–90 period, due to natural population growth and the sale of land. The average availability of land for Olkinos ranchers was reduced from 8.4 ha/person in 1986 to 7.1 ha/person in 1990, a 16% drop.

I investigated the outcome of the group ranch subdivision in Kajiado District with respect to transfers of land and improvements made in the 1986–1990 period. The main conclusions drawn from an analysis of the fragmentation, sale or mortgage of the subdivided areas are as follows:

- No distinctive set of characteristics could be assigned to the group of sellers. However, former group ranch committee members, who were relatively rich, were foremost among the group of non-sellers.
- Maasai pastoralists prefer to sell rather than mortgage (part of) their newly acquired parcels. Their rationale is that the selling of land is a much faster, less obstructive and more comprehensible way of obtaining money than mortgaging one's plot. Financial institutions also seem to be somewhat reluctant to provide loans, except in the case that the prospective borrower has been formally educated to a high level and has other more important sources of income besides livestock keeping.
- The proceeds of the sale of land, as well as loans, were mainly invested in infrastructural improvements of a non-productive nature such as the building of a modern house, fences, sprayraces and water pans. Repaying a debt as an instigation for selling land should also be mentioned.
- Land sales were also used to purchase more livestock. The higher stocking densities resulted theoretically in unviable ecological ranch sizes. This outcome questions the postulation that individual land ownership will lead to a destocking of the ranch at carrying capacity levels. In general, no significant change in natural resource management occurred. The fact that land could still be used in a communal way is most likely the main explanation for this observation. Most recently, however, the influx of outsiders involved in other land use activities created obstacles for the accessibility of pastures and mobility of herding patterns. The renting of land is also increasing, especially during times of stress.
- The group of land buyers is foremost composed of non-Maasai. (Inter)national companies, rich business men, politicians, civil servants, as well as less wealthy people are among this group. Maasai buyers were mainly well-off former com-

mittee-members, individual ranchers, civil servants, and politicians. Another group of informal buyers are Indians who practice strip gypsum mining in the Embolioi group ranch. Buyers of land, in contrast to the original Maasai owners, showed a preference for mortgaging land rather than reselling it.

Land Tenure Changes and Food Security: Concluding Remarks

If their strategies of risk aversion fail, pastoral households are faced with a temporary food shortage or transitory food insecurity (Salih, 1994:6). All pastoral societies have been living for centuries with this temporarily phenomenon. However, pastoralism seems to have become increasingly marginalized and many pastoral households are at present facing a kind of chronic food insecurity. It becomes more and more evident according to Schwartz and Schwartz (1985:5) that nomadic societies 'show a decreasing selfreliance in terms of food production'. Pastoralists are leaving the livestock economy as a result. Reports from all over Africa indicate the loss of pastoral areas for a growing group of pastoralists. This loss of land threatens to undermine access to and control over a resource which is crucial in securing food for many pastoral households.

Maasai pastoralists have experienced a drastic reduction in their land availability in the last century, foremost as the result of the arrival of the European colonizers. Moreover, African land laws were overruled and a closed Reserve was created. This enlarged control over mobility patterns hindered the Maasai's ability to market livestock in exchange for grains and ended the Maasai tradition of acquiring improved breeds from northern Kenya. The spread of European rule also meant a barrier to Maasai pastoralists in their bid to recapture herd losses after periods of crisis by raiding neighboring groups. Finally, the creation of a Reserve indulged the taxation of the Maasai to an extent that they became the heaviest taxed group in Kenya (Rutten, 1992). The colonial government also interfered with the Maasai land use patterns by introducing (unfit) grazing regulations and preserving certain (high potential) areas for game parks and agriculture.

After independence the Kenyan government and international donors introduced the group ranch concept. Land came into private hands, though still communally owned, except for some parts of the district where an educated elite had been able to grab large individual ranches. Corrupt committees, misuse of group ranch grazing by neighboring individual ranchers, and a growing population paved the way towards individual land ownership in the whole of Maasailand.

Land tenure changes might directly result in a change in the control over, access to and availability of this fundamental resource. *Access* denotes the ability to get and use a resource while *control* means power that goes beyond use (Sigot, 1995:6). Negative as well as positive effects can be distinguished from the individualization of landownership in Maasailand.

DIRECT NEGATIVE EFFECTS OF LAND INDIVIDUALIZATION

The viability of the system is undermined by removing (high potential) land out of the pastoral migration cycle. The loss of dry season grazing areas (*e.g.* Amboseli, Tsavo, the slopes of Kilimanjaro and Ngong) forces Maasai pastoralists to concentrate in less favorable places, such as the Chyulu hills, or to trek far into Tanzania. The risk of losing animals due to tsetse fly and exhaustion are much higher as a result.

The individualization of land ownership further reduces access to land due to landsales, fencing of private land and the request for grazing payments. Several cases are reported of Maasai households in disarray after the selling of land, with or without approval of the family members. Some Maasai men subsequently started living a luxurious life, including alcohol abuse. Rutten and Tonkei (1995) state that some of these husbands have fled their homes, leaving women and children in extreme poverty and hunger. Sons are taking their fathers to court to stop them from selling the remaining piece of land. Most of these 'young' Maasai, some in their thirties having families of their own, were not allowed to register as a group ranch member and were left out of the subdivision process. Conflicts over land allocation resulted in a large number of costly court cases, draining away much money and leaving some individuals bankrupt.

Officially the Land Control Board will not allow the selling of land without the consent of the wife, but daily practice differs. The gender biased access and control over land has acquired a profound negative effect after the commoditization of this basic resource. After subdivision of the group ranches only a few Maasai widows were given a title deed. Kenyan law does not bar women from ownership or control of land (Thrupp & Green, 1995:112), but Maasai women have limited hereditary rights and resources for acquiring land. Usually non-Maasai women (groups and rich individuals) are able to buy land.

In the past Maasai destitutes could always try to rebuild their herds by engaging in wage labor or through help from stock-friends. The loss of the land, however being of a more structural nature, is a direct negative effect of the commoditization of land in a arid to semi-arid region.

INDIRECT NEGATIVE EFFECTS OF LAND INDIVIDUALIZATION

Households are splitting up as a result of subdivision, in order to settle on their own plot. Labor availability for herding is decreasing (Rutten, 1992). One of the most important investments by land sellers is the purchase of livestock, which has often resulted in higher stocking densities on already overstocked farms (Rutten, 1995a).

Moreover, some of the new activities threaten the environment and thus the carrying capacity of the area. For example, high water use for flower production has lowered water tables, causing rivers and shallow wells to become dry (Rutten, 1995b). The use of pesticides and the application of agricultural techniques unfit for semi-arid regions cause erosion and threaten Maasai food sources. Influx of outsiders may also cause tensions between ethnic groups. Political patrons sometimes manipulate these tensions by instigating clashes.

The sudden loss of a secure access to land is pushing many Maasai into new directions, such as wage employment. However, they are usually not well prepared to compete in this market as a result of the lack of formal education. The process of marginalization and stratification has a detrimental effect on the environment (*e.g.* charcoal burning) and puts additional strain on cooperation. In the past, Maasai used to steal animals from neighboring groups to rebuild a herd. Today structural impoverishment has led Maasai to steal animals from fellow (poor) Maasai.

DIRECT POSITIVE EFFECTS OF LAND INDIVIDUALIZATION

A direct positive effect of the individualization of group ranch land is the ability of individuals to restrict misuse of grazing land by neighboring individual ranchers. They used to distribute their animals to relatives registered in the group ranch during the wet season. In this way they were able to rest their own individual ranches until the onset of the dry season when the animals returned to them. Nowadays wealthy herdowners are renting land from poor pastoralists which enables the latter to invest the profits in rebuilding their own herd.

Another positive outcome of land individualization: misuse of loans by corrupt committees is no longer possible. Individuals are now free to select their own friends and invest in (physical) improvements such as wells. Flexibility in pasture management is in a sense reintroduced with the dissolution of the group ranches.

INDIRECT POSITIVE EFFECTS OF LAND INDIVIDUALIZATION

The commoditization of land has resulted in an increase of non-Maasai entering the district and a subsequent boom in non-livestock activities. Agricultural groups have moved the cultivation frontier further into pastoral areas, which increased the availability of agricultural products such as maize and beans. However, the chances of a successful crop in rainfed agriculture are modest, especially in times of drought. Maasai can barter livestock and livestock products for maize. The purchase of cereals using the money obtained from selling animals is a lucrative trade for pastoral groups as the benefit in caloric terms is 7:1 (Zaal & Dietz, 1995). The rationale trading maize when the exchange rate is best from the livestock keeper point of view is slowly advancing. More Maasai buy and store some food. They also increasingly try to cultivate by themselves. Moreover, fields can be grazed in case of crop failure, or as stubble after harvesting. The influx of outsiders has also brought more job opportunities in the mining and flower industries. However, the participation by Maasai in these activities is minimal.

Another indirect positive effect of group ranch subdivision is the recorded increase in (informal) small and effective self help groups who invest money and labor in shallow wells, steer fattening, breed improvements and the like without interference by corrupt group ranch committees (Mwangi & Rutten, 1995). The investments in water facilities, *e.g.* rain water catchment from the roof of modern houses and shallow wells in the compound allows (kitchen-garden) cultivation, which boosts food availability. These investments also save time, money and labor to collect water from a distant borehole or river. Still more could be done such as the growing of napier grass as extra fodder for animals. This could increase milk production substantially.

The subdivision of group ranches threatens the access for wildlife to dispersal areas in case the new ranches would be fenced. The Kenyan Government tries to prevent such fencing by offering money to the Maasai living nearby Amboseli National Park to be used for the benefit of the community. Therefore, the Maasai profit more directly from the large sums of money earned by the tourist industry. They also participate themselves, as some tent camps have been built outside the park to earn money from tourists. Thus (the threat of) individualization of land might also result in an improvement of food security.

The main conclusion, however, is that the livelihood of individual Maasai families has become more vulnerable to changing property rights and the commoditization of the land resource. Access and control over grazing land—which allows for a flexible mobility pattern—is less assured at present. Water resources are depleted or diverted away from pastoral production by newly introduced activities operated by outsiders or elite Maasai. Stress-coping and survival strategies are under threat especially for the less fortunate within Maasai society. Levels of wealth are no longer merely expressed in ownership of cattle but foremost in the size of land holdings. This land reform process is solidifying Maasai society in a more permanent arrangement of poor and rich members.

One might argue that this is an unavoidable process. However, if no creative solutions are implemented one might encounter the destruction of these fragile environments to the detriment of the local human population. In the end the Kenyan society at large might be less well off. There is a need to strengthen ways of resource management intensification and diversification for both livestock rearing and agricultural development, preferably starting from initiatives taken by the Maasai, themselves, as shown in the field of water development. Moreover, labor-intensive job alternatives need to be developed, based on the livestock sector or upon other opportunities.

Notes

1. Behnke (1994) presented an excellent outline about the management characteristics of these non-equilibrium environments as compared to the equilibrium grazing systems in more temperate zones. The latter are characterized by high levels of climatic stability resulting in rather constant levels of primary production.

References

Århem, K. (1989) Maasai food symbolism. The cultural connotations of milk, meat and blood in the pastoral Maasai diet. *Anthropos* 84:1–23.

Behnke, R.H. (1994) *Natural Resource Management in Pastoral Africa.* Overseas Development Institute (ODI), International Institute for Environment and Development, Common Wealth Secretariat, London.

Bekure, S., P.N. de Leeuw, B.E. Grandin, and P.J.H. Neate (1991) Maasai herding. An analysis of the livestock production system of Maasai pastoralists in eastern Kajiado District, Kenya International Livestock Centre for Africa, ILCA Systems Study 4, Addis Abeba.

Besteman, C. (1994) Individualisation and the assault on customary tenure in Africa: Title registration programs and the case of Somalia. *Africa* 64(4):484–515.

Brouwer, I.D. (1994) Food and fuel: A hidden dimension in human nutrition. A study on the relationship between nutrition security and fuelwood availability in Ntcheu District, Malawi. Ph.D. thesis, Wageningen Agricultural University, Wageningen, The Netherlands.

Dietz, A. (1991) Crisis survival strategies: A summary of concepts and an example from the semi-pastoral Pokot in Kenya/Uganda. *In* J.C. Stone (ed.) *Pastoral Economies in Africa and Long Term Responses to Drought.* Aberdeen University African Studies Group, pp. 86–108.

Dietz, A. and M.M.E.M. Rutten (eds.) (1989) The future of Maasai pastoralists in Kajiado District, Kenya. Integrated proceedings of a conference in Brackenhurst Conference Centre, Limuru, May 28–31, 1989, ASAL Kajiado.

Dufour, J. (1971) The problem of collectively owned land in Tunesia. *FAO Land Reform, Land Settlement and Cooperatives* 2:38–51.

Eicher, C. (1984) Facing Up to Africa's Food Crisis. *In* C. Eicher and J.M. Staatz (eds.) *Agricultural Development in the Third World.* The Johns Hopkins University Press, Baltimore and London, pp. 453–479.

Food and Agriculture Organization/World Health Organization (1992) Improving household food security. Major issues for nutrition strategies, Theme paper no. 1. International Conference on Nutrition, FAO, Geneva.

Frankenberger, T., A. Pena-Montenegro, S. Tilakaratna, N. Velarde, and W.B. Eide (1993) Rural poverty alleviation and nutrition: IFAD's evolving experiences. International Fund for Agricultural Development, Rome.

Gower, R.H. (1948) The effect of a change of diet on Masai schoolboys. *Tanganyika Notes and Records* No. 26:77–78.

Hesseling, G. (1994) Legal and institutional conditions for local management of natural resources: Mali. *In* R.J. Bakema (ed.) *Land Tenure and Sustainable Land Use.* Royal Tropical Institute, Bulletin 332, Amsterdam, pp. 30–46.

Homewood, K.M. (1992) Development and the ecology of Maasai food and nutrition. *Ecology of Food and Nutrition* 29:61–80.

IFPRI (1995) A 2020 vision for food, agriculture and the environment—The vision, challenge, and recommended Action. International Food Policy Research Institute, Washington, D.C.

KDAR (several years) *Kajiado District Annual Report.* Kenya National Archives.

KLC (1934) *Kenya Land Commission: Evidence and Memoranda.* Government Printer, Nairobi.

Lawren, W.L. (1968) Masai and Kikuyu: An historical analysis of culture transmission. *Journal of African History* 9:571–583.

Maxwell (1995) Land access, tenure security, resource conservation, and food security, unpublished paper. Land Tenure Center, Wisconsin.

Mwangi, M. and M.M.E.M. Rutten (1995) Water Development in Maasailand: Participating without Participants. Manuscript, Nijmegen.

Nestel, P.S. (1989) Food intake and growth among the Maasai. *Ecology of Food and Nutrition* 23:17–30.

Platteau, J.-P. (1988) The food crisis in Africa: A comparative structural analysis. World Institute for Development Economics Research of the United Nations University, Working Paper 44.

Platteau, J.-P. (1992) Land reform and structural adjustment in sub-Saharan Africa: controversies and guidelines. FAO Economic and Social Development Paper 107, Rome.

Rutten, M.M.E.M. (1992) *Selling Wealth to Buy Poverty—The Process of the Individualization of Landownership Among the Maasai Pastoralists of Kajiado District, Kenya, 1890–1990.* Ph.D. thesis, University of Nijmegen, Verlag Breitenbach Publishers, Saarbrücken and Fort Lauderdale.

Rutten, M.M.E.M. (1995a) The tragedy of individualizing the commons, paper presented at the IASCP Fifth Common Property Conference 'Reinventing the Commons' 24–28 May 1995 Bodf, Norway.

Rutten, M.M.E.M. (1995b) Environmental problems in semi-arid southern Kenya—Blaming the wrong people, paper presented at the 'Regional Workshop on Public Interest Environmental Law and Community-Based Legal Incentives for Sustainable Natural Resources Management in East Africa' August 23–26, 1995, Jinja, Uganda, World Resources Institute, Washington DC.

Rutten, M.M.E.M. (1995c) The Maasai Concept of Drought. *Kajiado Focus* 8:1–3.

Rutten, M.M.E.M. and T. Tonkei (1995) Maasai intra household food allocation. Unpublished working paper no. 2 Netherlands-Israel Research Programme, African Studies Centre, Leiden, The Netherlands.

Rutten, M.M.E.M., D. Kitasho, Th. Kasaine, and W. Saruni (1996) Maasai and food relief. Unpublished working paper no. 3 Netherlands-Israel Research Programme, African Studies Centre, Leiden, The Netherlands.

Salih, S.A. (1994) Food Security in East and Southern Africa. *Nordic Journal of African Studies* 3:3–28.

Schwartz, S. and H.J. Schwartz (1985) Nomadic pastoralism in Kenya—Still a viable production system? *Quarterly Journal of International Agriculture* 24:5–21.

Sigot, A. (1995) Discourse on gender and natural resource management. *In* A. Sigot, L.A. Thrupp and J. Green (eds.) *Towards Common Ground—Gender and Natural Resource Management in Africa.* ACTS Press, Nairobi & World Resources Institute, Washington, D.C., pp. 1–11.

Thrupp, L.A. and J. Green (1995) Reflections on gender in natural resource management. *In* A. Sigot, L.A. Thrupp and J. Green (eds.) *Towards Common Ground—Gender and Natural Resource Management in Africa.* ACTS Press, Nairobi & World Resources Institute, Washington, D.C., pp. 103–114.

WCARRD (1988) The impact of development strategies on the rural poor. World Conference on Agrarian Reform and Rural Development, 1979–1989, 10 years of follow up, FAO, Rome.

Zaal, F. and A. Dietz (1995) Of markets, meat, maize and milk—Pastoral commoditization as a necessary but risky livelihood strategy. Paper for the Workshop 'The Poverties and Prosperities of East African Pastoralism, Uppsala Sweden, September 15–17, 1995.

ABOUT THE AUTHOR

Marcel Rutten is a geographer and research fellow at the Africa Studies Centre, Leiden, the Netherlands. He has occupied positions at the Centre for Environmental Studies of the Wageningen Agricultural University and at the Department of Policy Sciences, University of Nijmegen. He conducted research in Kenya among Turkana and Maasai pastoralists. His main interest is natural resource management in dry lands. He is the author of a book on the individualization of land-ownership among Maasai pastoralists, a process he continues to monitor. Currently he is also involved in research on sustainable water projects (*i.e.* shallow wells), combining indigenous and modern technical knowledge, as well as research about drought planning strategies among Maasai, Turkana and Bedouin pastoralists.

Dr. Marcel M.E.M. Rutten
African Studies Center
P.O. Box 9555
2300 RB Leiden
The Netherlands

Chapter 11

Gestion de Terroir at the Frontier: Village Land Management of Peasants and Pastoralists in Benin

Leo J. De Haan

Introduction

Growing concern about land degradation and deforestation in Africa in the 1980s has led to increased interest in land management. The 1985 U.S. National Research Council Conference on Common Property Resource Management (BOSTID, 1985) marks an important shift in this respect. Greater appreciation of traditional land management systems, such as communal property regimes, accompanied this shift.

A new development strategy emerged in the same period in French-speaking West Africa, in order to cope with the failures of existing rural development programs (Painter, 1993). Population pressure, drought and mismanagement by national authorities encouraged policy makers and donors, primarily the French development cooperation, the dominant donor in this region, to stimulate local initiative and to build up local capacity in terms of income generation and the exploitation and management of local resources.

This has become known as the 'Gestion de Terroir' (Land Management) approach. However, there are different definitions of the concept and there is no overall conceptual framework. The emphasis in the Gestion de Terroir approach strongly focused on peasants and crop cultivation, while pastoralists, as well as cattle raising by peasants, were neglected (Marty, 1993; Bonnet, 1990). The raising of cattle is of vital importance in the Sahel and Sudan zones. Its integration, therefore, in the Gestion de Terroir needs to be investigated, including the conditions for successful implementation of resource management which takes cattle-breeding into account.

The North Borgou in northern Benin is a meeting zone of peasants and pastoralists. Environmental degradation has already become a serious problem in the region due to population growth, increased commercialization and the immi-

H. J. Bruins and H. Lithwick (eds.), The Arid Frontier, 209–227.

gration of pastoralists from the Sahel. Conservation measures are being promoted among peasants and pastoralists in order to arrive at more sustainable land use. However, these interventions have had little success because most farmers tend to escape from environmental degradation, rather than control it. They migrate to sparsely-populated and uncultivated areas: 'the frontier region' in this chapter. However, signs of over-exploitation are already appearing in this frontier area, which is also increasingly used by pastoralists. Only a Gestion de Terroir approach, encompassing both modes of existence, may result in a more sustainable exploitation of the area's resources.

The Research Group on Livelihood and Environment of the University of Amsterdam conducted investigations in 1992 in collaboration with the University of Benin. In four survey areas in the Borgou Province (from south to north: Kalale, Kandi, Banikoara and Karimama, representing a cross-section of the province) 100 peasants and 100 herdsmen were interviewed in two villages.

This chapter focuses on North Borgou, encompassing the zones of Kandi, Banikoara and Karimama (Figure 1). Peasants and pastoralists occupy a frontier area, that includes a national park and forest reserves, situated in between these densely populated and cultivated zones. If present land management in this frontier area continues, environmental degradation will develop as in the surrounding, densely-populated zones. Only a Gestion de Terroir approach which includes both peasants and pastoralists, can lead to a more sustainable land use in this frontier area. Some scenario's for future land management, aiming at greater sustainability, will be developed in this chapter.

Modes of Existence: Peasants and Pastoralists

North Borgou lies in the north of Benin, between latitudes 11° and 12°30′N from the equator. It borders Burkina Faso, Niger and Nigeria, and it encompasses the districts of Karimama, Malanville, Banikoara and Kandi. The North Borgou area comprises roughly 27,000 km^2, being inhabited by more than 340,000 people. The population densities areas are low, on average 12 ppkm^2, but exceeding 60 ppkm^2 within the cultivated zones.

North Borgou is a transition zone of the Sudano and Sahelian climatic regions. The rainy season lasts from May to mid-October. Two main agro-ecological zones can be distinguished (see Figure 1). The first zone, in which the Karimama survey area is situated, has a Sudano-Sahelian climate with relatively little precipitation (700–900 mm). Ferruginous soils have been developed in sandy-clay sediments. Near the Niger river, relatively fertile, acid crystalline soils can be found, which have hydromorphic (pseudo-gley) properties. The main food-crops are millet, sorghum and beans. Cash crops are mainly onions and tomatoes, with some cotton and peanuts.

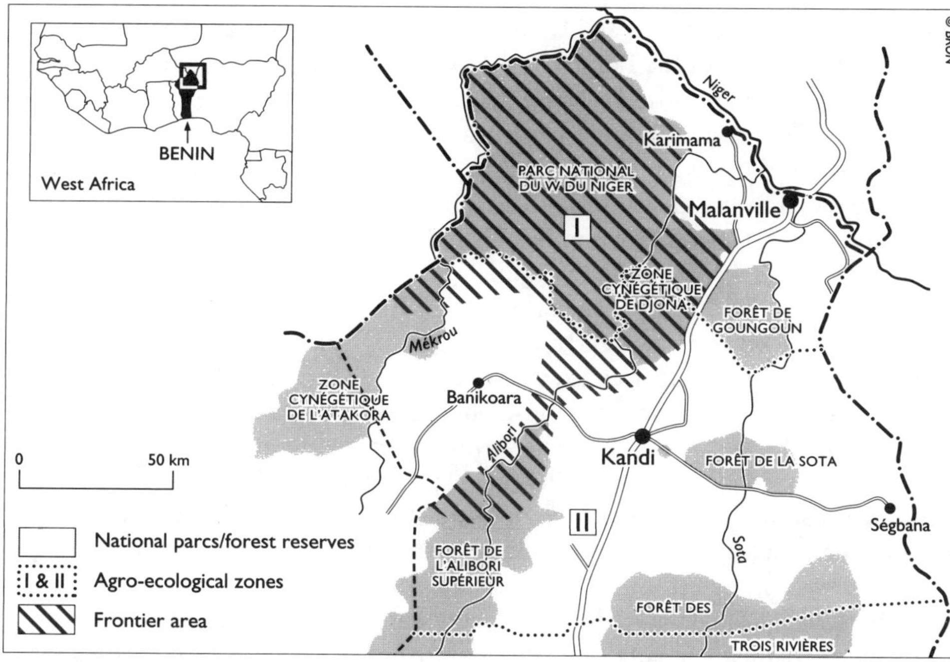

Figure 1. The peasant-pastoralist frontier area in northern Benin.

The National Park, the Cygenetic (Wildlife and Hunting) Zone of Djona and the Forest Reserve of Goungoun are situated in this first zone. Farming, herding and housing are officially prohibited in this uninterrupted zone of protected areas. Vegetation in these territories is mainly characterized by bush savannah, with gallery forests along the tributaries of the Niger river. These territories, together with the bush savannah adjacent to it in the south, make up the sparsely populated frontier area treated in this chapter.

Precipitation in the second zone, in which the Kandi and Banikoara survey areas are situated, is between 800–1,200 mm. Ferruginous soils have been developed on mainly crystalline basement rock; a high proportion of these acid crystalline soils are relatively deep. Vegetation is a bush savannah. The main food-crops are sorghum, maize and beans. Cotton is a very important cash crop.

This chapter deals with the cultivated zones of Kandi, Banikoara and Karimama and the sparsely populated frontier area in between these zones. The frontier area includes officially restricted zones such as the National Park, the Cygenetic Zone of Djona and the Forest Reserve of Goungoun.

The pastoral Fulani people migrated to the North Borgou more than a century ago, in search of grazing lands. The Bariba, the Boko and the Dendi had settled in this region a little earlier. They had developed farming (Bariba, Boko) or farming combined with fishing (Dendi). Since the peasants had the longest-standing rights, the Fulani had to ask them for permission to use the land. This principle still holds today. Gourmanche peasants and Dendi peasant-fishermen in the Karimama area, in the extreme north of the Borgou, used to grow millet, sorghum and beans. The area is hardly commercialized at present, although groundnuts are cultivated commercially, while irrigated market gardening is becoming increasingly important. This type of land use is very important in the neighboring district of Malanville. Bariba peasants in the areas of Kandi and Banikoara cultivate cotton, maize, sorghum and beans. Cotton growing has turned these areas into the most commercialized agricultural zones in the Borgou.

A symbiotic relationship between peasants and pastoralists, including the bartering of goods, services, as well as ways of sharing space, ensured the best living conditions in the fairly capricious, sub-humid and semi-arid climate. The Fulani settled in relatively permanent camps, where food crops were grown during the rainy season. The men left with the herds on transhumance in the driest months to areas where grass and water were still available. The remainder of the families stayed behind with some (milk) cows. The herds returned at the beginning of the rainy season, keeping a distance from the fields to avoid damage to crops, preferring grazing land a few kilometers outside the villages. After the harvest, the cattle browsed the stubble-fields. Flocks stayed overnight on request of a peasant on certain fields, depositing large quantities of dung, thus manuring the field. The peasant paid some cereals in return for this service. However, such 'manure contracts' were exceptions rather than the rule in the Borgou.

The herds went again on transhumance as soon as the stubble and water pools were finished. Transhumance herding is an extensive, but satisfactory response to climatic instability. The seasonal drought forces herds to leave in search for water. They only come back after the first rains and thus do not overburden the land.

Peasants used to contract Fulani to look after their cattle, which was also taken to the transhumance pastures. The Fulani herdsmen kept the milk and some of the calves in exchange. These 'keeper contracts' occurred regularly and are still practiced, although to a lesser extent. Circumstances changed since the 1980s, as tension between the two groups increased and the symbiotic relationship waned. This tension was partly caused by a period of relative drought. Increased cattle death rates had a negative effect on peasants' trust in Fulani herdsmen (personal communication A. Van Driel). Moreover, the number of foreign pastoralists that immigrated from the Sahel with their flocks increased considerably. Successful livestock health programs of the government of Benin and the European Development Fund also resulted in an increase in cattle. The overall increase of cattle caused competition between peasants and pastoralists about resources.

The peasants considerably extended the acreage under cotton, groundnut and irrigated market gardening, due to the introduction of draught animals and motorized pumps. Agricultural expansion thus reduced the area available for grazing. Because of the urgent need for grazing space, the Fulani now use areas like the National Park and the forest reserves (De Haan, Van Driel & Kruithof, 1990; De Haan, 1992; Breukers & De Hon, 1988). A new frontier has emerged.

Environmental Degradation

LAND DEGRADATION

Our survey revealed three important processes of land degradation that have occurred in recent years, *i.e.* soil erosion, topsoil deterioration and depletion of nutrients.

Soil Erosion

The most important types of soil erosion in the North Borgou are sheet erosion and gully erosion. Sheet erosion results in a loss of the topsoil and therefore in a loss of nutrients. It is caused both by farming and grazing and is found all over the region. Sheet erosion is not very serious but will increase if land use continues to intensify without improving the organic matter content of the soil. Gully erosion, caused by heavy rains at the beginning of the rainy season, washes away unprotected soils. It occurs only locally on steep slopes in deforested farming areas, along roads and, especially, along rivers. Gully erosion can also be found along rivers, caused by cattle trails.

Agricultural commercialization and cotton production have stimulated the use of the plough and the application of chemical fertilizers. Ploughing results in an increase of the area under cultivation per farmer, but also causes a more extensive clearance of trees as compared to fields cultivated by the hoe. This results in a large surface without trees, which leads to more runoff.

Fertilizer increases the quantity of nutrients in the soil and permits the farmer to cultivate a field for a longer period without diminishing returns. However, an increase of organic matter content does not occur. The prolonged cultivation of a field through the use of fertilizers, makes it more vulnerable to erosion.

Topsoil Deterioration

Topsoil deterioration in the North Borgou is mainly characterized by crust formation. A mixture of factors underlie this process. The first heavy rains in the beginning of the rainy season, directly followed by sun and wind, harden the top soil, if the soils lacks sufficient organic matter content. This process is aggravated by human activities like vegetation clearing, grazing or deliberate fires in the fields just before the rains start, thus exposing uncovered soils to heavy rains. Rigorous pruning of trees during the dry season by herdsmen to provide cattle with fodder (called aerial grazing) also contributes to this process. The topsoil structure deteriorates as a result, which is exacerbated by the action of cattle hoofs along trails.

Crusts are found everywhere in the North Borgou: on agricultural fields, fallow fields, and even in the National Park. The latter is a protected area, but nevertheless used intensively by Fulani for grazing. Soil crusts occur on areas of more than several square kilometers, inhibiting infiltration of rainfall and stimulating sheet erosion through increased runoff. Less water is available for plants as a result, while the ground-water level drops. Runoff causes regular inundations in the lower parts of the area.

Depletion of Nutrients

Crop harvesting without nutrient replenishment causes soil nutrient depletion. Even fallow periods of up to twenty years did not restore sufficient nutrients to the soil. The results of our soil survey show that all soils in the North Borgou are in fact exhausted. Organic matter content is low with an average of 1.6% (weight). Nitrogen content is very low with 15–50 ppm (parts per million), even in fallow fields. A quantity of 60 ppm nitrogen is sufficient for crop production. Phosphorus content is usually low with less than 15 ppm However, one manured field (no chemical fertilizer) reached 180 ppm. A quantity of 50 ppm phosphorus is sufficient for crop production. Potassium content is usually low, being less than 70 ppm. A manured field contained 340 ppm, which is well above the required level of 160 ppm.

Bush fires cause volatilization of nutrients, due to very high temperatures, especially at the end of the dry season when vegetation is very dry. The soil structure deteriorates, preventing quick regrowth of plants and facilitating crust formation.

DEGRADATION OF VEGETATION

Vegetation in the North Borgou is rapidly declining, partly due to increased cultivation. The vegetation of the large uncultivated areas, including the National Park and the Forest Reserves, deteriorates. Tree crowns cover maximally 20–50% of the soil in the Borgou Province, but less than 20% in the Kandi, Banikoara and Karimama survey areas. Real forest areas, with more than 50% soil cover, are rare.

Increased grazing is an important factor contributing to the degradation of vegetation. The growing population of pastoralists in the Borgou, a better veterinary service and increasing numbers of peasants with cattle, has augmented the total cattle population considerably.

Analysis of SPOT satellite images (Goossens, 1992) showed rapid deterioration of tree cover, especially in the National Park. Only 40% of the observed area of the park did not degrade in the period 1986–1990. Grass savannah has conquered large areas, because extensive bush fires caused bad germination, while roaming cattle prevented regrowth of trees. The National Park has become the most important refuge for Fulani herdsmen. The Fulani usually prune trees at the end of the dry season, using the leaves as fodder. However, part of the Fulani now prune trees during the dry season, while some cut all edible parts, thereby destroying the tree.

Bush fires are widely practiced by herdsmen and farmers at the beginning of the dry season to stimulate the growth of young offshoots, which are more palatable and contain more nutrients for the cattle. At that time of the year living plant material is still omnipresent, preventing very high temperatures. Such bush fires are considered less harmful than those at the end of the dry season. The latter fires are meant to clean the grazing areas from old herbs and to facilitate new grass to grow when the rains start. Both peasants and herdsmen also start bush fires for other purposes such as hunting and vegetation clearing.

Modern Conservation Practices

FARMING

Traditional farming and conservation techniques in the North Borgou were based on a system of shifting cultivation. The outer fields were cultivated for a number of years, followed by a period of fallow, during which the bush savannah restored fertility. The home gardens were cultivated permanently, while fertility was maintained by depositing garbage and dung. Different varieties of crops were used to accommodate various soil characteristics. Occasionally, bunds and plant rows were designed to slow down sheet erosion or to facilitate runoff in order to improve drainage. These traditional techniques are not adequate at present to secure a sustainable land use. Population growth and increased commercialization require a higher level of resource exploitation.

The Ministry of Agriculture is responsible for the large scale agricultural commercialization of the past two decades in the Borgou, through its regional agency CARDER (Centre d'Action Régionale pour le Développement Rural) and its field assistants in the villages. By offering seeds, pesticides, fertilizer and ox-drawn ploughs on credit to the farmers the CARDER has stimulated cotton production in particular. However, cotton growing and modern production methods are held responsible for part of the land degradation. Hence the CARDER also considers the introduction of conservation methods such as rotation, tree planting and the prevention of bush fires. Cotton production, nevertheless, remains the CARDER's most important field of interest.

The CARDER started a successful program in the late 1970s to promote ox-drawn ploughing, financed by the United Nations Development Programme

(UNDP) and the World Bank. Thus many peasants own a plough today: 84% in Kandi, 63% in Banikoara and 45% in Karimama. Others borrow or hire that equipment from relatives or neighbors, so that ploughs are used by almost 90% of the peasants in Banikoara and Kandi, and by 64% in Karimama.

The CARDER promotes an improved *rotation* system of crops, in which cotton is followed by a leguminous crop, then cereals and finally fallow. Each crop should be on the same field for only one year. Only 15% of the peasants interviewed in Karimama practiced some kind of rotation, including traditional patterns. Rotation use was much higher (>70%) in Kandi and Banikoara.

TABLE 1. Percentage of farmers per type of rotation in the survey areas

		Poor rotation	*Good rotation*	*Very good rotation*
Kandi	(n = 71)	15.5%	63.4%	21.1%
Banikoara	(n = 77)	28.6%	37.6%	33.8%
Karimama	(n = 15)	40.0%	13.3%	46.7%

Source: Survey results 1992.

However, not every type of rotation is sufficient to counterbalance declining soil fertility and erosion. A distinction has been made between three different types of rotation in Table 1. A rotation pattern is considered poor when a field is cultivated for a period of three years or more with one crop or a combination of crops from the same family, like millet and sorghum (both cereals). Every rotation period should be followed by a fallow period. A rotation pattern between two fallow periods is considered good when two years of cotton or a leguminous crop are followed by one or two years of cereals. A rotation pattern is considered very good if these crops alternate each year. The table shows that the rotation patterns are satisfactory in Kandi and Banikoara. These areas are the most important cotton producing zones in the Borgou. Peasants here seem to follow increasingly the advice of the CARDER on rotation methods.

Table 2 shows the percentage of farmers that have cultivated a field for more than five years. This is considered the maximum cultivation period in the North Borgou, after which fallow should follow. Comparing both tables it becomes clear that rotation is well practiced in both Kandi and Banikoara, but the cultivation periods in Banikoara are often too long. Extensive cultivation periods are caused by land shortage on the one hand and application of chemical fertilizer on the other. The use of fertilizer allows fields to be cultivated longer without decreasing yields. However, this results, on the negative side, in an increased danger of erosion, since the organic matter content of the soil is reduced. Note that only 224 out of 300 farmers practiced fallow and that in most cases the fallow period is too short for a sufficient soil recovery.

TABLE 2. Percentage of farmers cultivating fields less and more than 5 years before fallow in the survey areas

		Less than 5 years	*5 years or more*
Kandi	(n = 82)	68.3%	31.7%
Banikoara	(n = 54)	38.9%	61.1%
Karimama	(n = 88)	67.0%	33.0%

Source: Survey results 1992.

A more sustainable agricultural system should include increased use of manure. Only 40% (Kandi, Karimama) to 50% (Banikoara) of the farmers use some manure, while hardly anything is done to promote its use.

CARDER field assistants are now trying to stimulate farmers to save as many trees as possible when clearing new fields, in order to prevent excessive soil erosion. The CARDER aims at a coverage of some 40 trees/ha. Since many farmers cultivate their fields with ox-drawn ploughs, they are reluctant to follow this advice because of ploughing difficulties in fields with many trees. Alternatively, some village groups have planted rows of trees along the fields which may be considered as a first step towards alley-cropping. The planted trees in these examples are mainly fodder trees like *Leucaena leucocephala*. Some farmers plant trees for their fruits. However, these are rather exceptional cases.

HERDING

Whereas Borgou peasants have a broad indigenous knowledge of soil properties and crop varieties, Fulani herdsmen are famous for their knowledge of different types of vegetation and their potential use as fodder. The drying up of watering places was traditionally the impetus for the alternating use of pastures. If edible grasses were no longer available at the end of the rainy season, herders began pruning certain varieties of trees. Transhumance was even an exception in places were water and vegetation were abundant, such as in Karimama on the Niger river (personal communication, A. Van Driel).

Scientific opinion about traditional range management in Africa changed dramatically in the last decade. Success or failure of traditional range management was initially judged by means of botanic indices, measuring the state of the vegetation. However, this type of assessment is questionable with regard to Africa (Behnke, Scoones & Kerven, 1993; Scoones, 1994; Behnke & Kerven, 1994). The dominant factor in climatically unstable environments, influencing changes in vegetation, is rainfall, which lies outside the control of the herders. The only option for range managers, who cannot control the environment, is to adapt to it. This is called 'opportunistic range management', which is characterized by the objective to maintain large and productive herds if rainfall and vegetation permit, while destocking as quickly and profitably as possible when circumstances dictate (Behnke & Kerven, 1994).

Although the North Borgou is not as dry as the Sahel, the focus of the above debate, opportunistic grazing is also dominant in the former region. This is not merely caused by climatic factors, such as the decrease in precipitation during the last two decades. For many Fulani in the North Borgou, opportunistic range management is customary as they are transhumants from Sahelian areas in Niger and Burkina Faso, or immigrated to the North Borgou the past two decades. Moreover, the autochthonous Fulani also originate from the Sahel.

Moreover, the Fulani in the Borgou do not have clearly defined grazing rights, since they are relative newcomers. The lack of secure rights usually does not promote conservation strategies. There is no communal range management among the Fulani in this part of Africa. They tend to operate in small family groups and lack a clear, hierarchical power structure. With respect to the debate on common property regimes, their pastures can be regarded as open access areas, which favor individual free-rider behavior.

The herdsmen interviewed in the North Borgou showed little interest in the condition of the savannah or fallow land where they tend their flocks. Although they assess the quality of the grazing areas by means of the species present, they do not think about activities to maintain its quality. They are well aware of the negative effects of rigorous aerial grazing on the regeneration of the vegetation. The same can be said about the disastrous effects of bush fires at the end of the dry season. Nevertheless, they continue their present behavior because uncertainty with respect to climate and land tenure precludes other options.

The CARDER installed demonstration plots in various villages to show the negative impact of bush fires on soil and vegetation. However, these plots are oriented more towards peasants than pastoralists. Starting bush fires was officially forbidden in Karimama and this rule has been obeyed for a number of years. However, in 1992 the whole area burned again (personnel communication A. Van Driel; see also De Haan, 1995). Bush fires in the other zones are only allowed at the beginning of the dry season. Nevertheless, many herders, especially in the frontier area, including the National Park, set fire to the bush. Farmers still clear new fields with the help of fire and burn crop residues on their fields. The latter practice produces extremely high temperatures in the topsoil with the same negative effects as bush fires.

There is only one important project targeted at environmental conservation by Fulani and peasant cattle-owners (Projet de Développement d'Elevage dans le Borgou-Est; PDEBE), financed by the UNDP and the FAO, and situated in the South Borgou near Kalale. This project has constructed several small dams to serve as watering places for cattle, which are a starting point for the construction of numerous other small pools in the Borgou. These small water reservoirs should remove herding from the watering places in the forest reserves and other restricted areas. Committees of cattle-owners were established to manage the dam near Kalale. Each committee employs a guard who looks after the dam and the pool. He is paid from the collective account. Cattle are not allowed on the dam. Bush fires as well as crop cultivation around the pool are forbidden. Men and women are not allowed to wash or bath in the pool, which has only one entrance. After the rainy season, villagers must clean the area and repair the fence. Finally, farmers ploughing in the catchment area of the pool should follow the contour lines to diminish soil erosion in order to slow down silting up of the pool (De Haan, 1995).

ESCAPE TO THE FRONTIER

The most important mechanism to avoid excessive environmental pressure in both modes of existence, *i.e.* farming and herding, has traditionally been the mobility of both fields and animals.

Although this mechanism may have worked satisfactorily in the past, this is not the case any longer in the densely populated zones of Kandi, Banikoara and Karimama. Fields are cultivated longer and fallow periods are shortened. Soil fertility is not even restored after long periods of fallow, because bush fires and grazing cattle deplete the fallow fields. The mobility of the pastoralists' herds is hindered by expanding cultivation, conflicts with peasants about damage to crops, as well as disappearing passages to grazing areas and watering places.

An increasing number of peasants decided to migrate to the frontier zone, while keeping their plots in the home village. Numerous isolated farms and even hamlets have already been established. New fields are being cleared by means of fire. Cotton as well as food crops are cultivated. The settlers mainly concentrate in the transit zone, and along the road, between Kandi and Banikoara. They usually stay out of the protected areas because their fields are easily spotted by the forest guards.

The frontier area is also being used by Fulani and conflicts about crop damage by cattle, common in the densely populated zones, are already reported. The Alibori river, often bearing water throughout the dry season, is a pole of attraction to the Fulani. Since pastoralists are more mobile than farmers, the Fulani can be found throughout the National Park, the Cygenetic Zone of Djona and the Forest Reserve of Goungoun, risking both legal and illicit fines from the forest guards. Cattle density is highest in the frontier area during the dry season, because of the presence of transhumants from adjacent zones in Niger and Burkina Faso. Numerous Fulani have more or less settled permanently in this area.

Colonization at the frontier temporarily relieves environmental pressure in Kandi, Banikoara and Karimama. This probably reduces people's motivation to change farming and herding practices. However, the relief is only temporary and peasants and pastoralists will soon run into more serious environmental problems.

Conditions for a Successful Gestion de Terroir

Gestion de Terroir is more an applied strategy than a theoretical concept, as it grew from experiences in development projects. This is reflected in the current World Bank definition of PNGT (Programme National de Gestion des Terroirs) projects in West Africa: 'a multisectoral, decentralized and participatory methodology, that uses the concept of a village territory, managed by a responsible village group with the objective of using the natural resources in such a way as to assure their sustainability' (Guéye & Laban, 1992:12).

Important experiences for a Gestion de Terroir approach were accumulated during the so-called community forestry projects in the late 1970s and 1980s. These small-scale activities promoted self-reliance, sustainable land use and community responsibility for the management of natural resources within a community's territory. Moreover, these micro-projects proved to be feasible in the local context, as long as active participation, collective action and participation of women were guaranteed. This process-oriented and bottom-up approach, promoted and diffused through a network of rural extension agents from various government and non-government agencies, like the forestry departments, the cotton boards and NGO's, proved to be quite successful (Hijkoop *et al.*, 1991).

Integrated management of natural resources at village level became incorporated in integrated rural development projects as a result and were given political weight through its adoption by important funding agencies like the World Bank (Guéye & Laban, 1992). Gestion de Terroir implies on paper the definition of a territory (terroir), the establishment of sustainable production systems and the establishment of a development contract with the local population (Barrier, 1990:36). However, from an inventory of Gestion de Terroir programs (Réseau Recherche-Développement, 1993) we learn that in practice the definition of territory is vague and large scale disparities may occur. In addition, natural resource management is not always the primary goal; intensification of cotton production or food security are principal objectives too.

Three different views on Gestion de Terroir can be distinguished. Firstly, Gestion de Terroir is considered a suitable spatial planning concept. It serves as an instrument to partition the village territory into different sectors with particular resource use priorities (fields, woodland, *etc.*). It is hoped that conflicting resource use can be avoided by drawing clear boundaries (Bonnet, 1990). The second point of view tries to work out clear legislation on the issue of land rights in order to avoid conflicts when the Gestion de Terroir is applied to larger areas. They recognize the complexity of the land rights issue and they insist, therefore, on a solid legal base as a starting point for land use planning schemes (Kinz, 1992; Hesseling & Ba, 1994). Thirdly, political scientists and development agencies plead for more decentralization, to be implemented through Gestion de Terroir. They advocate a transfer of decision making power to local organizations based on territorial units (GRED, 1993:11).

These views assume that, at the local level, a kind of permanent consensus about the use of resources can be established. However, this does not reflect the nature of social relations in rural West Africa, which are characterized by many conflicting interests with respect to natural resources. Livestock keeping, an important type of resource use, has hitherto received little attention in the Gestion de Terroir (Bonnet, 1990; Barrier, 1990; GRED, 1992; Marty, 1993). Moreover, the livestock sector is becoming increasingly complex in West Africa: herders settle and are engaged in crop cultivation, peasants pick up animal husbandry and even the national elite invest in cattle. Thus the real issue at stake in the Gestion de Terroir debate is its underlying assumption that communal management of natural resources is still a viable option in rural West Africa.

Peasant-Pastoralist Relations in the North Borgou

The two modes of existence in the North Borgou do not function independently from one and another, but are interrelated in many ways. Despite conflicting land use there is still a sufficient degree of interdependence, like keepers and manure contracts, and thus a communal basis for spatial planning of resource use.

The shortage of farming land, pastures and water has led to conflicts between peasants and pastoralists, about crop damage by cattle and blocking of cattle passages by peasants. Table 3 indicates that 20 to 30% of the peasants suffer from crop damage. It is maintained in the Borgou that this is caused by 'irresponsible' foreign herders from Niger and Burkina Faso. However, our survey found out that crop damage by cattle belonging to foreign herders is less important than damage by autochtonous herders. Peasants and herders solve the problem between themselves in the majority of the cases (>70%), *i.e.* the herders pay for the damage or the peasants forgive them. Local authorities have to intervene in 18% of the cases, especially when a valuable crop like cotton is concerned. Occasionally violence, casualties and even deaths result.

Herders make use of passages to move from one pasture or watering place to another in cultivated areas. These passages have to be at least 100 meters wide to

TABLE 3. Percentage of farmers experiencing crop damage and origin of the cattle responsible

			Origin of Herd			
		Crop damage	Autoch	Alloch	Peasant	Unknown
Kandi	(n = 100)	33	64	6	–	30
Banikoara	(n = 100)	20	45	–	5	50
Karimama	(n = 99)	33	64	6	–	30

Source: Survey results 1992.

prevent damage by cattle to the surrounding fields. However, the passages tend to become narrower because of land pressure. Most passages are still open in Kandi and Banikoara (Table 4), but in Karimama, 65% of the herders are confronted with passages blocked by fields. These are mainly passages used in the dry season to go to the river Niger. Irrigated market gardening and cultivation on the river banks hinder the herders to reach the river.

TABLE 4. Percentage of herders confronted with blocked passages

		Blocked passage
Kandi	(n = 99)	3.0
Banikoara	(n = 97)	3.1
Karimama	(n = 99)	65.6

Source: Survey results 1992.

Fulani who arrived and settled in the North Borgou during the last two decades have no permanent camps as yet, but move their camps with their cattle to new pastures. Only 2% of the Fulani in Kandi and Banikoara had difficulties with peasants about finding a new site, but in Karimama 63% had encountered problems. Peasants in the North Borgou accumulate their wealth in cattle, entrusted in secret (in order to conceal their wealth to relatives and neighbors) to Fulani. The Fulani friend thus became the treasury of the peasant. The keepers contract gave the Fulani the right to use the milk of the cows and the ownership of part of the calves.

However, the decrease of rainfall and the decline in pastures forced herders to go farther away on transhumance, which resulted in higher mortality rates of cattle. At the same time, the introduction of plough-oxen, which stayed at the farm the whole year around, acquainted peasants with cattle keeping. A tendency to reduce the keeper contracts has emerged in addition to the growth of conflicts about damage to crop, particularly in Kandi and Karimama (Table 5). Today many farmers hire a Fulani herder on a seasonal basis to look after their cattle, which are combined in a village flock.

Manure contracts were formerly not very common in the North Borgou. Today, 2% of the farmers in Banikoara, 7% in Kandi and 9% in Karimama have manure contracts. The mutual advantage is that the farmer has his fields manured, as the

TABLE 5. Percentage of farmers having a keeper contract

		Presently	*Stopped*	*Before*
Kandi	(n = 99)	31.3	18.2	49.5
Banikoara	(n = 99)	31.1	1.0	32.4
Karimama	(n = 99)	13.1	25.3	38.4

Source: Survey results 1992.

herder locate the night kraal on his land, being allowed to feed his cattle with the stubble. Nevertheless, cattle browse the stubble fields even without a contract. Manuring is done less systematically in such cases, because the farmer has no influence on the location of the kraal. At present, peasants start collecting the stubble themselves to feed it to their own cattle or to sell it to the Fulani afterwards.

Although interdependence between peasants and pastoralists is diminishing and polarization is rising, there is still a broad communal understanding between both groups. Conflicts still involve only a minority of the people, and more important, most of the time they are solved privately. Polarization is more severe only in Karimama, squeezed between the National Park with its guards and the cultivated bands of the river, with many Fulani immigrants and transhumants from Niger. An approach towards communal resource management in Kandi and Banikoara may easily start with small groups of individual peasants and pastoralists, but in Karimama this process should be conducted by representatives of both groups.

Gestion de Terroir Scenarios for the Frontier

A successful Gestion de Terroir, being a communal management regime of natural resources, needs a sound communal basis. This is still a viable option in the North Borgou. Gestion de Terroir scenarios for the frontier area in the North Borgou are being proposed below.

The availability of water for men and animals during the dry season is the limiting factor in exploiting the frontier area. Human settlements and herds are, therefore, being attracted by the Alibori river. Interesting experiences, involving the creation of small ponds, exist in the South Borgou. Several small dams were constructed to create watering places for cattle. Joint committees of peasants and pastoralists to regulate the use of water and land around the pond function as starting points for communal resource management in the surrounding territory.

These kind of small ponds will also be interesting to the peasants and pastoralists in the frontier area. The ponds can be a watering place for cattle of both herders and peasants. The water may also be used for gardening and domestic purposes. A well distributed location of ponds may enable access to new pastures and may contribute to controlling environmental variability within the broader perspective of the debate about opportunistic range management.

Small ponds can be created upstream in the catchment areas of the tributaries of the larger rivers Alibori, Sota and Mekrou. These locations are suited for small earth dams, that can be constructed without heavy equipment and expensive concrete, but with substantial labor input of the local population. The small scale of the enterprise, the low costs, the local labor input and the limited water carrying capacity of the pond make it manageable for the local community. Collaboration between peasants and pastoralists can still be expected on the local level, thus providing a platform for communal management. As the water capacity of the pond is limited, few herds from outside the local community will be attracted to the place. Overgrazing by outsiders, so common around large wells and ponds in the Sahel, will thus be prevented.

These small ponds will carry water during the dry season, enabling the local population to control environmental variability, rendering transhumance superfluous. This has to be accompanied by other measures, such as the gathering or cultivation of fodder, the prevention of bush-fires, the improvement of pastures and fallow land, as well as the rotation of pastures combined with the allocation of land use types to certain zones within the territory of the community. All these measures contribute to more sedentary livestock keeping and the integration of crop cultivation and animal husbandry. This will reduce environmental variability and contribute to a shift from opportunistic range management towards more stable, integrated land management.

An interesting question in this respect is what role the National Park, the Cygenetic Zone of Djona and the Forest Reserve of Goungoun should play. These protected zones make up a large part of the frontier area and are in use by pastoralists. Any scenario for arriving at sustainable land use in the frontier area should also include these protected zones. Two major types of scenarios can be distinguished. The first is characterized by preservation and the second by utilization. Within each type, some sub-types can be identified.

THE PRESERVATION SCENARIOS

The first preservation scenario represents a complete ban on utilization of the resources in the protected areas by men and animals, foreigners and nationals. It requires large investments in personnel and equipment to enforce such a ban. The costs would by far exceed the investment capacity of the Beninese government. It is not yet clear to what extent foreign donors would be willing to supply the necessary funds. Pressure on resources in the frontier area outside the protected areas will increase as a result. This would be objectionable in the Karimama zone where polarization is most severe. Here a system of small ponds will probably not be sufficient. For the frontier area south of the National Park, between Kandi and Banikoara, the first preservation scenario would prevent the influx of Karimama and foreign transhumants and thus create a breathing space for the implementation of Gestion de Terroir around a system of small ponds.

The second preservation scenario would create passages from north to south through the National Park, the Cygenetic Zone of Djona and the Forest Reserve of Goungoun in order to alleviate pressure on the Karimama district. These passages should have a width of several kilometers in order to allow cattle to graze during the passage and to prevent overgrazing. Thus the frontier zone south of the National Park would serve as an overflow for the Karimama zone. Transhumants from outside the community should be accommodated within the local communities of herders and peasants. This might erode the local bearing surface for communal resource management. On the other hand, herders from Karimama already stay in this area during the dry season. Their integration into local resource management is in any event a prerequisite.

THE UTILIZATION SCENARIOS

The first utilization scenario implies the lifting of all restrictions on the exploitation of resources by peasants and pastoralists in the protected areas. This would most certainly relief pressure in Karimama and in general will make the frontier area even more attractive than at present. However, within ten or fifteen years at the most, peasants and pastoralists will encounter a situation similar to today and the need for an environmental management strategy will again emerge. One may argue that such a period is needed to arrive at any viable Gestion de Terroir scenario at the frontier. Only such logic would make the first utilization scenario acceptable. However, one may also argue that a certain degree of resource scarcity is needed to motivate peasants and pastoralists to participate in Gestion de Terroir. Full utilization of the protected areas may than be considered as a waste of time and resources.

Nevertheless, some temporary or occasional utilization of the National Park, the Cygenetic Zone of Djona and the Forest Reserve of Goungoun could be desirable during the implementation phase of Gestion de Terroir in the frontier area outside these protected territories. In fact, this second utilization scenario is the most promising for the pastoralists. It can be expected, during the implementation of the program focused on a Gestion de Terroir approach around small ponds, that the pastoralists are not capable of abandoning opportunistic range management practices at once, because environmental variability will remain important in the initial phase and land rights will not have been established immediately. Mobility of herds must be guaranteed in that case. Utilization of natural watering places and pastures in the protected areas may then serve as safety valves and contribute to a successful implementation of Gestion de Terroir.

Conclusion

A Gestion de Terroir approach organized around small ponds and combined with north-south passages through the National Park, the Cygenetic Zone of Djona and the Forest Reserve of Goungoun, coupled with their occasional utilization, might be the most promising scenario for frontier area development in the North Borgou aimed at more sustainable land use. Today, there is still a mutual understanding between peasants and pastoralists on which communal resource management can be based. However, foreign herders from Burkina Faso and Niger constitute the uncertainty factor in this scenario. On the one hand, they already make use of the frontier area and thus will be interested in participation in the Gestion de Terroir, if advantages are clear. On the other hand, their weak relations with local peasants and pastoralists might prevent such participation.

Acknowledgements

This chapter is based on a survey in Benin carried out under the author's supervision by the University of Amsterdam and the University of Benin in 1992. The project was financed by the Commission of the European Community. The author, accepting full responsibility for the chapter, gratefully acknowledges the following persons: (in Benin) M. Igué, H. Adegbidi, M. Adam, G. Biaou, J. Fanou, A. Adegbidi, A. Mouph-Taou, G. Otto, R. Tossou, T. Djedjebi. M. Houinato, E. Quenum and A. Zinsé; (in the Netherlands) P. Douma, A. Van Driel, J. Dirven, A. Mahieu, S. Paris, V. Coenen, D. Niemeyer, J. Berkhout, A. Maas and E. Goossens (Belgium). The author thanks Antje Van Driel, who investigates peasant-pastoralists conflicts in Karimama, for her valuable comments on an earlier draft of this paper, as well as the editors of this book for constructively reviewing and improving the text.

References

Barrier, C. (1990) Développement Rural en Afrique de l'Ouest Soudana Sahelienne, Premier bilan sur l'approche Gestion de Terroirs Villageois. *Les Cahiers de la Recherche Developpement* 25:33–42.

Behnke, R., I. Scoones, and C. Kerven (1993) *Range Ecology at Disequilibrium. New Models of Natural Variability and Pastoral Adaptation in African Savannas.* Overseas Development Institute (ODI) / International Institute for Environment and Development, London.

Behnke, R., and C. Kerven (1994) *Redesigning for Risk: Tracking and Buffering Environmental Variability in Africa's Rangelands.* Natural Resource Perspectives 1. Overseas Development Institute, London.

Bonnet, B. (1990) Elevage et Gestion de Terroirs en zone Soudanienne. *Les Cahiers de la Recherche Developpement* 25:43–67.

BOSTID (1986) Common property resource management. *Proceedings* of a Conference prepared by the Panel on Common Property Resource Management. Board on Science and Technology for International Development, National Research Council. National Academy Press, Washington.

Breukers, G. and M. De Hon (1988) Monographie Régionale de la Province du Borgou, République Populaire du Bénin. Programme de Coopération Universitaire au Développement NUFFIC/PUO: Amsterdam: Projet Economie Rurale UNB/UvA/ER, Université Nationale du Bénin/Université d'Amsterdam.

De Haan, L. (1995) Local environmental management in north Benin: Bright spots in the local impasse. *In* J. Van Den Breemer, C. Drijver and L. Venema (eds.) *Local Resource Management in Africa.* John Wiley & Sons, Chichester, pp. 129–146.

De Haan, L. (ed.) (1992) *Rapport entre Agriculteurs et Eleveurs au Nord Bénin: Ecologie et Interdépendence transformée.* Tome I et II. Université d'Amsterdam/Université Nationale du Bénin/Commission des Communautées Européennes, Amsterdam.

De Haan, L., A. Kruithof, and A. Van Driel (1990) From symbiosis to polarization? Peasants and pastoralists in Northern Benin. *The Indian Geographical Journal* 65:51–65.

Goossens, E. (1992) Étude de la Dégradation environnementale dans le Parc régional W au Nord Bénin. *In* De Haan, L. (ed.) *Rapport entre Agriculteurs et Eleveurs au Nord Bénin: Ecologie et Interdépendence transformée.* Tome II. Université d'Amsterdam/Université Nationale du Bénin/Commission des Communautées Européennes, Amsterdam, pp. 1–14.

GRED (1992) *Gestion des Terroirs et Elevage.* GRED, Paris.

Guèye, I., and P. Laban (1992) From woodlots to Village Land Management in the Sahel. ODI Paper 35. International Institute for Environment and Development, London.

Hesseling, G., and B. Ba (1994) *Le foncier et la Gestion des Resources Naturelles au Sahel.* CILLS-Club du Sahel, Paris.

Hijkoop, J., P. Van Der Poel, and B. Kaya (1991) *Une lutte de longue haleine.* Institut d'Economie Rural/Institut Royal des Tropiques, Bamako/Amsterdam.

Kintz, D. (1992) *Le Foncier Rural au Mali.* Caisse Centrale de Cooperation Economique, Paris.

Marty, A. (1993) La Gestion des Terroirs et les Eleveurs; un outil d'exclusion ou de negociation? *Revue Tiers Monde* 34:327–344.

Painter, T. (1993) Getting it right. Linking Concept and Action for Improving the Use of Natural Resources in Sahelian West Africa. Dryland Network Programme Issues Paper 40. International Institute for Environment and Development, London.

Reseau Recherche Developpement (1993) Gestion de Terroirs; Problemes identifies par les operateurs de terrain en Afrique et a Madagascar et des Pistes et Quelques solutions. DSA GIRAD, Paris.

Scoones, I. (1994) *Living with Uncertainty. New Directions in Pastoral Development in Africa.* Intermediate Technology Publications. International Institute for Environment and Development, London.

ABOUT THE AUTHOR

Leo De Haan (born 1950 in Dordrecht, The Netherlands) is Associate Professor in Human Geography of Developing countries at the University of Amsterdam, The Netherlands. After a few years of employment in the Caribbean, he did his doctorate research in the rural North of Togo on the integration of the local mode of existence in the regional and national economy, focusing in particular on market and labor integration. His current research mainly concerns West Africa, covering themes like food security, cattle marketing and border trade of cereals and cattle, as well as land management regimes by peasants and in particular pastoralists, against the background of their livelihood strategies. He is scientific coordinator of a sustainable development cooperation project between Dutch universities and the National University in Benin.

Dr. Leo J. De Haan
Research Group Livelihood & Environment
Faculty of Environmental Sciences
University of Amsterdam
Nieuwe Prinsengracht 130
1018 VZ Amsterdam
The Netherlands (e-mail ldehaan@isg.frw.uva.nl)

Chapter 12

Wild Plants as Food Security in Namibia and Senegal

Patrick Van Damme

Introduction

Traditional communities in arid and semi-arid areas often depend on wild plants and their products for survival during tiding-over periods, when agricultural food stocks have almost been depleted. These food sources can be used as hunger or famine foods to alleviate temporary food shortages. In periods of limited food stress such foods may be eaten only occasionally and more often by children and poorer sectors of society with only a few direct or exchange entitlements. The seasonal variability of wild food supply from different sources influences food acquisition strategies of pastoralists (the Fulani throughout West Africa, Bernus, 1988), crop growing communities (the Serere in Senegal, Rosetta, 1986; the Toucouleur of Senegal, personal observations), fisherfolk and hunters, and hunter-gatherers (Bushmen or !Kung San[1] of Botswana, Lee & de Vore, 1976).

Documentation of famines in Africa shows the importance of wild foods in sustaining livelihoods. During the 1973 famine in Sudan, the Berti of western Sudan collected wild grass seeds and survived off these (Holy, 1980). This pattern was repeated during the 1984–1985 famine, during which grass seeds and tree fruits (*Boscia senegalensis*, Capparidaceae, among others) were collected in order to survive (De Waal, 1989). A similar strategy was reported from Wollo, Ethiopia, where leafy plants (e.g. *Portulaca oleracea*, Portulacaceae; *Amaranthus* spp., Amaranthaceae), and plants with berries and fruits (*Rosa abyssinica*, Rosaceae; *Ficus sur*, Moraceae), seeds and roots were incorporated into the diet during food shortage periods of the early 1980s (Rahmato, 1988). In the Senegalese Ferlo (a pastoral region in the north), the seeds of *Combretum aculeatum* (Combretaceae) and several grasses are consumed only in extreme conditions.

Hunger food products usually constitute only a part of the normal diet of numerous traditional communities, and only become the sole source of food during stress or in frontier situations at other times. Famine foods include wild cereals,

H. J. Bruins and H. Lithwick (eds.), The Arid Frontier, 229–247.

tubers and roots, vegetables, berries, nuts, fruits, but also insects and wild animals. Africa has more than 2,000 of these little-known plant species, according to BOSTID (1996). Indigenous grains such as *Eragrostis teff* (tef) or *Digitaria exilis* (fonio), although known to many Africans, are only used in limited areas, where they are grown. Some other (wild) grass species like *Panicum laetum* (wild fonio), *Cenchrus biflorus* (cram-cram) or *Echinochloa colona* are an important part of the pastoral production system and provide hunger food in times of scarcity. *P. laetum* even yields 400–1,100 kg/ha in the most productive, low-lying valley parts of the Malian Gourma, which is higher than local yields of millet (*Pennisetum americanum*, Poaceae).

The movement patterns of the Tuareg of northeast Burkina Faso are influenced by the siting of particularly good patches of wild *P. laetum* (Barral, 1977). In Mali, poorer households may combine the use of porridge from *Boscia senegalensis* fruits with their limited supplies of stored millet (Martin, 1985). In northern Nigeria, leafy vegetables and other bush foods are collected as daily supplements to relishes and soups and/or used as famine food in times of food stress (Longhurst, 1986). In Swaziland, a wide range of wild foods is used; its use increases with staple food shortages. Species used include *Bidens pilosa* and *Sonchus oleraceus* (both Asteraceae) and *Commelina* spp. (Commelinaceae) which are considered weed species and are commonly found in fields, on fallow land, pastures and along paths. Mushrooms, bulbs and flowers are also collected (Ogle and Grivetti, 1985). A similar pattern is shown in Ghana (Asibey & Beeko, 1989; Abbiw, 1990).

A diversified resource basis increases the options for maintaining food security. Different products complement each other with seasonally different patterns of availability, the potential for storage for hunger season use, or for use in times of extreme food stress (Scoones *et al.*, 1992). In the Miombo woodlands of Tanzania, a few tree species provide food for the Sandawe throughout the year (Newman, 1975). In Zimbabwe, indigenous wild fruits seasonally complement exotic fruits as food sources in the communal areas, with indigenous trees providing an important dry season supplement (Gumbo *et al.*, 1990); within this context, *Sclerocarya birrea* (vern. name: marula or maroola plum; Anacardiaceae) holds a very prominent position and is in the process of being domesticated.

Wild foods available for consumption at times when other foods are in short supply are particularly important. In dryland farming areas of north India, *Ziziphus* spp. (Rhamnaceae) fruits provide food between February and April, when other foods are scarce. Likewise, locust bean (*Parkia biglobosa*, Fabaceae) provides dry season food in West Africa where the ground, brown seeds are prepared into a fermented dish (Campbell-Platt, 1980) called *soumbala*; the yellow powder in the pod is consumed raw or as porridge. In the Kalahari, nuts of the mongongo tree (*Ricinodendron rautanenii*, Euphorbiaceae) are a dry season staple food (Lee, 1973). Many wild foods can be collected and stored for later use: wild grass grains are stored along with crops; nuts and berries can be processed and stored; wild vegetable leaves may be dried and insects or wild meat can be preserved for later consumption.

In this chapter, case studies are presented of two types of traditional societies: hunter-gatherers in the Namib desert and farmers in a deciduous subtropical forest area in southern Senegal.

The Topnaar Society in the Namib Desert[2]

The Topnaar, formerly known as Hottentots, constitute a relatively small group numbering 800–1,000. They belong to the Nama people who form part of the Khoi-Khoin race now basically confined to Namibia. The Nama people are divided into nine tribes, including the Topnaar of the lower Kuiseb valley and the Topnaar of Sesfontein (Figure 1). Both areas are situated in the Namib desert. The climate of

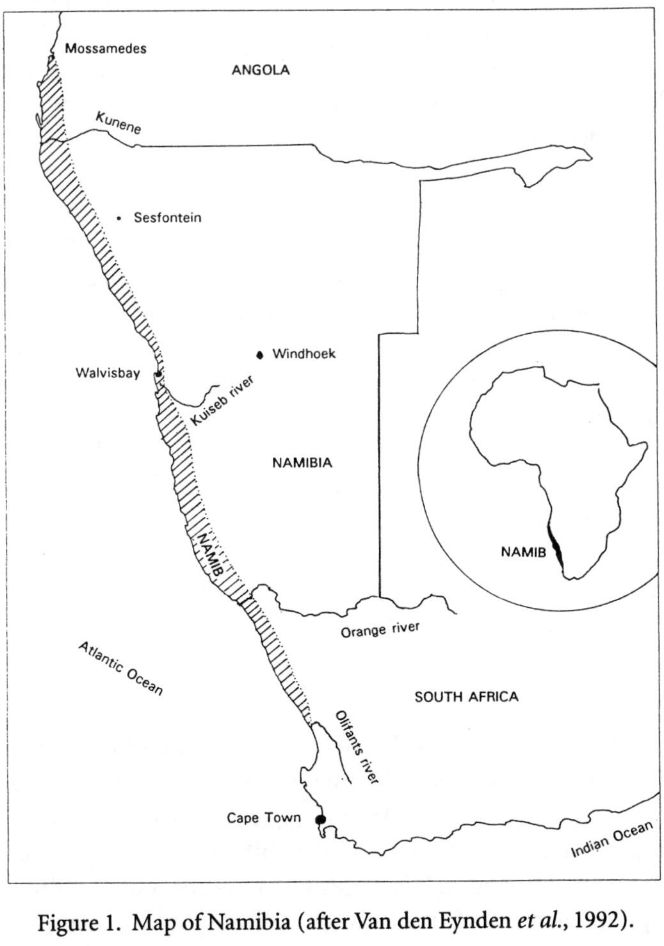

Figure 1. Map of Namibia (after Van den Eynden *et al.*, 1992).

the Namib is characterized by a low and very variable rainfall, nowhere exceeding 25 mm/year in this hyper-arid part. Insolation is high; temperatures range from cool to extremely high (more than 40°C) and there are frequent occurrences of fog and stratus clouds. The coastal areas have a nearly continuous high relative humidity created by the cold Benguela Current that flows northwards along the coast.

The Topnaar are among the oldest inhabitants of Namibia, who according to oral tradition, came from the north—exact area not specified—prior to occupying the Walvisbay territory on the coast. They were among the first traders in that area, bartering beef, goats, milk, *!nara* and fresh water with European sailors as early as 1677 in exchange for general supplies, clothes, weapons and liquor. The Topnaar's territory was much larger in the past than it is now. They were driven from their original area by the Herero, who migrated from the north and by other Nama people who came from the south.

The Topnaar are hunter-gatherers by tradition, and still practise this lifestyle, although there are some differences between the Kuiseb and Sesfontein Topnaar.

The Kuiseb Topnaar live near their original immigration site. However, it is not clear why the only other concentration of traditional Topnaar lives in Sesfontein, situated 500 km north of the Kuiseb in the Kaokoveld area. Many, primarily young Topnaar individuals have migrated into urban and/or industrial areas over the last decades, finding more stable income sources through formal labor opportunities, which are not available in their traditional environment.

Gathering or collecting is the most direct method of obtaining plant products. It may include regular or irregular harvesting of uncultivated plants. It is thus dependent on a diverse source of products that can offset seasonal and interannual variability in wild food production. Hunting or fishing usually go hand in hand with collecting. Most Topnaar depend almost exclusively upon gathering and, to a lesser degree, on hunting to provide themselves with subsistence food. However, the Sesfontein Topnaar have gradually become more or less sedentary farmers. The possibilities for the Kuiseb Topnaar to hunt were severely limited when their area was proclaimed a nature reserve. This decision has influenced and changed their traditional lifestyle tremendously, *inter alia* confining them to hunting zones far away. As a result, they have gradually moved into small stock raising, especially goats, to cover their animal protein needs.

THE TOPNAAR OF THE KUISEB VALLEY

The Topnaar of the lower Kuiseb valley live in 14 small communities each consisting of only a few families. Their shacks are made of driftwood or other scrap material. They live from cattle herding, gardening, and gathering !nara (Acanthosicyos horridus, Cucurbitaceae). The latter is a cucurbit which grows in the wild and has, until now, never been cultivated. It is endemic to the coast of Namibia and provides food to the Topnaar. Archaeological studies indicate that the !nara has been a food for humans living in the Namib desert for the past 8,000 years. It can be considered the traditional staple of the Topnaar, also known as the !Naranin, the people living off the !nara. They depended rather heavily on its fruits for their survival, before they had extensive contacts with (white) outsiders. The !nara grows abundantly in the sand dunes near Walvisbay, where subterranean water is present. Each family possesses a number of !nara bushes. This ownership is unlimited in time and was approved by Queen Victoria herself during the last century. Each family can only harvest from its own !nara bushes. The property rights are hereditary: if the parents die, the !nara field is divided among those children interested in the !nara. The Topnaar's chief and his council may rule in disputes if necessary. During the harvesting season of the !nara, whole families move down to the coastal !nara fields and stay there until the end of the harvest. Only a few people stay behind in the "villages" to attend livestock.

The harvesting season of the *!nara* lasts from November till May. When the pumpkin-like fruits are ripe, they are collected and buried in the soil or put in the sun for a few days to make them softer. Traditionally, fruits are peeled and boiled for some hours, until the pips are released from the pulp, which then becomes deeply orange and thick. Then the pips are separated from the pulp by sieving. They are dried in the sun for a few days and thereafter stored in bags. They are eaten like nuts or ground and added to dishes. The pips are highly nutritious: they contain up to 57% oil (which has a high percentage of poly-unsaturated fatty acids) and 31% protein. The boiled pulp is poured on the sand or on bags and dried in the sun for a few days. This dried pulp forms flat cakes which can be chewed or added to porridge. It can be stored for months and is eaten the whole year round. The fresh fruit can also be eaten raw and has a pleasant fruity taste (but contains cucurbitacins which can irritate the mouth). The pips, separated from the pulp by rubbing them in the sand, are eaten raw or roasted.

At present, it is rather difficult to assess whether the *!nara* is still a true staple or whether it has become a hunger food. Day-to-day food supply of Topnaar families living in the desert but having relatives in nearby towns consists of "mealy meal" (*i.e.*, maize flour) and to a certain extent *!nara*. In the absence of maize flour traditional communities can still fall back on the *!nara* cakes which are prepared and stocked by each family. No hard figures are available on the relative importance of either, nor is there information about the exact amounts of other food sources that are currently available to them. Traditional knowledge about hunger food sources is still very much alive, and most Topnaar communities use them to some extent. Table 1 lists the species of wild plants most commonly used in the Kuiseb area.

Date palm (*Phoenix dactylifera*, Arecaceae) was originally introduced by German soldiers who used the date fruits as a source of energy-rich food during their campaigns. Some seeds were later planted in a missionary's garden from which new plants spontaneously multiplied and spread into the nearby (dry) river beds, forming oasis-like spots. The Topnaar eat the dates fresh or dried and use the roots in a decoction drunk to treat tuberculosis.

The Asclepiadaceae or milkweed family comprises a few typical desert or semi-arid species endemic to the area. The succulent stems of *Hoodia currori* and *Trichocaulon pedicellatum* are eaten raw after removal of the outer skin and thorns. They resemble small cucumbers, and are consumed as such by the Topnaar. Pieces of the

TABLE 1. Species used by the Kuiseb Topnaar as hunger food (taxonomical classification according to Cronquist, 1988)

Species	*Family*	*Species*	*Family*
Leaves			
Pergularia daemia	Asclepiadaceae	*Tagetes minuta*	Asteraceae
Euclea pseudebenus	Ebenaceae	*Salvadora persica*	Salvadoraceae
Cullen obtusifolia	Fabaceae	*Acacia albida*	Fabaceae
Monsonia sp.	Geraniaceae	*Acacia erioloba*	Fabaceae
Myrothamnus flabellifolius	Myrothamnaceae		
Fruits			
Phoenix dactylifera	Arecaceae	*Capparis hereroensis*	Capparaceae
Orthanthera albida	Asclepiadaceae	*Boscia albitrunca*	Capparaceae
Euclea pseudebenus	Ebenaceae	*Boscia foetida* ssp. *foetida*	Capparaceae
Salvadora persica	Salvadoraceae	*Ficus sycomorus*	Moraceae
Lycium cinereum	Solanaceae		
Gum			
Acacia albida	Fabaceae	*Acacia tortilis* ssp. *heteracantha*	Fabaceae
Prosopis glandulosa	Fabaceae		
Stems			
Hoodia currori	Asclepiadaceae	*Myrothamnus flabellifolius*	Myrothamnaceae
Trichocaulon pedicellatum	Asclepiadaceae		
Roots			
Phoenix dactylifera	Arecaceae	*Aspilia eenii*	Asteraceae
Orthanthera albida	Asclepiadaceae	*Senecio marlothianus*	Asteraceae
Capparis hereroensis	Capparaceae	*Acanthosicyos horridus*	Cucurbitaceae
Boscia albitrunca	Capparaceae	*Cullen obtusifolia*	Fabaceae
Boscia foetida ssp. *foetida*	Capparaceae	*Tephrosia dregeana*	Fabaceae
Pods			
Acacia albida	Fabaceae	*Acacia tortilis* ssp. *heteracantha*	Fabaceae
Acacia erioloba	Fabaceae		
Prosopis glandulosa	Fabaceae		
Seeds			
Parkinsonia africana	Fabaceae	*Citrullus ecirrhosus*	Cucurbitaceae
Acacia albida	Fabaceae	*Monsonia* sp.	Geraniaceae
Acacia erioloba	Fabaceae		

stem added to sugarwater give a refreshing drink. The fruits of *Orthanthera albida*, which belongs to the same family, are eaten mainly by children: young fruits are eaten completely; in the case of old fruits only the inner part of the peel is eaten—the outer, somewhat fibrous part and the seeds are removed. The roots of the smae plant are put in beer to improve its flavor. *Pergularia daemia* var. *daemia*, and *P. daemia* var. *leiocarpa*, are apparently not consumed by the Topnaar, but are known in Botswana and South Africa (Zululand) where the leaves are eaten as a sort of "wild" spinach.

Aspilia eenii (Asteraceae or daisy family) has a root which can be put in milk to induce curdling and to improve its flavor. This yields a sort of yoghurt, that lasts longer than fresh milk. *Senecio marlothianus*, which belongs to the same family, yields roots that can be added to flavor home-brewed beer. Its root is wrapped in a cloth, put in the sun, crushed and then added to the beer. The leaves of *Tagetes minuta* have a strong to pungent smell; they are used to add flavor to tea. The roasted and ground seeds of *Parkinsonia africana* (Fabaceae, Caesalpinioideae or cassia (sub-) family) can be added to coffee to improve its taste.

The Capparaceae or caper family comprises a few endemics which yield edible fruits: *Capparis hereroensis*, *Boscia albitrunca* and *B. foetida* ssp. *foetida*. The latter two, however, are not used as such by the Topnaar, who are more interested by their medicinal properties. They are better known by the Bushmen (Botswana and Namibia) who eat their berries and roots. Throughout southern Africa, the berries of *B. albitrunca* are eaten fresh or are made into a sweet drink by mixing them with water or milk. Young roasted and ground roots are used as a coffee substitute. The roots are also prepared as a food in several ways, and are used in the fermentation of beer and the curdling of milk. The Bushmen first pound and cook the roots of *B. foetida* ssp. *foetida* into a porridge before eating it. The root is also used as a chicory substitute.

The cucumber family (Cucurbitaceae) comprises, besides the *!nara* root, yet another interesting plant, the tsama melon or bitter apple[3] *(Citrullus ecirrhosus)*, roasted and eaten by the Topnaar. These seeds are also eaten by the Bushmen of the Kalahari. The Ovambos of northern Namibia mix the oil from the seed with red ochre to make a cosmetic.

The fruits of *Euclea pseudebenus* (black Cape or false ebony; Ebenaceae, ebony family) are edible but quite astringent and thus only rarely consumed. The leaves are browsed by livestock and the berries can be fed to chickens, thus having an indirect, rather than a direct effect on the Topnaar's food situation. The toothbrush tree or salt bush (*Salvadora persica*; Salvadoraceae, the mustard tree family) has similar properties and uses; the fruits smell like cress, are edible, but cause diarrhea. It is only in a few exceptional cases that people will turn to them as a source of food.

Cullen obtusifolia (Fabaceae, pea family) is a small perennial herb. The leaves or the whole plant can be added to tea for flavoring. Adding the plant's roots to milk enhances its taste and curdles it into a kind of yoghurt. The roots of *Tephrosia dregeana* (Fabaceae) are used in the same manner. The same family also comprises a few Acacia tree species which are rather famous as a hunger food source. These acacias are among the most prominent trees in the Kuiseb river bed and are therefore known and appreciated for their value by all Topnaar communities. *Acacia albida* (ana tree; syn. *Faidherbia albida*) yields seeds that are eaten by people in Zimbabwe during times of famine. The Kuiseb Topnaar use pods and seeds especially as fodder for goats and cattle. The leaves are also browsed. The tree yields an edible gum. The pods and leaves of *A. erioloba* (camel thorn) are used in the same manner. In periods of food scarcity, the pods' pulp is a source of food for the Kuiseb Topnaar. Its seeds are used as a coffee substitute all over southern Africa. Less important is *A. tortilis* ssp. *heteracantha* (umbrella thorn; *basterkameeldoring, krulpeul* or *haak-en-steek*): its pods are a nutritious fodder, and its gum is edible. The pods of the related *Prosopis glandulosa* are eaten by both people and livestock. The pods of the same species (or a close relative) are used by American Indians to make a stew or an alcoholic beverage. The gum exuded by the tree has been used as a substitute for arabic gum in the manufacture of adhesives and drops.

The geranium or cranesbill family contains a few *Monsonia* species that are also used by the Kuiseb Topnaar. The seeds of the plants are collected by ants. The Topnaar gather these seeds from the ant nests and add them to tea, or roast and grind them and then add them to coffee for a better flavor. The leaves are used to improve the flavor of tea. Baked seeds are eaten.

Ficus sycomorus (sycamore fig; Moraceae, the fig and mulberry family) is very prominent at certain places in the Kuiseb valley. The figs are smaller than the cultivated equivalent: they are 1.5–2.5 cm in diameter and grow profusely in panicles on the main branches and trunk. The figs are covered with soft hair and turn yellow when ripe. They are eaten fresh or dried. The raw fruits can be put in a jar with sugar, which makes a kind of jam. The dried and ground fruits are used as a substitute for coffee. The fruits are used in spirit distillation in Zambia and northern Namibia.

The resurrection bush or *Myrothamnus flabellifolius* (Myrothamnaceae) is a botanical oddity, as it basically looks like a dead plant throughout the dry part of the year. However, the plants come alive with the first good rains and unfold their small leaves that turn green. Leaves and stems are added to tea to flavor it. They are also used as a spice.

The fruitlets of *Lycium cinereum* (*bok-* or *kraaldoring, slangbessie;* Solanaceae, nightshade family) can be eaten when ripe (as in South Africa), but this practice is apparently unknown in the Kuiseb.

THE TOPNAAR OF SESFONTEIN

The natural conditions in the residential area of the Sesfontein Topnaar differ completely from those in the Kuiseb. The Sesfontein Topnaar community is situated on an alluvial plain, north of the Hoanib river, on the fringes of the northern Namib. The climate is more humid than that in the Kuiseb, resulting in a more abundant vegetation with a higher variety of useful plants. The Sesfontein Topnaar have integrated agricultural practices in their gathering lifestyle: almost all families have a garden in which they grow wheat (Triticum spp.) and corn *(Zea mays),* using irrigation. However, wild food plants are still being gathered from the environment. Different seeds are used for the preparation of porridge. Edible tubers and bulbs are collected, as are a number of wild fruits.

The *!nara* which is so important for the Kuiseb Topnaar is completely absent from the Sesfontein Topnaar's diet. The only cucurbit worth mentioning is a *Coccinia* species which yields a tuberous root that can be eaten roasted.

The beautiful red flowers of *Ruellia diversifolia* (Acanthaceae) contain a lot of tasty nectar, rich in energy, and which is sucked out. The leaves and stems of *Amaranthus dinteri* ssp. *dinteri*[4] (pigweed; Amaranthaceae) are cooked and eaten as vegetables. They can also be fried with onions. The plant grows in humid soils, and is a weed in Sesfontein's crops.

The Sesfontein Topnaar eat the fruits (sweet fibrous mesocarp) from the date palm, and also the fruits of the makalani, ivory, dum or real fan palm (*Hyphaene petersiana*; Arecaceae). Other people in Zimbabwe, Zambia and northern Namibia use this palm extensively: The sap tapped from the trunk is fermented into palm wine and the heart of the palm is/was consumed as a vegetable, which is now forbidden since this practice kills the tree.

Hoodia currori, Orthanthera albida and both *Pergularia daemia* species are used as in the Kuiseb. There is another interesting member of the milkweed family in Sesfontein which yields tubers with a high water content, that are eaten roasted (*Fockea angustifolia*, water root kambroo).

A typical shrub for the Sesfontein river beds is *Cordia gharaf* (Boraginaceae; heliotrope or forget-me-not family). It carries orange elliptic fruits of about 1 cm long which are eaten fresh.

Combretum imberbe (leadwood; Combretaceae) yields an edible gum of minor importance.

The sedge family (Cyperaceae) comprises a few species which are common as weeds in the crop fields in the Sesfontein area. *C. fulgens* (or maybe *C. usitatus*; *uintjie* in Afrikaans which means small onion) yields tubers which can be eaten raw or boiled.

The area supports some *Strychnos* (Loganiaceae or strychnos family; the exact species could not be identified) and *Ximenia americana* trees (sour plum, *kleinsuurpruim* or *doring pruim;* Olacaceae, the sour plum family). The pulp of the spherical, fleshy, many-seeded strychnos fruit is eaten raw. The seeds are extremely toxic since they contain strychnine. Sour plums are ellipsoid to ovoid in shape,

TABLE 2. Species used by the Sesfontein Topnaar as hunger food (taxonomical classification according to Cronquist, 1988)

Species	*Family*	*Species*	*Family*
Leaves			
Amaranthus dinteri		*Monsonia* sp.	Geraniaceae
ssp. *dinteri*	Amaranthaceae	*Myrothamnus flabellifolius*	Myrothamnaceae
Orthanthera albida	Asclepiadaceae	*Pergularia daemia*	Asclepiadaceae
Cullen obtusifolia	Fabaceae	*Salvadora persica*	Salvadoraceae
Fruits			
Hyphaene petersiana	Areceae	*Cordia gharaf*	Boraginaceae
Ficus sycomorus	Moraceae	*Strychnos* sp.	Loganiaceae
Ximenia americana	Olacaceae	*Berchemia discolor*	Rhamnaceae
Ziziphus mucronata	Rhamnaceae	*Salvadora persica*	Salvadoraceae
Grewia tenax	Tiliaceae	*Cyphostemma* sp.	Vitaceae
Gum			
Acacia tortilis		*Combretum imberbe*	Combretaceae
ssp. *heteracantha*	Fabaceae		
Stems			
Amaranthus dinteri		*Hoodia currori*	Asclepiadaceae
ssp. *dinteri*	Amaranthaceae		
Myrothamnus flabellifolius	Myrothamnaceae		
Roots			
Orthanthera albida	Asclepiadaceae		
Cullen obtusifolia	Fabaceae	*Coccinia* sp.	Cucurbitaceae
Pods			
Acacia tortilis			
ssp. *heteracantha*	Fabaceae		
Seeds			
Monsonia sp.	Geraniaceae	*Sesbania sphaerosperma*	Fabaceae
Setaria verticillata	Poaceae		
Nectar			
Ruellia diversifolia	Acanthaceae		
Trunk			
Hyphaene petersiana	Arecaceae		
Tubers			
Fockea angustifolia	Asclepiadaceae	*Cyperus fulgens/usitatus*	Cyperaceae

yellow to red when ripe and some 3 cm long. They are eaten raw in the Sesfontein area, but in South Africa they are used to make lemonade or beer.

Monsonia spp. (Geraniaceae), *Ficus sycomorus* (Moraceae), *Myrothamnus flabellifolius* (Myrothamnaceae) and *Cullen obtusifolia* (Fabaceae) are used as in the Kuiseb area. Among the other legumes, the seeds of *Sesbania sphaerosperma* are roasted and ground and prepared into a coffee substitute. *Acacia tortilis* ssp. *heteracantha* is much more used in Sesfontein than in the Kuiseb area: the Topnaar grind its pods, remove the seeds and boil the remaining pulp with milk, blood or water into an edible porridge (corn or wheat can also be added); its gum is also eaten.

Berchemia discolor (bird plum or wild date; Rhamnaceae, buffalo-thorn family) is another source of plum-like, edible fruits which are eaten fresh or dried. They are also cooked in milk and the consumed. The tree is common at the highlands of southern Africa. Elsewhere in the area, the fruits are fermented into alcohol, or cakes are prepared from the dried, ground fruits. *Ziziphus mucronata* (buffalo thorn or shiny leaf) produces bright red to red-brown, round fleshy fruits, 1–2 cm in diameter, with a single hard pip inside; they are eaten raw or boiled. Other people in neighboring areas use it in different ways: the Ovambo brew alcohol from it; pounded fruits are roasted and ground as a coffee substitute throughout southern Africa.

Salvadora persica is used in the same way as in the Kuiseb area but fruits are also dried, rehydrated and rolled into balls. Such a preparation is a knapsack food for travellers.

The moist soil conditions near the Sesfontein springs allow *Setaria verticillata* plants to grow. The annual bur-bristle grass (Poaceae) is the only "wild" member of its family that is being used: the seeds are ground and boiled in water, milk or fat to make a porridge. The seeds are used as a source of alcohol in South Africa.

The fruits of *Grewia tenax* (Tiliaceae, jute family) and *Cyphostemma* spp. (cobas tree; Vitaceae, grape-vine family) are eaten raw. With this last species, the fruits' juice can be squeezed out and drunk.

SUMMARY FOR NAMIBIA

Many species listed here can be considered as complementary or additional food sources, supplementing the diet under normal conditions. However, in times of famine these plants become a real source of hunger food, yielding useful products, even during periods of drought.

The *!nara* clearly occupies a very special place among the species mentioned, as it is a true staple and a source of juice and medicine. The Kuiseb Topnaar know some 20 other species that can be used as hunger food. Some of them yield edible fruits used in times of scarcity *(Acacia erioloba* or *Salvadora persica).* Plant parts of at least 10 species are used as a coffee substitute or to make tea. Several species can curdle milk into a kind of yoghurt (or curd).

Forty percent of the mentioned species are endemic to the Namib desert *(sensu lato).* Some of the remaining species are confined to southern Africa and have only a regional spread and importance. Many of the endemic species are either succulents *(Trichocaulon, Hoodia)* or xerophytes with no leaves *(Acanthosicyos)* or reduced leaf size *(Orthanthera),* or with succulent leaves *(Boscia, Capparis, Salvadora).* The most extreme case is that of the resurrection plant which can survive complete dehydration. The above-mentioned species remain small, except for *Boscia.* Several bushes and trees several have spines to protect them from excessive browsing *(Acacia, Acanthosicyos).*

The wild plants used by the Sesfontein Topnaar show a somewhat greater variety of possibilities and uses. The climatic conditions of this area, combined with the abundant presence of spring water, are much more favorable for plant growth than those in the Kuiseb area. This explains a relatively greater presence of more cosmopolitic species, *i.e.* plant species that occur in a wide variety of ecologies all over the world. However, there are only a few more species used in Sesfontein than in the Kuiseb, while most species are also equipped with arid condition features. Nine species used in the Kuiseb area are also used in Sesfontein.

It is quite noteworthy, in comparison to other regions, that grasses (1 species) and root or tuber plants (3 species) are hardly present at all in the list of food plants. Most species mentioned contain a number of secondary compounds which make them sources of medicine and/or veterinary products as well.

The Sub-humid Environment of Southern Senegal

Senegal is partly situated in the Sahel zone. Rainfall is often erratic. People throughout the Sahel have learned to cope with a precarious food security situation, by broadening their traditional, agricultural food basis with the addition of wild plant products. These include several grass species (Poaceae) in the northern semi-arid to arid areas which, under "normal" conditions, would never be used for food. The grasses are basically a source of edible seeds, or grains, which are processed much in the way of the cultivated cereal grains, such as millet and sorghum, resp. *Pennisetum americanum* and *Sorghum bicolor,* and are thus a source of energy (carbohydrates). Wild rice *(Oryza barthii)* is a semi-arid zone species that is a very close relative of upland rice, *O. glaberrima.* Other interesting wild grasses used to make cakes or porridge, belong to a number of widely known genera: *Brachiaria* (with *B. distichophylla, B. jubata,* and *B. stigmatista*), *Cenchrus* (*C. biflorus,* eaten or used in preparing drinks), *Dactyloctenium* (*D. aegyptium,* for porridge), *Digitaria (D. debilis, D. horizontalis* and *D. ciliaris), Echinochloa (E. colona* known as jungle rice or shama millet; *E. pyramidalis,* antelope grass, and *E. stagnina*), *Eleusine* (*E. indica* which has a high protein content), *Eragrostis* (*E. cilianensis* or stink grass, *E. ciliaris, E. gangetica, E. pilosa,* and *E. tremula*), *Panicum* (*P. pansum, P. fluviicola, P. subalbidum, and P. laetum* which are considered to be among the best wild cereals), *Paspalum* (*P. orbiculare* whereby the arid land variety is better than the one growing on swampy soils), *Pennisetum, Sorghum (S. arundinaceum), Setaria, Sporobolus* (for a more extensive list and comments: see Abbiw, 1990).

People also rely on a number of tuber plants and wild fruit trees for survival. People living in the semi-arid to sub-humid zone also have a thorough knowledge of possible hunger food sources, albeit not well documented. In this last chapter data is presented concerning the hunger food knowledge of tribes living in the Kolda area of southern Senegal (Casamance province).

CASAMANCE

The Casamance area and more specifically the Moyenne Casamance (Kolda region, Figure 2) is inhabited by a number of tribes, of which the Mandingue, Fulani and Wolof are the most important. The Fulani are traditionally nomadic cattle raisers, who have gradually become sedentary in their area. They grow the same crops as

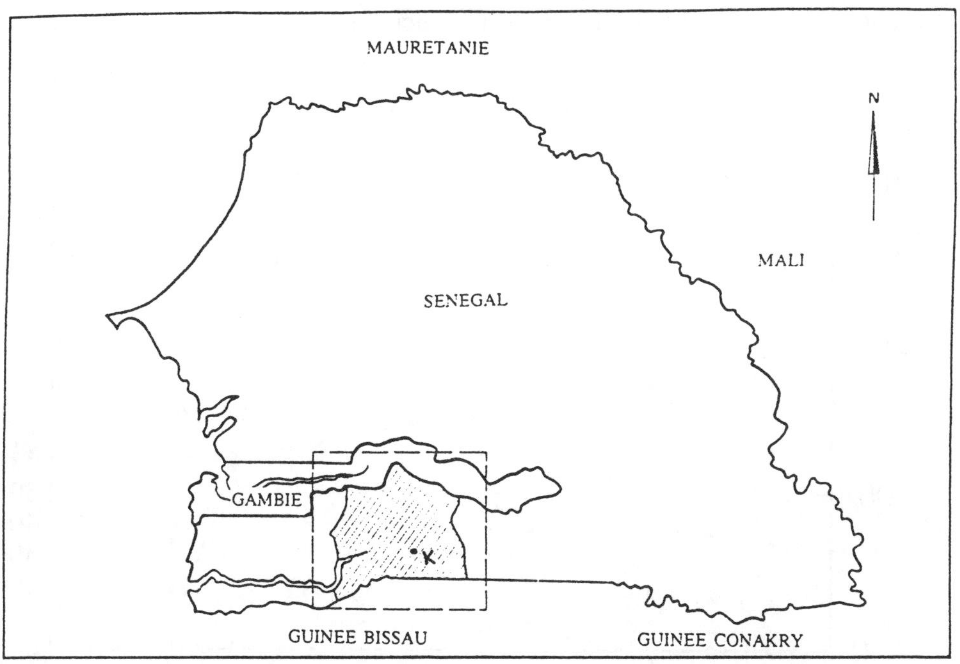

Figure 2. Map of Senegal with Moyenne Casamance area and Kolda (K) highlighted in the south (after De Wolf & Van Damme, 1994).

the two other tribes which traditionally have been more familiar with agriculture. The Mandingue grow fonio *(Digitaria exilis)*, sorghum and several millets, rice *(Oryza sativa)*, maize *(Zea mays)*, groundnut *(Arachis hypogaea)* and several vegetables. The Wolof also grow a mixture of millets and sorghum, cassave *(Manihot esculenta)*, cotton *(Gossypium hirsutum)* and groundnut. The three tribes have gradually intensified their agricultural system, resulting in almost permanent occupation of plots, leading to soil degradation. Land clearing is usually done through slash and burning of the original, natural vegetation. There is little use of fertilizers or other inputs.

The Casamance has a Soudano-Sahelian climate. The rainy season starts in June and ends in October. Average yearly rainfall is 800–1,200 mm distributed over 60–90 rainy days. Temperatures vary between 7 and 45°C (Figure 3). The local flora is a source of numerous products used for food, protection (fibres for clothing and

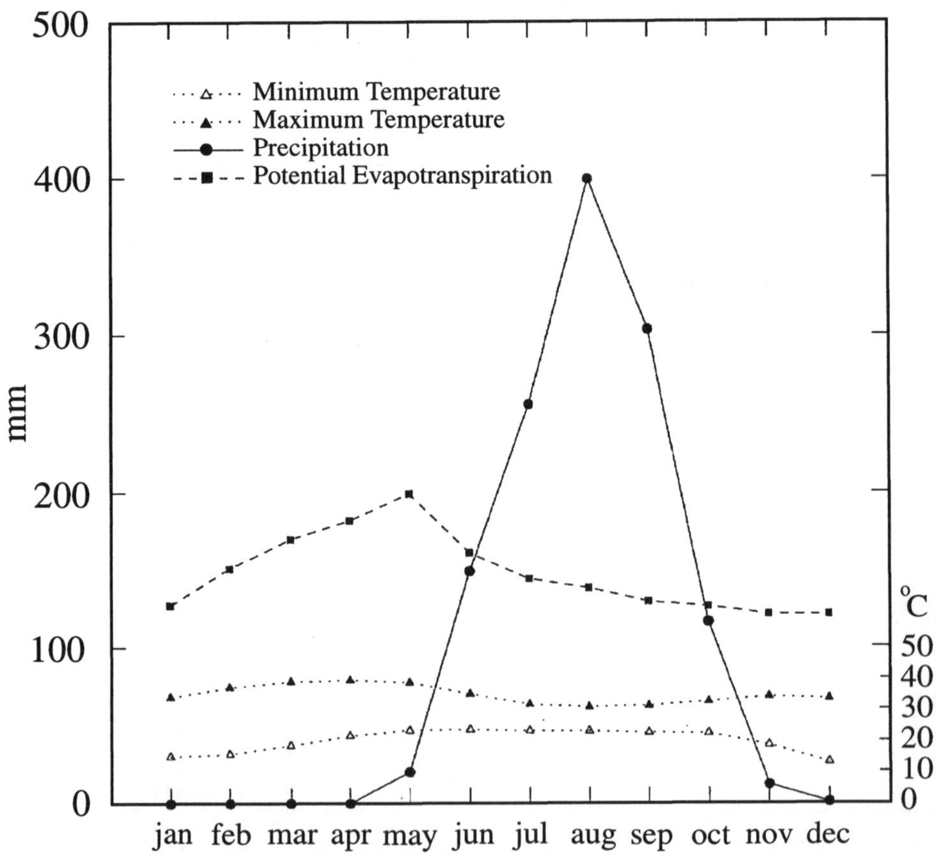

Figure 3. Climate data for Kolda, Casamance, Senegal (after De Wolf & Van Damme, 1994).

thatching), construction, and medicine. The local pharmacopoeia is very well developed and plants are used in remedies to cure human and animal diseases (for more details consult Van den Eynden & Van Damme, 1994).

Table 3 presents the 40 species which are used for their fruit. Most of these plants are trees and shrubs. Fruits are usually eaten by children as an addition to the normal diet. Older people will only eat them in a limited number of cases. A number of species presented in Table 3 were introduced from abroad, but have become part of the natural vegetation through natural dispersion processes (*e.g. Anacardium occidentale*, cashew). Some species, such as *Mangifera indica* (mango) or *Carica papaya* (pawpaw), and also cashew, are still being planted as part of reforestation programs. Most listed fruit plants are naturally occurring species that grow in the forests. When land is cleared for agriculture some of these are actually left uncut or allowed to coppice during the cropping cycle (*e.g. Cordyla pinnata* or *Annona senegalensis*). Some species are very much appreciated and even harvested to be sold on the local markets (*e.g. Detarium senegalense* and *D. microcarpum*, *Landolphia heudelotii*, *Saba senegalensis*, *Vitex madiensis*, *Ziziphus mauritiana*, or *Neocarya macrophylla*).

TABLE 3. Trees and shrubs used for fruit (* indicates: seeds used seperately)

Species	*Family*	*Species*	*Family*
Anacardium occidentale	Anacardiaceae	*Neocarya macrophylla* *	Chrysobalanaceae
Lannea acida	Anacardiaceae	*Parinari curatellifolia* *	Chrysobalanaceae
Lannea microcarpa	Anacardiaceae	*Diospyros mespiliformis*	Ebenaceae
Lannea velutina	Anacardiaceae	*Icacina senegalensis*	Icacinaceae
Mangifera indica	Anacardiaceae	*Strychnos spinosa*	Loggganiaceae
Sclerocarya birrea	Anacardiaceae	*Azadirachta indica*	Meliaceae
Spondias mombin	Anacardiaceae	*Ficus sycomorus*	Moraceae
Spondias purpurea	Anacardiaceae	*Ximenia americana*	Olacaceae
Annona senegalensis	Annonaceae	*Borassus aethiopum*	Arecaceae
Hexalobus monopetalus	Annonaceae	*Elaeis guineensis*	Arecaceae
Landolphia heudelotii	Apocynaceae	*Ziziphus mauritiana*	Rhamnaceae
Saba senegalensis	Apocynaceae	*Gardenia erubescens*	Rubiaceae
Adansonia digitata	Bombacaceae	*Sarcocephalus latifolius*	Rubiaceae
Cordyla pinnata	Fabaceae	*Lepisanthes senegalensis*	Sapinaceae
Detarium senegalense	Fabaceae	*Quassia undulata*	Simaroubaceae
Dialium guineense	Fabaceae	*Cola cordifolia*	Sterculiaceae
Tamarindus indica	Fabaceae	*Grewia lasiodiscus*	Tiliaceae
Andira inermis	Fabaceae	*Vitex doniana*	Verbenaceae
Parkia biglobosa *	Fabaceae	*Vitex madienses*	Verbenaceae
Carica papaya	Caricaceae	*Cissus waterlotii*	Vitaceae

TABLE 4. Ingredients used in sauce preparation

Species	*Family*	*Species*	*Family*
Leaves			
Tylophora sylvatica	*Apocynaceae*	*Strychnos spinosa*	Loganiaceae
Adansonia digitata	*Bombacaceae*	*Ficus thonningii*	Moraceae
Cassia tora	*Fabaceae*	*Moringa oleifera*	Moringaceae
Flowers			
Bombax costatum	*Bombacaceae*	*Moringa oleifera*	Moringaceae
Bark			
Grewia mollis	*Tiliaceae*		
Fruits			
Cordyla pinnata	*Fabaceae*	*Cissus rufescens*	Vitaceae
Tamarindus indic	*Fabaceae*		
Gum			
Sterculia setigera	*Sterculiaceae*		

TABLE 5. Ingredients used in tea preparation

Species	Family	Species	Family
Terminalia avicennnioides	Combretaceae	*Grewia mollis*	Tiliaceae
Pterocarpus erinaceus	Fabaceae		

Tables 4 and 5 present species which are used in sauces and for making tea. Although they are not to be considered as hunger food, they supplement the daily diet with essential vitamins and proteins.

Combretum nigricans (Combretaceae) yields a resin which is occasionally processed into sweets.

The bulbs of *Raphionacme brownii* and *R. daronii* (Asclepiadaceae) are eaten fresh. The yam roots of *Dioscorea prehensilis* (Dioscoreaceae) are eaten cooked or roasted. The aerial bulbils of *Dioscorea bulbifera* must first be cooked before they can be eaten, because they are slightly poisonous. These four species seem to be the only carbohydrate sources that are sometimes used by the three tribes that were investigated. It should also be noted that there are no cereal grains among the species listed here. The few Poaceae species they collect are only used for medicinal purposes. This could indicate that the local food situation is perceived as being not too precarious.

The local people know the wild plant species that can be used for a variety of purposes. This knowledge is important for times of famine.

Concluding Remarks

Gathering of (hunger) food plants is obviously one of the major activities among hunter-gatherers. Their livelihood and survival clearly depend on their knowledge about the use of wild plant species. Sedentary farmer communities occasionally gather wild food plants to supplement dietary deficiencies, caused by a drop in food production or food availability. The perception of 'hunger food' is clearly different, depending upon the setting and the activities of a certain population. In both cases, however, major famine foods are able to cover energy needs and sustain people through famines, or at least through the first stages until outside (food) help arrives.

The natural environment of an area influences the absolute number of species to choose from when looking for (hunger) food. The wetter Casamance area yields more useful species than arid Namibia. Some 400 odd plant species occur in the Namib desert, whereas twice or thrice this number is found in Casamance. The greater variety allows for better coping strategies when living conditions get worse and food becomes scarce.

Most species presented in this text are perennials which are better adapted to stressed environments. Although the aboveground plant parts may, in some cases, not be present throughout the year, regrowth starts each year from underground reserve material, that is sometimes used for food *(Cyperus, Dioscorea, Raphionacme).*

In Namibia we are dealing with species which have features allowing them to survive, grow and develop in very harsh arid conditions. These are interesting species which could be introduced in arid frontier zones where people are confronted with similar climatic constraints.

Hunger food plant species can be the basis for further agricultural research. A number of the mentioned species are already used in other regions. Endemic species with a limited distribution area may have more than just a local vocation. Further research into their specific properties and possibilities for improvement and cropping may make them interesting candidates for agricultural status in other environments as well. The same reasoning can be applied to other species which have a broader distribution area, and are already more widely used, but have never been formally developed into an agricultural crop.

The case of the buffalo gourd *(Cucurbita foetidissima)* perfectly illustrates this point. Originally it was only known from its use in Botswana by some local tribes. However, research work into its physiology, biology, and agronomic properties, developed the plant into an agricultural crop which is now yielding commercial fruit in the *U.S.A.* under marginal arid conditions.

Notes

1. The !Kung San and Topnaar, as well as a number of other people in southern Africa are known to use click languages. The four clicks are presented through special signs that by convention correspond to a specific sound.
2. Most of the information in this chapter has been taken from Van den Eynden *et al.,* 1992.
3. Whenever possible we have added the English vernacular name to the scientific Latin name of the species presented here.
4. The Latin species epitheton *dinteri* refers to Dinter, the first botanist of imperial South West Africa who published a book on the country's vegetable *veldkos* or food from the field, so that one can say that the species name was well chosen.

References

Abbiw, D.K. (1990) *Useful Plants of Ghana.* Intermediate Technology Publications and The Royal Botanic Gardens, Kew.

Asibey, E.O.A. and C.Y.A. Beeko (1989) Forest and food security. Symposium on Ghana's forest policy, 3–7 April, 1989, Greenhill, Accra, Ghana.

Barral, H. (1977) Les populations de l'Oudalan et leur espace pastoral. *Travaux et documents de l'ORSTOM,* 77, France.

Bernus, E. (1988) Seasonality, climatic fluctuations, and food supplies (Sahelian nomadic pastoralist societies). *In* de Garine, I. and Harrison, G.A. (eds.) *Coping with Uncertainty in Food Supply.* Clarendon Press, Oxford, pp. 318–336.

BOSTID (1996) *Lost Crops of Africa,* vol. 1, *Grains.* National Academy Press, Washington.

Campbell-Platt, G. (1980) African locust bean (*Parkia* spp.) and its West African fermented food product, dawadawa. *Ecology of Food and Nutrition* 9:123–132.

Cronquist, A. (1988) *The Evolution and Classification of Flowering Plants.* New York Botanical Garden, New York.

De Waal, A. (1988) Famine early warning systems and the use of socio-economic data. *Disasters* 12:81–91.

De Wolf, J. and P. Van Damme (1994) Inventaire et modelage de la gestion du couvert végétal pérenne dans une zone forestière du Sud Sénégal. Rapport final, partie A: Étude phytosociologique, RUG, Coupure Links 653, B 9000 Gent, 82 pp.

Gumbo, D., B. Mukamuri, M. Muzondo, and I. Scoones (1990) Indigenous and exotic fruit trees: Why do people want to grow them? *In* Prinsley, R.T. (ed.) *Agroforestry for Sustainable Production: Economic Implications.* CSC, London, pp. 185–214.

Holy, L. (1980) Drought and change in a tribal economy: The Berti of Northern Darfur. *Disasters* 4:65–72.

Lee, R.B. (1973) Mongongo: the ethnography of a major wild food resource. *Ecology of Food and Nutrition* 2:307–321.

Lee, R.B. and I. de Vore, (1976) *Kalahari Hunter-Gatherers: Studies of the !Kung San and Their Neighbors.* Harvard University Press, Cambridge, Mass.

Longhurst, R. (1986) Household food strategies in response to seasonality and famine. *IDS Bulletin* 17:27–35.

Martin, M. (1985) Design of a food intake study in two Bambara villages in the Segon region of Mali with preliminary findings. *In* Hill, A. (ed.) *Population, Health and Nutrition in the Sahel.* Routledge Kegan Paul, London.

Newman, J.L. (1975) Dimensions of Sandawe diet. *Ecology of Food and Nutrition* 4:33–39.

Ogle, B.M. and L.E. Grivetti (1985) Legacy of the chameleon: Edible wild plants in the kingdom of Swaziland, Southern Africa. A cultural, ecological, nutritional study. Part II—Demographics, species availability and dietary use, analysis by ecological zone. *Ecology of Food and Nutrition* 17:1–30.

Rahmato, D. (1988) Peasant survival strategies in Ethiopia. *Disasters* 12:326–344.

Rosetta, L. (1986) Seasonal variations in food consumption by Serere families in Senegal. *Ecology of Food and Nutrition* 20:275–286.

Scoones, I., M. Melnyk, and J.N. Pretty (1992) *The Hidden Harvest. Wild Foods and Agricultural Systems. A Literature Review and Annotated Bibliography.* IIED, London, 256 pp.

Van den Eynden,V., P. Vernemmen and P. Van Damme (1992) *The Ethnobotany of the Topnaar.* RUG, Coupure Links 653, B 9000 Gent, Belgium, 145 pp.

Van den Eynden, V. and P. Van Damme (1994) Inventaire et modelage de la gestion du couvert végétal pérenne dans une zone forestière du Sud Sénégal. Rapport final, partie C: Étude Etnobotanique, RUG, Coupure Links 653, B 9000 Gent, Belgium, 102 pp.

ABOUT THE AUTHOR

Patrick Van Damme (born 1956) is professor of Tropical and Subtropical Agronomy, Ethnobotany and Rural Development Economics in the Department of Plant Production, University of Gent, Belgium. He obtained his Ph.D. in stress physiology working with a desert plant *(Euphorbia tirucalli).* Before starting a university career, he worked in Nigeria for the International Institute of Tropical Agriculture (IITA) and for the Food and Agricultural Organization of the United Nations (FAO) in Senegal. His current research is focused on agriculture and botany in arid and semi-arid zones, blending plant taxonomy and anthropology into ethnobotany. Recently, pistachio and olive have been added to the list of species earmarked for further research and development. The author is regularly consulted for development project planning and evaluation by the World Bank, the International Fund for Agricultural Development (IFAD), and the European Union.

Prof. Patrick Van Damme
University of Gent
Faculty of Agricultural and Applied Biological Sciences
Department of Plant Production
Lab. of Tropical and Subtropical Agronomy and Ethnobotany
Coupure Links 653
B-9000 Gent
Belgium

Part Five

Management of Water Resources

Chapter 13

Irrigation Development: Proactive Planning and Interactive Management

Leslie Shanan

Introduction

About 240 million ha (17% of the world's cropland) are irrigated, producing one-third of the world's food. As nearly 75% of this area is in developing countries, irrigation will remain a crucial component of future agricultural programs. However, it must be noted that in the course of the last three decades, there has been a sharp decline in the growth rate of new irrigation areas, particularly in developing countries. During the 1960s and the 1970s for instance, the annual growth rate of the irrigated areas exceeded 2%, an increase of more than 20% every decade. Annual rates sometimes exceeded 4%. On the other hand, in the 1980s the annual growth rate dropped to less than 0.8%. A further decrease in growth is expected to continue in the foreseeable future.

The following factors have been identified as contributing to this changing scenario:

1. Low cost irrigation projects have generally been exhausted and only expensive options remain.
2. Investments in agricultural and irrigation development projects are declining as economic constraints limit national budgets. The industrial, transportation, housing and defense sectors are being given higher priorities.
3. Projects with land capable of long-term sustainable production next to sustainable water supplies have all been completed. New projects require the transport of water over large distances, generally trans-basin diversions.
4. The rates of return on investments in irrigation have usually been significantly below their anticipated levels.
5. Most irrigation projects in developing countries operate at low efficiencies, sometimes as low as 10-15%. The result is that as much as 85-90% of the water delivered to the project at the headworks often runs to waste in the drainage

H. J. Bruins and H. Lithwick (eds.), The Arid Frontier, 251–276.

networks, in seepage losses from canals and ditches and in over-irrigation of many fields, particularly in the headstream sections of the project areas. The water is not only wasted, but it invariably causes waterlogging of extensive areas in the project. In the waterlogged areas, water tables rise and soil salinity problems appear as the salts accumulate in the soil profile and in the aquifers. In the water-logged areas, land either goes out of production, or yields are reduced significantly. The term 'twin menace' is commonly used to describe this irreversible process, because soil and aquifer salinity problems invariably follow waterlogging. Consequently, the returns both in terms of finance as well as yield have been disappointing.

6. Finally, the world community has become aware of the urgent need to solve a wide range of environmental problems that include *inter alia*, the prevention of accelerated rates of pollution of our natural resources, especially land and water. The challenge presented to the planner is no longer *high productivity* but *long-term sustainability.* The irrigation engineer must discard many of his old concepts and technologies and replace them with new ones that will minimize these hazards. It is now recognized that irrigation *per se* is a dynamic process that disturbs the water and salt balance of a region and may cause irreversible damage to an ecosystem.

Most, if not all existing irrigation projects will have to be rehabilitated in order to meet the demands of the 21st century. The upgrading of projects is estimated to cost $1500-$2000 per ha. As new projects cost $5000-15000 per ha (depending on whether the scheme includes a multipurpose dam or trans-basin diversion), the rehabilitation of existing projects will undoubtedly be the preferred alternative of the future.

Understanding the reasons for the poor performance of irrigation projects in the past is a prerequisite to proposing feasible ways of improving the situation. In most of the systems, water-deliveries in the main networks, from the headworks down to the government outlets, is fluctuating and unassured. During the last three decades, numerous attempts have been made to improve the situation in developing countries. These covered a wide range of comprehensive programs composed of a varying mix of components such as:

1. Improving water management at the farm level.
2. Land consolidation programs.
3. Introducing water distribution policies and schedules.
4. Increasing water charges.
5. Funding credit programs for introducing high yielding varieties, fertilizers and pest control measures.
6. Establishing research and extension services.

7. Reducing network losses by lining canals.
8. Transferring management from government to private or farmers' agencies.

Unfortunately, the results of rehabilitation and improvement schemes often fell far below the planners' expectations. Some projects are now in their second and third cycle of rehabilitation.

Irrigation at the Crossroads: The Present Dichotomy

The major and overriding shortcoming of these programs has been that they did not succeed to tackle the primary cause of the problem, namely, the mismanagement and unreliable delivery and distribution of the water in the network. Prior to these programs, water deliveries were often controlled by the strong upstream farmers who had become sufficiently powerful politically and economically to corrupt the operation staff, ensuring that their fields received more than their fair share of the available supplies. The rehabilitation programs did not change the situation. Farmers (generally downstream and at tail-ends) who in the past had received unassured, and untimely deliveries continued to receive inequitable supplies, remaining the *poorest of the poor*, living below the poverty line.

In the 1990s, the diversity of proposed solutions has merged into a sharp dichotomy of two principal schools of thought. The one school, led by the International Irrigation Management Institute centered in Sri Lanka, maintains that the low performance of the existing system is due to the fact that they are government controlled and the problems can only be solved by the transfer of the management responsibility from the government to locally managed non-government institutions. They have recently rectified the title of their conceptual approach by calling it *Privatization and Self Management of Irrigation* (Vermillion, 1996). This modification is more than just semantics, since the term *privatization* is an economic theory with political connotations. It raises the basic issue of the 'public *versus* private goods' debate. Should 'water' be a *public good*, like 'national defense' and 'primary education' or a *'private good'* like 'shoes' and 'cars'? Are farmers' water rights less important than other human rights—the right of free speech, education, health services *etc.*? There is certainly no consensus on this issue. Even in the United States of America, we find wide disagreement on the role that government should play in the free market system. If we compare the U.S.A. today with the same country a hundred years ago, we see major differences in the extent to which goods are public or private. There has been a gradual encroachment of government intervention into to the free market. Medical care, regulated industries, regulated banking and interest rates, state educational institutions, national roads, highways and transportation systems are just some of the sectors that today have major elements of public choice compared to the 19th century. Part of the political controversy in the U.S.A. (as is the case also in many secular democratic countries in Europe) has been the attempt by some sections to roll back the scope of public choice, where such an approach is thought to be inessential, while other sectors have tried to further the extension of public choice.

The privatization and free market concept holds that the common good is best served by the uninhibited pursuit of self interest. The question "What kind of society do we want?" often receives the response "Let the free market decide!" Hence the farmers must be the dominating and controlling voice in the management organizations of privatized irrigation projects, in order to ensure that their demands will be met. Consequently, at every take-off point in the rehabilitated network, the water flows are regulated by constructing adjustable (usually manual) control gates to meet the continually varying water demands, as determined by the perception of the farmers. Theoretically, the numerous regulating structures make possible exact and efficient control of the deliveries to meet a wide range of irrigation schedules and cropping patterns. Project production is recorded as yield per unit of land.

The second school of thought, of which the author is one of its protagonists, is based on a *weltanschauung* that maintains there is something wrong with making *the survival of the fittest* a guiding principal of a civilized society, particularly when it is dealing with a section of the community that lives at or below the poverty line. Cooperation, mutual aid and government intervention should be as much a part of the system as competition. Survival of the fittest distorts and corrupts.

Societies derive their cohesion from shared values, and a common historical memory that is rooted in culture, religion and tradition. To survive, society must be based on the rule of law that safeguards the rights of the individual. In relating this concept to irrigation, for the farmer, his water rights are more valuable to him than any other right, and unless these are guaranteed and assured, they are worthless to him. Unpredictable fluctuations in rainfall make rainfed agriculture a risky business. The purpose of irrigation is to reduce the risks. If deliveries are unreliable, the 'irrigation' farmer is worse off than before. It therefore follows that only *after* a government has shown it has the political will and ability to assure a seasonal supply to every farmer, delivering the water to him according to a timely predetermined schedule that guarantees these rights, can we expect any improvements in equity and efficiency of irrigation projects.

This school of thought proposes, once farmers' water rights are assured and can be enforced by law, that the rehabilitation of the project must be based first on a simple operating and management system, and secondly, allocation policies and seasonal delivery schedules of all canals and government outlets must be promulgated by the government at least one month prior to the irrigation season. These principles have led to the concept of *structured networks*. A rehabilitation plan must take into account the negative effects of the prevailing social, political and economic structures in the project area (particularly corruption, poor law enforcement and powerful upstream farmers). The network must not only be simple to operate, but difficult to corrupt. The design of the network ensures that in the distributaries and minors delivering to the government outlets, the flows are constant and not fluctuating. In these sections of the network, the canals are either run full or not at all (*i.e.* it is an ON/OFF system). Either all outlets receive their full flow, or no flow at all. Water is supplied to the government outlets at full supply levels and hence water supplies are easy to monitor volumetrically.

The seasonal water delivery schedules are planned to meet the critical growth stage of the predominant crops and these structured networks do not supply water in order to meet individual farmers' demands. Project production is measured in terms of yield per unit of water, and not yield per unit of land. In the sections of the network that deliver on ON/OFF schedules, control structures are regulated by proportional divisor devices which assure that the flows at all take-off points and at outlets are automatically divided proportionally (Horst, 1983). These structures consist of fixed overflow weirs, having widths proportional to the areas irrigated. The advantages of the proportional division are:

1. Operation is simplified significantly. Standard regulating structures may have to be constructed at the head of secondary canals, but downstream, the fixed proportional divisors require no regulating control. Operational manpower can be reduced by as much as 80%.
2. If less or more water is supplied in the ON/OFF section depending on the availability of supplies, the shortage or surplus will be spread proportionally over all the blocks and shared by all the farmers.
3. The downstream divisor structures are cheaper to construct and maintain because they have no moveable parts.
4. The distribution of water is simple and understandable, not only for the operational staff, but also to the farmer. The system fosters the farmers' confidence in the network and in his water right, and also enables him to monitor the deliveries, on a time basis, because the flows are constant and do not fluctuate. Below the government outlet, the blocks and deliveries to each farm should be locally managed, based on a rotational rotary system.
5. This system gives little opportunity for mismanagement and corruption above the government outlet. Mismanagement will be clearly visible. Consequently, equitable water distribution is enhanced. The irrigation system performance depends on how the different elements work together, and not on how they work separately.

The structured network concept believes that in developing countries, it is better to produce a reliable subdivision of the water to the farmer on a predetermined delivery schedule and so enable him to use his skills, resources and enterprises for the optimum exploitation of his two main capital resources—his land and his assured water right—than to try and regulate the water supply to meet the individual needs of the individual farmers, as is usually done in the advanced industrialized developed countries like the U.S.A.

The fundamental difficulties that face the irrigation planner in search for solutions to this sharp dichotomy—Privatization *versus* Structured Systems, can be found in case studies in *Planning and Management of Irrigation Systems in Developing Countries* (Shanan, 1992) and in *The Privatization and Self Management of Irrigation* (Vermillion, 1996).

Environmental Hazards in Irrigation Development

Several environmental problems may occur in the wake of irrigation development, as a result of bad planning and management:

1. Sedimentation
2. Waterlogging
3. Soil salinization
4. Sea-water intrusion
5. Land subsidence

The latter four environmental topics, including the twin menaces of waterlogging and salinization, are extensively discussed in a previous article by the author (Shanan, 1987). The problem of sedimentation and its control through proactive irrigation planning is presented in the following section.

SEDIMENTATION PROBLEMS IN IRRIGATION PROJECTS

Erosion, transportation and deposition of sediment are processes that have occurred throughout geological times. Man's interference with the balance in nature has accelerated the rates of erosion taking place on our planet, in many cases to a hazardous degree. Sediment production in watersheds commonly range from 50 ton/km^2 to 1,000 ton/km^2 (1 metric ton = 1,000 kg), depending on the geology, climate, physiography, vegetation and other conditions of the watershed. The concomitant sedimentation problems that arise in irrigation projects depend on whether the project receives its water supply from a river where an upstream storage dam controls the river flow, or whether it is a *run-of-the river* project where a take-off structure diverts all or part of the river flow to the project area.

Sedimentation Control in Projects with Upstream Storage

Sedimentation occurs in reservoirs upstream of a dam constructed on a natural watercourse. The problem of concern is the rate of sedimentation in the reservoir, and the period of time that will elapse before the usefulness of the storage dam is seriously reduced. The annual loss of storage in reservoirs due to sedimentation ranges from a low rate of about 0.07% (giving the reservoir a half life of about 700 years), to an excessive rate of 5.0% (resulting in a reservoir half life of only 10 years). In the U.S.A., the replacement cost of this lost storage amounts to hundreds of millions of dollars annually. The engineer, in designing the capacity of the reservoir, must take the sediment problem into account. He must anticipate changes that might occur upstream of the reservoir (such as deforestation, urbanization, and industrialization), that would increase the rates of sedimentation and shorten the useful life of the project.

The construction of a dam *per se* alters the hydraulic characteristics of the river flow and the resultant sediment-transport capacity often creates serious problems both upstream and downstream from the reservoir (Gottschalk, 1964).

Deposition in the river bed above the reservoir gives rise to *aggradation* upstream that will reduce the upstream channel capacity. This often causes more frequent flooding of the adjoining plains as well as a rise in the local ground water tables. The extent of the aggradation depends on the river gradient, sizes of the sediment particles, and the degree of reservoir-storage fluctuation. Low gradient streams caused sediment to be deposited for almost 90 km upstream of the Imperial Dam, a comparatively low dam built in the Colorado River in the U.S.A. in 1938. In another typical case in the U.S.A., aggradation above the Elephant Butte Dam, during a thirty year period, raised the river valley and channel bed of the Rio Grande about 3 m above the high water level upstream of the reservoir for a distance of about 15 km upstream. As a result it became necessary to protect urban upstream areas from flood damages and from actual burial in the deposits. In addition, an extensive system of drains and ditches were required to control the drainage problems that arose in the flood plains.

The outflow from a reservoir is relatively free of sediment and consequently possesses a tractive force capable of initiating movement of materials in the downstream channel, causing *degradation* of channel bed. The rate of degradation depends on the type of material in the channel, as well as the hydraulic characteristics of the flow. Lowering of the main channel bed causes in a concomitant degradation of the tributary channels and may initiate a new cycle of head-ward erosion in the watersheds. In severe cases, ground water tables are lowered with resultant damage to agricultural land. This degradation of the channel bed, in many cases, undermines the foundation of bridge piers and other structures in the rivers.

Numerous cases of degradation have been reported from all over the world. The quantities of earth moved in the process often reach enormous proportions. For example, tremendous amounts of materials have been degraded from the river bed below the Hoover Dam on the Colorado River in the U.S.A. About 150 million m^3 were removed from the channel for a distance of about 150 km downstream of the dam during a 16 year period. The depth of the degradation ranged up to 5 meters. The process continued further downstream and in the next 45 km downstream, an additional 100 million m^3 were degraded.

Sediment Control in Run-of-the-River Projects

In these projects, the silt laden waters from the river are diverted directly into the main canal. These canals have smaller gradients than the river and because of their resultant low sediment transport capacities, large quantities of silt are deposited in the main canals and in branches of the network. The silt load entering the canal always jeopardizes the operation of the network because of the large quantities involved. On a 20,000 ha project, for example, the main canal inlet would divert about 10-20 m^3/sec of silt laden water from the river and the annual sediment load would amount to 50-200,000 ton/year. At least half of this amount would settle in the network and unless cleaned out every year, the canal would silt up completely.

Various types of equipment have been developed for removing silt from canals (dredges, drags, tractors, conveyer belt systems). The disposal of the excavated silt presents a serious problem. Unless waste land is available near the canal, the expense of transporting the material further increases the maintenance costs. Sediment removal from the network is an ongoing maintenance task in all projects and the annual costs of maintaining a network free of sediment problems ranges from \$20 to \$60 per ha per year.

The sediment load carried from the river into the canals presents a difficult problem to the engineer. First, he must ensure that the amount of silt entering the main canal is minimized and second, that the silt is not deposited near the headworks and so reduce the capacity of the entire network.

Furthermore, he must design stable canals that maintain their design sections. This is no easy task and a wide variety of approaches have been developed in an attempt to present practical solutions to the problem. Empirical solutions have been developed in several countries, mainly in Egypt, India and the U.S.A.; analytical solutions have been presented by scientists mostly in Europe and the U.S.A. The problem facing the irrigation engineer is to design a channel that, over time, is neither in a state of silting or scouring. The stability of the canal depends on a wide range of parameters that are continually in a dynamic process of change. These parameters include:

1. The nature and composition of the materials from which the canal has been constructed. These vary as the canal passes through ever changing soil profiles along its alignment. Further changes take place as the processes of aggradation and degradation occur in the canal bed itself.

2. The silt content (bed load, suspended load, sediment sizes, particle cohesiveness, surface drag in the canal bed and sides) varies widely from watershed to watershed. These parameters are also in a constant state of flux in any specific case and vary annually, seasonally, and even daily, depending on the nature of the flow in the main river.
3. The velocity of the flow of the canal depends on the canal's dimensions, shape, depth of the flow and hydraulic characteristics (slope, hydraulic radius, and roughness coefficient). These parameters are also changing constantly depending on the discharge of the canal as well as the seasonal nature of the vegetation growing on the side-slopes.
4. The stability of a canal is also affected by bends and curves in the alignment depending on the gradient, velocity, sediment load and discharge. When the canal section becomes unstable, local bank erosion and scour, and in extreme cases may lead to a break in the canal bank.

Because of the complicated dynamic nature of the interrelation of these factors, the solutions proposed both by the empirical and analytical approaches have varied widely. The design engineer must be experienced in choosing a suitable method for solving a specific case in order to ensure that the canal beds and side slopes will, with the passage of time, remain stable. He must also ensure that the canal head regulating structure is designed to minimize the amount of silt entering the canal (Punmia, 1981). The structure should include: (1) Undersluices upstream of the entrance to scour the silt deposited in the river above the approach channel. (2) An entrance crest-level to the regulator that is sufficiently high above the crest of the undersluices to prevent the heavy river bed loads from flowing into the canal. (3) A silt extractor (Punmia, 1981) which is a device to catch and eject part of the silt that has entered the canal. It is constructed at some distance downstream of the head regulator and consists of a horizontal diaphragm constructed above the canal bed that separates the top water from the bottom water carrying the heavier silt load. Tunnels below the diaphragm collect and transport some of the silt back to the river.

Downstream of the headworks regulatory structure, there is generally no control on the sediment flowing in the canal. The processes of aggradation and degradation give rise to deposition and erosion in the network. Unless the canals are maintained on a regular basis, the performance of the system is impaired, sometimes radically. In most networks, the sediments settle in the upstream sections and in those stretches where the velocity of flow is low due to low gradient or weed growth which block the section. When the deposits occur close to outlets, discharges through the outlets are reduced considerably and deliveries to distributaries and minor ditches fall below the anticipated design flow. In most projects in developing countries, the operation of deteriorating networks leads to mismanagement and corruption.

In order to overcome this problem and reduce the quantity of silt in the network, an ingenious outlet device called the Adjustable Proportional Module (APM) has been widely used in Punjab and Rajasthan in India (Punmia, 1981). Irrigation engineers who have not worked in the northwestern sector of India are seldom acquainted with the structure. The APM is designed to deliver a constant discharge from the parent channel, irrespective of the upstream level, within the working limits of the device. The main features of this outlet consist of (1) a horizontal crest near the upstream entrance of the structure and (2) a gradually expanding flume constructed at the downstream side of the structure. Flow through the outlet is super-critical and hence a hydraulic jump develops on the downstream side. The water level downstream therefore does not affect the discharge. The structures are sturdy, difficult to tamper with, and have proven satisfactory on an extensive scale. APMs not only divide the flow in fixed proportions, but also draw off form the parent channel *a fixed and fair share of silt.* This is an essential requirement for networks that carry significant silt loads and/or are designed as stable channels. Unless the outlets divert an equitable share of silt as well as of water, the silt loads in the system build up in the downstream reaches of the network, causing serious problems of sedimentation in the canals, as has occurred in numerous systems in many countries of the world.

THE GEZIRA PROJECT: A CASE STUDY IN MANAGEMENT AND SEDIMENTATION PROBLEMS

The Gezira project in Sudan highlights the problem of sedimentation and shows how the silting of the canals was a significant factor that considerably reduced the performance of a once successful large-scale project. It also shows how social, economic, political, health, population, management and environmental factors are all inter-related in irrigation projects (Plusquellec, 1990; Shanan, 1992).

The Gezira area in Sudan, lying between the Blue Nile and the White Nile, has a semi-arid climate with seasonal rainfall in summer. A major dam was constructed in 1925 to irrigate 400,000 ha of clay soils for growing cotton. Following the Independence of Sudan, the project was nationalized in 1950 and plans were initiated to expand the irrigated area to 840,000 ha. From 1960 to 1980, a program of agricultural intensification was implemented, but the net result was complicated by management difficulties, deterioration of the network, the rapid spread of cotton pests, critical labor shortages and the increase of waterborne diseases (malaria and bilharzia). Nevertheless, under a central management board, the scheme became the largest irrigated 'farm' in the world. In the mid 1970s, a program of diversification was introduced and summer crops, mainly sorghum and groundnuts, followed by wheat, were added to cotton that remained the major crop.

Agricultural labor requirements expanded as a result of increasing the irrigated area and extending the irrigation season. Population density increased from two persons per hectare in 1960 to three persons in 1980. Approximately 400,000-500,000 migrant laborers entered the area every year to participate in cotton picking and cultivation. The resulting labor shortages were aggravated by the spread of malaria in the mid 1970s: over 20% of the labor force was down with malaria in 1974. Increased silt and weed growth in the network became ideal snail habitats for the growth and transmission of bilharzia. Surveys in the early 1980s showed that 50-60% of the village population was affected by this debilitating disease.

At this point, the small farmers attempted to overcome the limitations imposed by the management of the project in operating the network. The centralized management board (Sudan Gezira Board) established operating rules, sanctioned the crops the farmers planted, and determined water allocations. The primary objective of the operating rules in the late 1970s was to foster cotton production in the project area. For this purpose, water was delivered on a fixed 14-day rotation schedule. Conflicts developed between the management's objectives and the farmers' interests. The farmers, instead of growing cotton as stipulated by the management, preferred to grow sorghum and generally ignored the government's rules. They found various ways of introducing modifications which enabled them to maximize their income but the bureaucratic management refused to recognize these innovations, that included giving high priority to irrigation of sorghum. Since labor was a serious constraint on production, the farmers developed cultivation and irrigation practices that minimized labor requirements not only during daylight hours but also at night—a practice that was contrary to the original project plan.

These innovations were the major factor in contributing to the intensification of agricultural production in the project, even though the management refused to recognize that farmers were not following the rules. Though water is the limiting factor in kToduction, farmers in spite of the management's plans, developed innovative practices in several important respects to cope with this constraint. Production in the project was quite good, but the top level of the bureaucracy was not informed of this, because the cropping system did not conform to the management's objectives.

In the original design of the project, the engineers used their ingenuity to store the night flows in the canal, so as to enable the farmers to irrigate only during the daylight hours. In most irrigation projects in developing countries, night irrigation is inefficient and large quantities of water are wasted. In the Gezira project, the canals with delivery capacities between 0.5 m^3/sec and 1.5 m^3/sec were over-dimensioned so that they could act also as night storage reservoirs. Bed widths were increased to 6 m, water depths to 1.8 m and regulator gates constructed at distances of 1 to 4 km. Below these reaches, water was delivered only during daylight hours. The night storage system operated successfully for more than four decades. In the mid-1960s, however, a backlog of silt and weed clearance started to clog many canals, particularly the nights storage reaches where velocities of flow were low. With the deterioration of the network, most farmers left the outlet gates open for 24 hours a day (instead of 12 hours) so as to compensate for the reduced deliveries.

Water distribution in the network became not only inequitable, but anarchy reigned. The farmers left field gates open in order to irrigate as large an area as possible. This continuous irrigation prevailed without farmers' attendance—particularly at night time. Because of the continuous irrigation practices, discharges were often well below full supply levels and deliveries 30-50% of design discharge. The original field outlet pipe gates discharging into 38 ha blocks were subject to theft and tampering. Most were replaced by drum bottoms, bags, local materials, or were not replaced at all. The infestation of the canals with weeds and the deposition of silt became the most serious problem of the project and the capacity of the canals was seriously reduced.

Water delivery in some areas was neither timely nor reliable and little water reached the tail-end farmers. Investigations of present system performance have revealed that the management information system is manipulated by the project staff and large discrepancies were found between the recorded water orders (indents) and actual deliveries. Further investigations also disclosed that there is little relationship between the observed values of gate openings and the values recorded by the project staff. Hired labor is still an important constraint to production—only 15% is supplied by family labor and the remainder comes from local or immigrant hired labor. Agricultural production is still disappointing. Average cropping intensity is about 60%—much lower than the 75% planned by the authorities. Yields of cotton, wheat, groundnut and sorghum are low and average about 20-30% of those achieved at research stations. The government now intends to embark on extensive weed eradication and silt clearance programs in order to improve equity and reliability in water distribution. These improvements would in turn increase agricultural production, but the farmers would still have little choice in deciding which crop they can grow. Lessons of the Gezira scheme include the inherent limitation of large scale irrigation projects, centralized authority and lack of response to markets, as well as sedimentation.

Since the mid-1980s a new stage has begun in the project area—privatization has begun to take over a large sectors of the project. Rich landowners have bought out the small 5 ha farmers and created private estates 5000 ha or more in size. The now landless small farmers have been replaced by cheap migrant laborers.

Planning and Management of Structured Irrigation Networks

The successful irrigation planner, if he wishes to bring about positive change in developing countries, must take fully into account the economic, political, institutional and human environment. His projects must be able to function in this environment, by recognizing that tradition, history, culture and religion define the social structure of the community. These attributes will only change gradually, if at all. The only effective innovation he can make, provided there is the political will for change in the government, is to foster the understanding that farmers must have inalienable water rights before they can develop confidence in themselves and in the irrigation projects. Only then will they invest in seeds, fertilizers, pest control measures and technologies and take the risks needed to move from traditional low productive peasant farming to high yielding efficient irrigated agriculture.

The case studies presented below, confined to India, have been taken from the author's book *Planning and Management of Irrigation Systems in Developing Countries* (Shanan, 1992). We have selected India because she has the largest and most ambitious irrigation program in the world on an area exceeding 55 million hectares (25% of all irrigated areas). She has been one of the leaders in irrigation science, particularly in developing countries, for the entire 20th Century. On gaining Independence in 1947, she set herself the target of establishing irrigation projects to protect the farmers from the ravages caused by recurring droughts that often devastated the entire rural population. By the 1980s, she had succeeded in meeting this challenge, and the menace of starvation is no longer the curse of India. The networks were designed so that the water could be delivered through thousands of kilometers of canals and ditches as rapidly as possible and spread over the greatest extent possible. Speed of delivery was the essence of the operation, not efficient water use. India has now set herself the target of rehabilitating many of these projects.

INDIA—A LEADER IN IRRIGATION TECHNOLOGY

Irrigation projects in India range in size from 'minor projects' with a command area of less than 2000 ha, to 'medium projects' from 2000 ha to 10,000 ha, and to 'major projects' serving areas greater than 10,000 ha. The command areas of many large projects exceed 200,000 ha and represent outstanding engineering achievements. Cost-benefit analysis of a proposed mega-project, the Sadar Sarovar Project in the State of Gujarat, is presented by El-Bihbety & Lithwick (Chapter 15, this volume). Huge and extensive projects have understandably given rise to many problematic issues, such as: high transmission and seepage losses in the canal delivery systems; difficulties in managing efficiently the distribution network; changes in local and regional water balances which have led to rising water tables, waterlogging and salinity. The construction of a major dam in some projects necessitated the government to relocate village communities to areas outside the planned reservoir lake.

The principal delivery systems of India can be distinguished as either 'northwestern systems', practiced in the States of Punjab, Haryana and Uttar Pradesh or as 'southern systems', characteristic for most of the rest of India. In both systems the government delivers the water to farmer service areas, 10-100 ha in size, and the farmers (5 to 100 in number) are responsible for distributing the water within this service area, locally called a 'chak'.

The northwestern systems were constructed to provide irrigation water from run-of-the-river canals primarily to protect farmers from long dry spells occurring during the wet monsoon season. These systems supply water intermittently to government outlets on rigid delivery schedules. They were designed to spread the available supplies 'thinly' over extensive areas by delivering minimum quantities of water to as large a number of farmers as possible. Irrigation intensities are low, approximately 30 % to 40% of a farmers land holding in each irrigation season.

Performance of the northwestern systems is comparatively high. The management is able to operate the networks according to the agreed objectives and the projects are characterized by relatively high efficiencies with equitable and assured deliveries. Water delivery schedules are fixed at the beginning of each season and the government implements them conscientiously. Farmers have introduced sophisticated cropping systems and, within the limits set by other constraints, use modern, advanced agricultural practices. Farmers are fully aware of their rights to a fair share of the assured supplies and defend these rights even if they have to use physical force against unauthorized water use. The conveyance systems are relatively simple, sections of the system are operated on a rotational intermittent basis, and outlets to farmer service areas are ungated (usually a proportional divisor which does not require adjustment). Within each block, a system of rotational water supply ('warabundi') ensures that each farmer receives his fair share of the available supply in proportion to the size of his land. The operating rules have been in practice for generations and are strictly maintained. Differential seepage losses in the minor network are taken into account in many projects to establish a farmer's allocated time, thus achieving more equitable allocations. The water available from the government supplies is not sufficient for all of a farmer's land and he is able to irrigate only approximately 30%-40% of his landholding each season. He is free to choose his crops and the proportion of his land on which he wishes to use his water; the irrigation administration does not interfere in his decision. Many farmers have developed local ground water resources to supplement the government supplies and can thus extend their irrigated areas and diversify their cropping patterns. These are essentially structured networks.

The southern systems ('shejpali'), on the other hand, are typical for most Indian States, and provide an interesting contrast to the northwestern systems. The southern systems are characterized by (1) large storage reservoirs upstream of the main headworks to store inter-annual and inter-seasonal river flows. (2) The canal systems are designed with numerous control and regulatory structures to deliver varying rates of flows in the supply network. (3) The projects are generally planned for relatively higher intensities of irrigation—often greater than 60% per season—and include a 'localized system', where particular areas are limited to specific crops such as paddy and sugar cane. Finally (4), the water distribution procedures are based on allocation rules of crop sanctioning where the authorities approve (in theory, at least) a farmer's cropping pattern which is supposed to conform to the overall plan of the project. The project management is supposed to deliver water according to the requirements of the approved cropping patterns. These are not structured systems. They are demand systems, theoretically aiming to supply the needs of individuals or groups of farmers.

Performance of the southern systems is usually low, in contrast to the northwestern systems. This is reflected in low delivery efficiencies (generally less than 30%), while the total irrigated area seldom exceeds 60% of the planned target command. Most of the available supplies are controlled by the headstream farmers, who use and waste large quantities of water. Consequently, tail-end holdings suffer from insecure deliveries.

During the 1970s the Government of India introduced a diversity of innovative components into many irrigation projects to improve project performance. These included programs for improved organization and maintenance, canal lining, land consolidation, drainage works and watercourse contruction. In many projects, rotational water supply programs below the government outlet were combined with other components. In most projects, however, inequality in water distribution and wastage of water remained central issues. Head-reach areas continued to receive more than their allocated supplies, while tail-end zones often remained unirrigated. In a series of follow-up programs, the government established several water management studies to evaluate the reasons for this poor performance and to propose improvements. Studies were conducted under the direction of the author in the Chambal River Project in Rajasthan and Madhya Pradesh, and in the Indira Gandhi Project (Shanan, 1992). The studies were planned and organized hy the World Bank for the United Nations Development Program (UNDP) and the field work was carried out by Government Irrigation Department staff. A review of one project is presented in the following section.

THE INDIRA GANDHI NAHAR PROJECT—A STRUCTURED NETWORK

The Indira Gandhi Nahar Project is an ambitious plan to settle a large population on new lands in a desert zone by irrigating about 750,000 ha on the northwest system principles. The main canal, some 600 km in length, delivers water from tributaries of the Indus River according to water allocations determined seasonally by an interstate Management Board. Construction of the project was divided into two phases, with Stage I starting in the 1950s to develop 540,000 ha. Project works initiated in 1974 concentrated on intensifying development of 200,000 ha in this area by promoting infrastructure development, on-farm works, watercourse lining, and road construction. Progress was satisfactory and there was a continuous improvement in yields, cropping intensities and production.

In 1979 Stage II of the project was initiated to extend development to an additional 200,000 ha. Although the network had been designed and constructed with a system layout suitable for distributing water according to the northwest practices, many inequities had developed. Stage II, therefore, included a Water Management Area (WMA) with the object of demonstrating, on an extensive area, ways in which optimum irrigation benefits could be achieved.

In 1980 the Narangdesar Distributary at the head of the project was selected as suitable for the Water Management Area. This distributary, about 40 km in length, serves an area of approximately 30,000 ha. In the first year the network was upgraded, some 180 ungated divisors were constructed, regulating structures improved, linings repaired, canals cleared of silt, and all outlet and control structures calibrated to deliver design discharges. The ungated divisors, known us Adjustable Proportional Modules (APMs), described previously, deliver constant discharge from a parent channel.

It was recognized that fluctuations in deliveries in the main canal affected discharges to the Narangdesar Distributary. In order to evaluate the effect of these variable flows in the past, historical cropping patterns were studied by stratifying the area served by the distributary into three sub-zones: head, middle and tail. The analysis showed that: (a) the number of delivery days to each zone decreased from head to tail, with head-reaches receiving 40% more flow days than tail-reaches; and (b) the decreasing availability of water was reflected in the cropping intensities, with head-reach intensities averaging 35% more than tail-reaches. A further analysis of the sub-zones showed that in each one, cropping intensities of upstream farms exceeded those in tail-end farms also by approximately 40%. In the past, tail-end farmers had received about half the quantity of water that was delivered to headstream farms, despite the fact that the entire network was lined. In order to improve water distribution in the Water Management Area (WMA), the following operational rules were specified and agreed upon:

1. Seasonal allocations to the area would be established 4 weeks before the start of each season, depending on availability of supplies and according to the estimates of the inter-state Management Board.
2. Seasonal allocations would be translated into weekly ON/OFF schedules, so that when the network was operating it would deliver at full supply level.

3. ON/OFF schedules would be finalized in consultation with the Department of Agriculture, which would notify the farmers, through the Extension Service, of the programmed delivery weeks.
4. The operation of the network would take into account filling and emptying times so that farmers would receive multiples of 7 days of deliveries.
5. Cross-regulators, regulating gates and flow-measuring flumes would be maintained in good working condition at all take-off points, to ensure that downstream reaches receive their equitable allocation.
6. All inlets to farmer service areas would be of the ungated APM type to minimize the size of the operating staff and also to ensure accurate delivery and minimum interference by farmers.

During 1982 and 1983 the Water Management Area (WMA) was operated successfully as a fully structured network, first on approximately 7000 ha and then on 22,000 ha. By 1984 it had been extended to 35,000 ha, beyond its original target of 30,000 ha. The main achievements of the WMA were as follows:

1. Fluctuations in the flows to farmers were reduced from 50% of design flow to an acceptable level of approximately 10%.
2. Seasonal water deliveries to the farmers were reduced by approximately 15% and made to conform to authorized project allocations; all farmers preferred the improved schedules to those that had been in operation previously.
3. The operating staff required to manage the Narangdesar Distributary included one gauge reader and one ditch-rider per 2500 ha, at an annual cost of less than $1 per ha, a reduction of 80% in staff.
4. Despite the reduction in seasonal water deliveries, yields and intensities increased significantly in the WMA; for example, cotton yields were approximately 30% higher in the managed areas than in a comparable non-managed area. At the same time, irrigated intensities were 65% in the WMA versus 47% in a non-management zone.
5. Annual net returns were estimated to be $300/ha higher in the WMA than in a non-management zone.
6. The operating staff quickly became adept at planning the delivery schedules and full cooperation developed among the Extension Service, the irrigation staff and the farmers.
7. Predetermined schedules proved popular with *all* farmers in the WMA and farmers outside this area began demanding similar services.
8. Equity in water deliveries was introduced to an extensive area without raising conflicts between tail-end and headstream farmers.

The WMA was a marked success due to the diligent work and cooperation of the engineers and the Extension Service. Because of its success and the demand of the farmers in other areas, the project management in 1984 began to apply the principles to an additional 40,000 ha. The methodology of delivering water according to predetermined schedules had proven simple, inexpensive and practical, and achieved equity and high levels of return to farmers. Predetermined schedules had been criticized originally on the supposition that they might set limits to farmers' options. However, a detailed review showed that cropping patterns in the WMA vary widely from farm to farm. This indicated that, although the predetermined schedules were *inflexible* during a particular season, farmers selected their cropping systems from a mix of many compatible crops. The 'fixed' schedules did not prevent them from adjusting their programs to their resource constraints (soil, capital, land, labor, equipment *etc.*). They were able to remain below their preferred levels of risk aversion and achieve high economic yields. Finally, the improved performance of the WMA area was not achieved at the expense of the rest of the system. The water delivered to the WMA conformed to the official seasonal allocation, and in no way infringed on the deliveries to the rest of the project area.

The Israeli Experience

The planning, management and operation of the water supply system and the irrigation network of Israel is an example of a successful attempt to establish a modern irrigated agricultural sector on a national scale and at the same time minimize that rate of development of deleterious environmental hazards resulting from this program. The policies, water laws, planning and operating criteria that enabled this to be achieved are reviewed briefly below.

The State of Israel was established in 1948 and covers about 21,000 km^2. The country started with a stunted industrial base and few raw materials. Agriculture was based on minimal financial investment, cheap manual labor and consisted mainly of rainfed hillside cultivation (orchards and vineyards), subsistence farming in the plains (grains, sunflowers) and some irrigated orange groves. The government decided to harness the enterprising spirit of its farmers and develop the potential of the new immigrants to push agriculture form its typical 'felaheen' (peasant) farming to a modern, high-investment and productive sector.

The construction of the Natural Water Carrier was completed in 1964, enabling agriculture to focus on new irrigation technologies combined with high yielding varieties, fertilizer use and pest control measures. Extensive irrigated orange groves, subtropical fruit orchards, cotton, commercial fish ponds, export flowers, vegetables and green-house agriculture are some of the outstanding features of present day agriculture in Israel (Shanan & Berkowitz, 1995).

THE 'WATER LAW' AND FARMERS' WATER RIGHTS

From the inception of the State of Israel, water was recognized as a scarce national resource. The first Prime Minister, David Ben-Gurion, considering that more than half of Israel is desert, declared: "*We must conquer the desert, lest the desert conquers us.*" Water would have to become a nationalized resource, owned and regulated by the government. The problems of security and the differences between the social and private costs justified the introduction of a comprehensive water supply system. The 'Water Law' was adopted in 1959 to grant the government the powers to control water development on a national scale. This law, which formed the basis for all irrigation development in Israel, stated: 'Israel's water sources are public property, controlled by the State and devoted to the needs of its residents and the country's development ... A person's right on land does not grant him rights on a water source located on that land or passing through it."

The Water Law thus enabled the government to control both surface water and groundwater and Israel became a forerunner in regulating these two sources in an integrated national plan. Control over existing private wells was considered necessary to avoid over-exploitation that would inevitably occur as private costs do not reflect the social costs of replenishment, aquifer pollution and the value of opportunities foregone elsewhere. People who had been users of water (both surface water and groundwater) prior to the promulgation of the Water Law were given rights to continue using these waters contingent on their beneficial use.

The major objective of the Water Law was to exploit the water resources by allocating water rights to consumers for beneficial use. Another objective of the Water Law was to establish water prices and maximize the economic contribution of irrigated agriculture to the national wealth. In line with the egalitarian approach of the government to the development of all regions of Israel, especially border areas and arid zones far from water sources, the Water Law provided for the establishment of an equalization fund "to reduce the difference between water charges in different parts of the country." The fund was financed by levies on water allocations to users. The main instrument for ensuring equitable allocations was based on granting non-transferable quotas to users. Under the law, a relatively low rate was charged for a particular quotum, but higher rates (up to triple the basic rate by using a step-function principle) were charged for the amount of water used in excess of the quotum. The higher rates served as a deterrent to non-compliance.

The Water Law established a Water Commissioner's Office responsible for implementing the policies. The National Water Carrier was designed as a pressure pipe system, so that all deliveries could be recorded with reliable standard water meters. They monitored bimonthly quotas, taking into account differential evapotranspiration rates. In this way, water use in agriculture was made a function of crop requirement and the penalties for exceeding the quotas were imposed bimonthly.

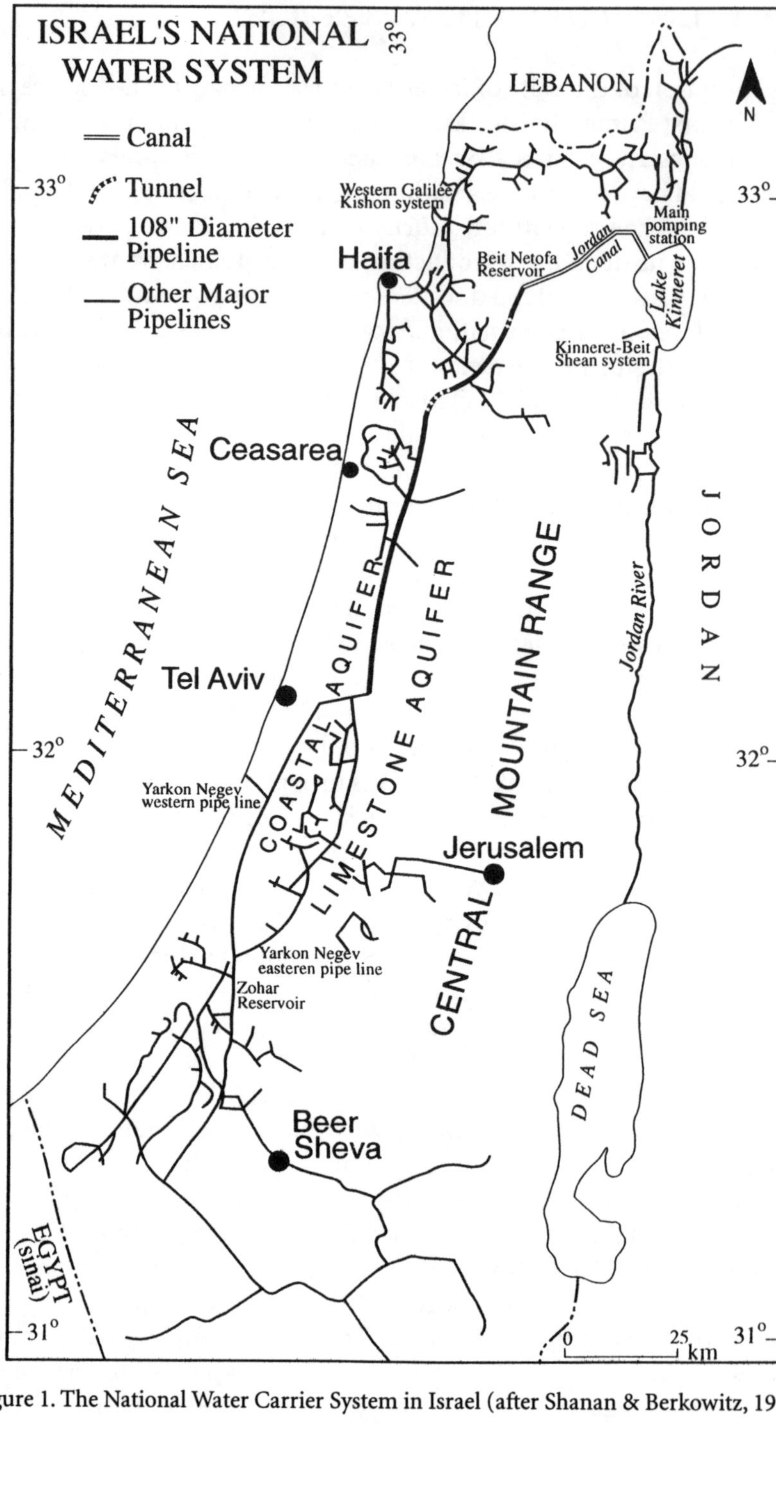

Figure 1. The National Water Carrier System in Israel (after Shanan & Berkowitz, 1995).

These policies acted as a strong incentive for efficient use of water particularly in the season of peak demand.

THE NATIONAL WATER CARRIER SYSTEM

The overall water plan for Israel integrated all the reosurces into a comprehensive scheme so that (a) water could be conveyed from regions of excess in the north to regions of scarcity in the south and (b) the system would have operational flexibility so that surface water and groundwater could be stored in aquifers and transferred interregionally. This storage transfer and distribution of water is carried out by the National Water Carrier (Figure 1). Part of this carrier consists of an open canal but the main networks comprise pre-stressed concrete pipes 70 to 108 inches in diameter. About 400 million cubic meters (MCM) of water are pumped annually from Lake Kinneret (210 m below Mediterranean Sea level) to an evaluation of 152 m from which point it flows by gravity to the Coastal Plain. The central mountain areas and regions further south receive water by additional pump lifts. In years of good rainfall, surplus water in Lake Kinneret is transferred to the south and stored in the Coastal and Mountain Aquifers. In years of below average rainfall, these recharged aquifers can be used to supplement shortages that may occur in the south. The carrier delivers water to local systems rather than supplying individual consumers directly from the main system. Its peak delivery is 20 m^3 per second. The project comprises pumping stations, pipelines, canals, tunnels and operational reservoirs. Construction commenced in 1953 and was completed in 1964. The length of the system's main conduit is 130 km, and the capital investment in the project was valued at the time at US$ 147 million (Shanan & Berkowitz, 1995).

WATER DEMAND AND SUPPLY

The short-term water potential from the present network is estimated to be about 2,020 MCM per year. Groundwater represents about 60% of this amount and the aquifers, together with Lake Kinneret, have an operating storage capacity equivalent to about a one-year supply. This allows for considerable interannual and interseasonal flexibility.

The short-term demand which can be met with the present engineering structures is about 2,090 MCM, supplied from the combined use of three sources: fresh water, saline water and the reuse of treated sewage. Agricultural demand is about 56% of the total demand, receiving about 63% of its requirents from fresh water sources, 27% from saline sources and the remaining 10% from treated sewage projects. However, as the population continues to grow, fresh water allocations to agriculture are continuously declining, while the use of treated wastewater increases. Between 1950 and 1990, while the amount of land under cultivation increased by 175% to reach about 437,100 ha, there was a concomitant 550% increase in the irrigated area which expanded to 205,700 ha. The overall average water allocation for the irrigated areas is about 575 mm per year, measured at the head of the system.

INTEGRATED PLANNING AND OPERATION OF THE SYSTEM

Optimum management strategies for the short-term and long-term use of water resources were determined by combining a systems analysis approach with modelling techniques. Political and economic policies and alternative water allocations are incorporated into the models in the form of targets and constraints. The benefits foregone by using a specific strategy by a particular sector instead of its next best alternative can be evaluated by a model. In this way, the economic costs of constraints are calculated and submitted to the political and economic decision makers (Frankiel & Goodall, 1978).

The model has undergone continual improvement so as to include new problems and aspects which were not considered during the earlier stages of the national water plan (such as water quality, salinity and groundwater pollution). The analysis also evaluates the costs of developing alternative resources, using aquifers for storage and controlling the growth of water demand. The objective function maximizes (at various probability levels) the present value of the net benefits resulting from a specific water supply condition. The model also simulates the effects of variable seasonal allocations to economic sectors and regions taking into acccount different water resources development levels. The economic costs of institutional constraints are determined so that proper weights can be given to them to simulate anticipated levels of production, employment and income in the rural regions.

RESEARCH AND EXTENSION

Agricultural research in Israel has been closely associated with the efficient use of water. The Agricultural Research Organization (ARO) of the Israel Ministry of Agriculture established field research stations in all the climatic regions of Israel, in order to deal with local problems. A well-organized and active extension service closely linked to these stations and to the farmers succeeded in:

1. Training new immigrants with no previous agricultural experience to become excellent farmers within a decade.
2. Developing new irrigation techniques suitable for intensive and extensive irrigation in the private sector and cooperative farming sector.
3. Introducing new high-yielding crop varieties which are efficient water users.
4. Encouraging the introduction of greenhouse agriculture, primarily for export flowers and vegetables.

The extension service developed a special methodology for training farmers, which later became known as the 'Training and Visit System' and which, over the last two decades, has been adopted either implicitly or explicitly by many developing countries on the recommendation of the World Bank. Dynamic changes have taken place in the extension service. In the 1950s, extension service officers, usually high school graduates with some general farming experience, could easily keep ahead of the less-educated immigrant farmers. By the 1980s, farmers had acquired considerable experience and many of the new generation held academic degrees. Consequently, the extension service is today staffed by highly qualified personnel with academic specialization as well as practical experience.

Greenhouse agriculture is an important aspect of Israel today and many farmers attain record yields. For example, tomato plants are produced in individual pots and can develop as creepers reaching 15 meters in length and producing up to 320 tons/ha. Arab farmers in Israel have also moved from traditional basin and furrow irrigation to sprinkler and drip systems, and many are also greenhouse farmers. It is worth noting that in the last decade, these advances have 'crossed the border' into the administered territories and sprinkler, drip and greenhouse schemes are today common features in these areas.

Water has always been a constraint limiting farmers' income and they endeavor to use the resource as efficiently as possible. The changes that have taken place in the use of water for cotton is a good example of significantly increased water use efficiency. In the 1960s, production efficiency was about 0.5 kg cotton per m^3 and by the 1980s it had reached 1.0 kg/ m^3. Similarly, in the commercial fish pond sector water use efficiency during the same two decades rose from 0.17 kg fish per m^3 to about 0.63 kg fish per m^3.

FUTURE PROBLEMS AND PROSPECTS

Israel, like other advancing industrialized countries, faces environmental problems related to agricultural development and urbanization. Water quality records show that Israel must manage its groundwaters skillfully so as to prevent hazardous consequences developing, particularly in the Coastal Plain. The combination of a shallow phreatic water table, sandy soils, dense populations, and excessive pumping have caused a rise in the salinity and pollution levels of this aquifer. Using the results of simulation models, groundwater abstraction has been controlled and groundwater has been recharged with surface water. The aquifer is being managed carefully so that water quality will conform to acceptable standards.

Economists hold that government policies must be motivated by two objectives —increasing national wealth and improving income distribution. Israel's development programs in the water resource and irrigation sectors have conformed to these aims by using a comprehensive planning process which took into account the limited resources, continually changing demands, pollution parameters, and social and political variables. Israel has demonstrated that with a democratic structure, planning combined with a political will allows vested interests to compete in a mixed economy and, despite limited resources, enables satisfactory rates of economic growth to be maintained. The Israel Export Institute has reported that all agricultural exports showed an increase from US$55 million in 1962 to US$1,500 million today. Gross National Product of Israel rose tenfold in 40 years to reach about US$ 16,000 per capita per year in 1995, equivalent to a mean annual rate of growth of about 6%, and it was achieved despite the hostile political situation prevailing in the region.

As the 21st century approaches, it is clear that Israel will have to make adjustments to its priorities in water use. With the anticipated increase in population and the expansion of industry, some of the fresh water supplies at present being used by agriculture will have to be allocated to domestic and industrial use. Agriculture will have to adjust its production targets by reducing the extent of the areas at present targeted for relatively high water users (fish ponds, cotton, bananas, *etc.*) and by replacing them with more efficient water users (sunflower, greenhouse agriculture, *etc.*) as well as by expanding the program for the reuse of sewage effluents. It is also expected that early in the 21st century, the price of desalinated sea and brackish water will allow it to become a new economic source of water for municipal and domestic supplies.

Towards the 21st Century

In looking to the future we must differentiate between the problems and the possible solutions for industrialized developed countries and the developing countries. In the developed countries, less than 8% of the population is engaged in agriculture and family farm net incomes are $10,000 per year or more. On the other hand, in developing countries, more than 65% of the population depend on agriculture for their livelihood, and net family farm incomes are often less than $500 per capita per year.

The successful Israeli experience followed basically the same pattern of strategies as the Newly Industrial Economies (NIE) in south-east Asia: Singapore, South-Korea, Hong-Kong and Taiwan. Israel inherited an agricultural base, typical of British Colonial rule: peasant agriculture with large landowners and tenant farmers. With Independence in 1948, Israel set her national priorities: education, equality for women in the work place, an egalitarian work ethic (especially for labor), controlled import and export policies, high levels of savings, subsidies for agriculture so as to replace low productive cheap labor farming with a much smaller but highly skilled and motivated educated agronomist (in both the private and the collective sectors). Israel thus fostered rapid advance into modern agriculture that incurred investments, credits, marketing and new technologies aimed at the export markets whenever possible. In Israel however, unlike the NIEs in south-east Asia, the U.S.A. was the role model, not Japan.

A tremendous responsibility now falls on the shoulders of the rich industrial countries and the International Agencies who must recognize that the 'losers' in developing countries are hampered by their present circumstances. Most depend on exporting food and raw materials, sugar, timber, coffee, bananas and even drugs. Much international aid goes into these developing countries. But when the economic balance-sheet is examined, we find that far more money flows out of the impoverished regions of Africa, Asia and South America into the rich countries, than is invested by the latter into the former. About $43 billion a year of capital is transferred back as interest payments, profits, capital flight, royalties and fees (Kennedy, 1993). Hopefully, the world will use its available resources more efficiently in the coming century.

The lessons learnt from the NIEs and from Israel *vis-à-vis* the improvement of irrigation projects and agriculture in developing countries are:

1. Irrigation farmers must have water rights that ensure secure and timely supplies.
2. Under the prevailing socio-economic conditions of developing countries, delivery networks must be simple to operate, and difficult to corrupt.
3. Management is only one link in the chain of improvement that must be introduced into irrigation projects in developing countries in order to raise performance. It is certainly not the first link in the chain. Management cannot replace 'mortar and bricks' and no amount of skill can improve the performance of an irrigation project if the supplies in the main network are unreliable, fluctuating, control gates mismanaged and corruption rife.
4. Networks must be structured physically so that they serve a hierarchy of administrative subdivisions (districts, villages) each with its specific project controlled delivery point, and the management structured with a devolution of responsibilities so that a manager is responsible for a *single* village or district that forms a legally established unit from the point of view of government administration and law enforcement.
5. A strong law enforcement policy must be supported by a government to ensure the 'transparency' of the management and the rights of the farmers.
6. Agriculture must be subsidized because farmers net incomes are generally below the poverty line.
7. The rate of migration from canal areas to the cities must be controlled and be a function of the rate of industrial development in the metropolitan areas.
8. The younger generation in the rural areas must be trained and educated to high technical levels so that their generation will in time be able to advance the agricultural practices from the traditional cropping methodologies to those appropriate for the 21st Century.

9. As population growth stabilizes and rural communities form smaller proportions of the society, farm sizes can be increased because it is an impossible mission to expect millions of farmers to make more than a subsistence living on small 1-2 ha irrigated plots.

References

Frankiel, F.M., and D.W. Goodall (eds.) (1975) *Simulation Modelling of Environmental Problems, Scope 9*. John Wiley & Sons, New York, pp. 1-21

Gottschalk, L.C. (1964) Reservoir sedimentation. *In* Ven Te Chow (ed.) *Handbook of Hydrology.* McGraw-Hill Book Company, Toronto, pp. 17.1-17.33.

Horst, L. (1983) Irrigation Systems - Alternative Design Concepts, Overseas Development Institute (ODI), Agricultural Administration Unit, Network Paper No. 7c, London, pp. 1-88.

Kennedy, P. (1993) *Preparing for the Twenty-First Century.* Harper Collins Publishers, London.

Plusquellec, H. (1990) The Gezira Irrigation Scheme in Sudan: Objectives, Design and Performance. World Bank Technical Paper No. 120. Washington, D.C., pp. 1-90.

Punmia, B.C. (1987) *Introductory Irrigation Engineering.* Standard Publishers Distributors, New Delhi.

Shanan, L. (1987) The impact of irrigation. *In* M.G. Wolman and F.G.A. Fourier (eds.) *Land Transformation in Agriculture.* SCOPE 32, John Wiley & Sons, New York, pp. 115-131.

Shanan, L. (1992) Planning and Management of Irrigation Systems in Developing Countries. *Special Issue: Agricultural Water Management Journal* 22(1,2). Elsevier Science Publishers B.V., Amsterdam.

Shanan, L., and S. Berkowitz (1995) The Context of Locally Managed Irrigation in Israel: Policies, Planning and Performance. *Short Report Series on Locally Managed Irrigation* 1D:1-21, International Irrigation Management Institute (IIMI), Colombo.

Vermillion, D. L. (ed.) (1996) The Privatization and Self Management of Irrigation. Final Report, International Irrigation Management Institute (IIMI), Colombo.

ABOUT THE AUTHOR

Leslie Shanan has been a consulting engineer since 1950 for water and agricultural development projects. In 1982, he was appointed Adjunct Professor of Hydrology in the Department of Geography of the Hebrew University of Jerusalem. Since 1970, he has been a consultant to numerous missions of The World Bank and other international agencies. His principal contributions to the ecology and hydrology of desert environments were summarized in *The Negev: The Challenge of a Desert,* co-authored with M. Evenari and N. Tadmor (Harvard University Press, 2nd edition, 1982). He presented his methodology for designing and evaluating irrigation projects in *The Planning and Management of Irrigation Systems in Developing Countries,* by L. Shanan (Elsevier Scientific Publishers, Amsterdam, a special issue of the *Agricultural Water Management Journal,* October 1992).

Professor Leslie Shanan
P. O. Box 2188
Savyon 56530
Israel

Chapter 14

Drought and Water Management in California

Maurice Roos

Introduction

The past 30 years have seen rapid economic growth in California with increasing demands for developed water supply. Population grew from nearly 16 million people in 1960 to an estimated 30 million in 1990 and to 32 million in 1995. Irrigated acreage was about 3.3 million ha in 1960 and rose to a peak of about 3.9 million ha in 1981, then declined to about 3.7 million ha in 1990. Many large reservoirs and water development projects were built during the 1960s and 1970s. Rising concern about the effects of growth on the natural environment led in the 1980s to a slowdown in water development, and seemed to eliminate new on-stream reservoirs as supply options. Public desires to restore some of the environmental losses led to an increase in demand for water for environmental purposes, especially fisheries and waterfowl, partially at the expense of previous agricultural and urban supplies. The six year drought from 1987–1992 brought further strains and probably accelerated the trend for more efficient water use (water conservation), interagency cooperation, and water marketing. The difficulty of developing new water supplies and the increasing competition for water by urban, agricultural, and environmental interests have placed a premium on increasing the efficiency of current water use through conservation, recycling, and conjunctive use of surface and ground water. However, these measures cannot meet all water needs; continuing population growth will require new water supplies for future needs.

Climate Factors

Most of California enjoys a Mediterranean climate, with cool, wet winters and warm, dry summers. The state, which extends across nearly 10 degrees of latitude from about 32.5° to 42° north, is near the southern margin of the prevailing westerly

H. J. Bruins and H. Lithwick (eds.), The Arid Frontier, 277–293.

wind belt, a region between approximately 30° and 60° north latitude where a continuing series of cyclonic storms progress from west to east producing periodic rainfall. To the south is a zone of semi-permanent high pressure areas with descending warm, dry air. The atmospheric high pressure area which affects California is known as the Pacific High. The global zones of weather shift with the seasons. Much of the year California is in the high pressure belt which accounts for the fair weather and lack of precipitation during the summer. During the winter season, the storm belt shifts southward to occasionally place the state under the influence of Pacific storms which bring vitally needed rain and snow.

Most of California's moisture originates in the Pacific Ocean to the west and southwest. Storms with a long southwesterly fetch generally produce more precipitation (sometimes floods), because these storms tap air with higher moisture content originating over warmer water. As moisture laden air is transported over mountain barriers, such as the Sierra Nevada, the air is lifted and drops additional rain or snow on the westerly slopes. The mountain induced (orographic) precipitation is very important for water supply. For example, the 1,600 m high Blue Canyon weather

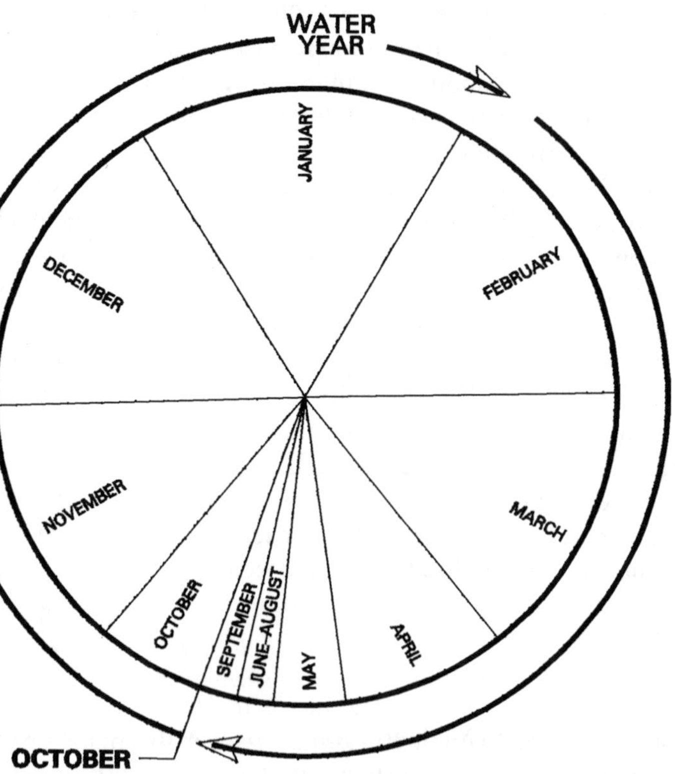

Figure 1. Monthly precipitation distribution in the Sierra Nevada.

station, situated about 100 km northeast of Sacramento, averages 1,600 mm of precipitation a year, about 3 1/2 times the 450 mm expected at Sacramento in the middle of the Central Valley.

The direction of orographic wind flow is important. The greatest amount of water is wrung out when wind flow is at right angles to the mountain barrier (that is, from the southwest for the Sierra Nevada). A more southerly wind, such as occurred frequently during water year 1992, the last year of the recent 1987–1992 drought, is not as productive.

Normally, some 5 to 7 major winter storms occur during the wet season. Each one drops 25 to 50 mm of rain in the Sacramento Valley and corresponding equivalents (80 to 160 mm) of rain and snow in the Sierra. A shortfall of a couple of major storms causes a dry year; conversely a couple of extra storms or an extended stormy period produce a wet runoff year. An unusually persistent Pacific High over California during the three wet mid-winter months (December through February) predisposes the year toward the dry side. On average, half the year's annual precipitation occurs during the December through February period. Three-fourths occurs during the five-month November through March period (see Figure 1).

Water Resources Management in California

The wetter North Coast and Sacramento River hydrologic regions in northern California contribute most of the runoff (Figure 2), but the need for water is greatest in the central and southern portions of the state. Three-fourths of the state's 87 billion m^3 of average natural runoff originates north of Sacramento, the state capital; about three-fourths of the need for urban and agricultural water is south of Sacramento. Agriculture is important in California with over 250 crop and livestock commodities. The total value of agricultural production is about one-tenth that of the entire United States. Around 3.7 million ha of land are irrigated (CA DWR, 1994). Irrigated acreage is expected to decline slightly, perhaps 5% by the year 2020, as population increases from 30 million in 1990 to a projected 49 million. Environmental demands of more freshwater for fisheries and wetlands are rising, partly at the expense of agriculture. Water conservation and other water management measures can stretch supplies (about 4% when existing reuse is taken into account) but these tools cannot solve the problems associated with major shortfalls.

Evaluation of true water savings by programs fostering water conservation requires a look at the entire hydrologic system. Fresh water is lost when tail water, deep percolation, or sewage goes to saline underground or surface water bodies. In most interior areas tail water or deep percolation is not lost but returns to usable surface or ground water supplies. In California, the future potential savings in net water demand has been estimated to be around 4% of total agricultural and urban water use (CA DWR, 1994).

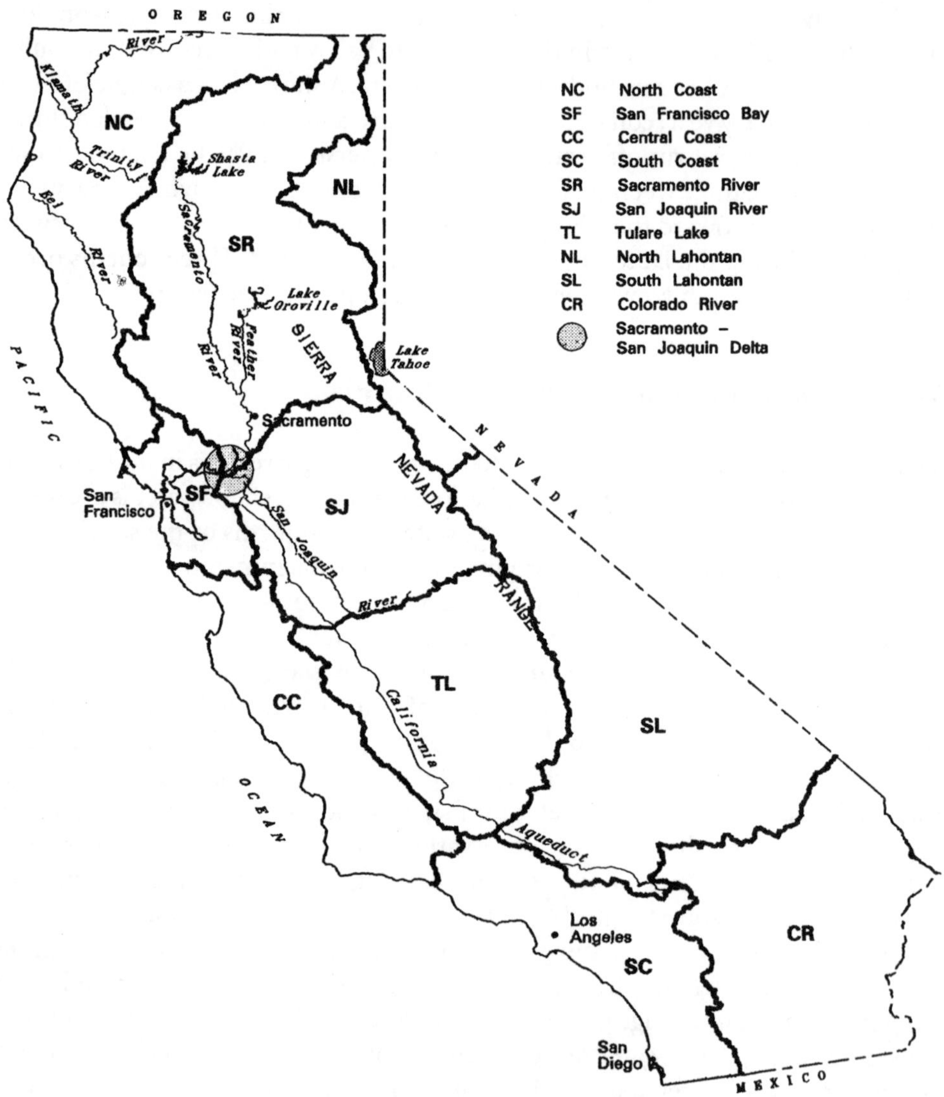

Figure 2. Hydrologic regions of California.

TABLE 1. Estimated California water demand in 1990 in billions of cubic meters ($m^3 x 10^9$)

Category of Use	*Average Year*	*Drought Year*
Urban	8.4	8.8
Agricultural	33.1	34.8
Environmental	35.0	20.2
Other*	1.8	1.8
Total	78.3	65.6

*Includes major conveyance facility losses, recreation uses, and energy production.

Estimated total net water demand[1] (water needed after allowing for reuse within service areas of return flow and deep percolation to usable ground water) within California in 1990 for average and drought conditions is shown on Table 1 (CA DWR, 1994).

Water demand is somewhat higher during drought because less rainfall is available to replenish soil moisture in the root zone during the rainy season. The drier soil means a greater demand for irrigation water during the growing season. A large portion of the environmental needs, about 22 billion m^3, is natural runoff to the ocean from North Coast streams which are designated wild and scenic rivers, left in their undeveloped state. That runoff is much reduced in a drought year and hence these water demands are less (by about 12 billion m^3). The wild and scenic riverflow doesn't impact reservoir storage. Other major environmental water components are freshwater outflow from the Sacramento-San Joaquin River Delta for salinity repulsion and fisheries, water for wildlife habitat areas, and instream fishery flows below dams.

Water development in California is based on the concept of storing water during the wet season and in wet years for use in drier periods with long distance conveyance to areas of need. Major water transfer projects have been built to convey water from areas of plenty to the drier central and southern portions of the state. The largest such facility is the California Aqueduct which extends some 700 km from the Sacramento-San Joaquin Delta to the South Coast region. The tidewater region of the Delta is the hub of water transfer; water quality there can be affected by salinity intrusion from the ocean via San Francisco Bay.

Large reservoirs in the north provide water storage and flood control. Water is then released downstream to be exported by large pumps in the southwestern corner of the Delta. The large reservoirs are multipurpose projects built to provide flood control during the wet winter season. Flood control requirements are eased during the spring, and if all goes well, the reservoirs fill from spring snowmelt. During low flow months, additional water is released to repel ocean salinity in the western Delta as a hydraulic barrier to preserve suitable freshwater quality within the estuary, to maintain good quality water for export, and to provide proper migrating fish habitat. Total export capacity of the combined federal and state water facilities is over 8 billion m^3 per year. However, exports have been limited to an average of 6 billion m^3 the past several years due to lack of regulated supply into the Delta and export pumping constraints to protect threatened or endangered fish species. This is approximately one-fifth of all the current (1995 level) water use in Central and Southern California, including the San Francisco Bay area.

Recent Droughts in California

For the second time this century, 6 years of drought gripped Northern and Central California during 1987–1992. The drought was broken in most parts of California by the wet year 1993, in which runoff was 125% of average. But a relapse occurred in 1994, which was again critically dry, raising fears that the drought had resumed. The formally designated 'drought watch' of 1994 was finally washed out to sea in the two large flood events (January and March) which made 1995 one of the wettest years this century and refilled all but a couple of the state's major reservoirs. The following pages present the hydrologic facts of the recent drought and include some comparisons with other droughts (Roos, 1995).

TABLE 2. Percentage of average precipitation

	Water year						
	1977	1987	1988	1989	1990	1991	1992
Statewide	45	61	82	86	69	76	86
Northern Sierra	38	57	70	101	72	65	72

PRECIPITATION AND RUNOFF DURING THE 1987–92 DROUGHT

Statewide precipitation was below average in each year of the above drought. Statewide and Northern Sierra percentages are listed in Table 2. Water year 1992 (which extended from October 1, 1991, through September 30, 1992) did produce well above-average precipitation across the southern third of California. However, amounts were light across the northern third of the state and especially in the Sierra Nevada. As a result, the statewide precipitation average was 86% and the runoff even lower at 43% of average.

In 1989, when Sacramento basin runoff (Table 4) was 80% of average, northern basins were near normal and the southern portion of the state was dry. Water year 1977, which was the driest single year of record, is also shown on the next four tables.

TABLE 3. Percentage of average April 1 snow water content

	Water year						
	1977	1987	1988	1989	1990	1991	1992
Statewide	25	55	30	75	40	75	60
Northern Sierra	25	55	20	80	35	65	50
Southern Sierra	20	50	25	70	45	80	60

TABLE 4. Percentage of average runoff

	Water year						
	1977	1987	1988	1989	1990	1991	1992
Statewide	20	48	48	72	45	43	43
Sacramento River	28	50	50	80	50	46	48
Sacramento River Amount in m^3x10^9	6.3	11.4	11.3	18.2	11.4	10.4	11.0

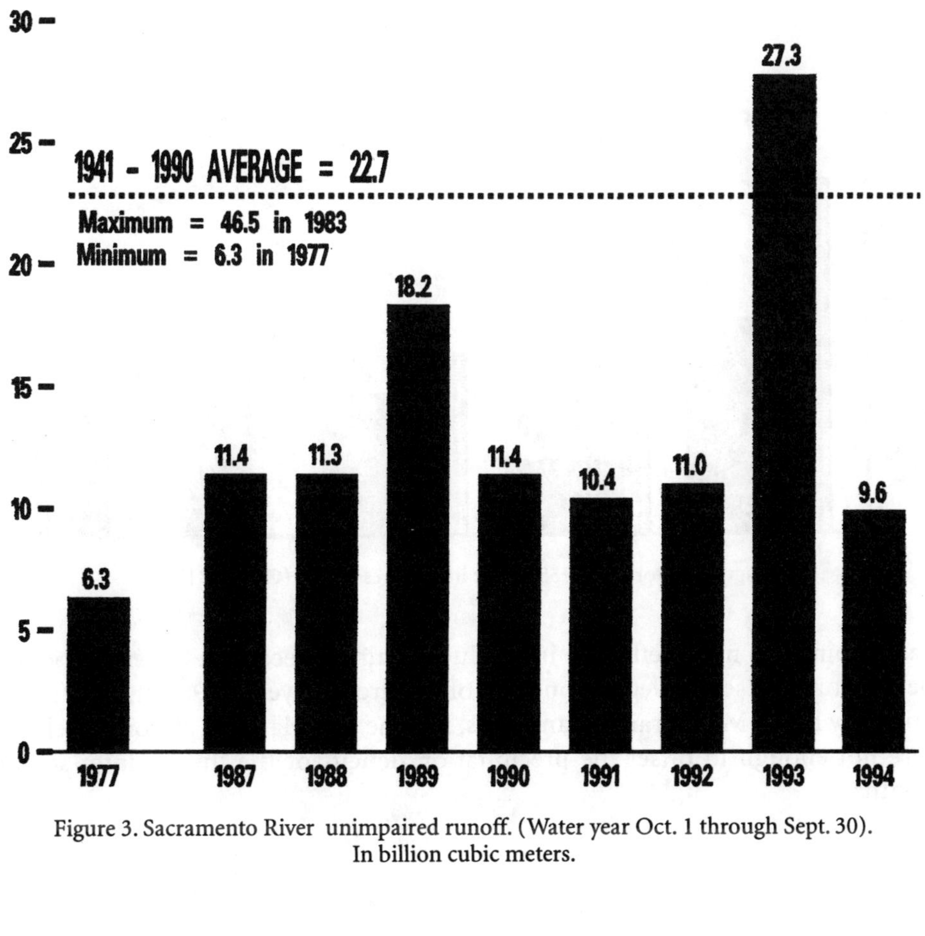

Figure 3. Sacramento River unimpaired runoff. (Water year Oct. 1 through Sept. 30). In billion cubic meters.

During the drought, the mountain snowpack mirrored the precipitation deficiency. Normally the peak snow accumulation occurs near the first of April. The April 1 snowpack, then, is an important measure of prospective water supplies. Table 3, derived from CA DWR Bulletin No. 120 reports, gives a comparison of snowpack water content. Table 4 shows runoff statistics.

Each drought is different. The recent drought in the Sacramento River basin is unique in that runoff in five of the years was similar, about half of average. Only 1989 was different. The Sacramento River runoff (the sum of unimpaired runoff of the four major rivers in the basin) is also shown on Figure 3.

Precipitation during the 6-year 1987–1992 period was about three-fourths of average. The deficit in precipitation was magnified in runoff, which was about half of average over the 6-year period. A portion of each rainy season's precipitation goes into wetting the ground before runoff can begin. Therefore, the impact of a shortfall in precipitation is amplified in runoff deficits. Likewise, early and late

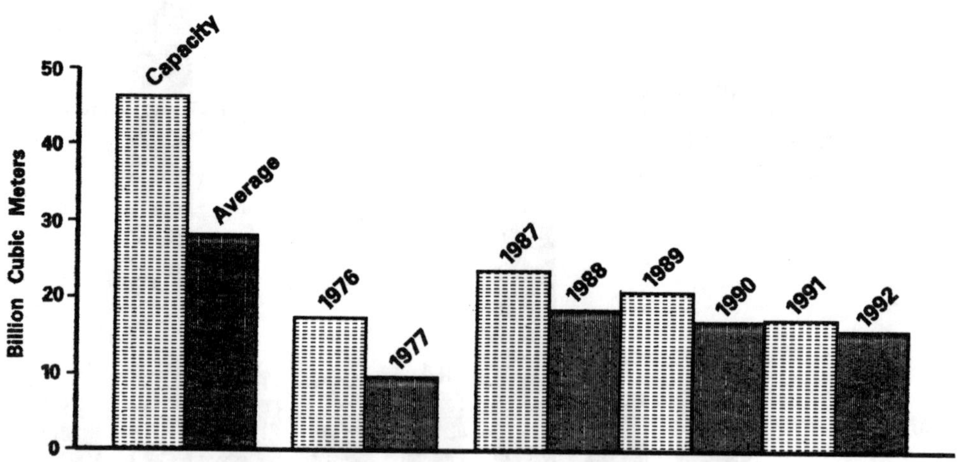

Figure 4. Storage in 155 major in-state reservoirs (October 1).

season rainfall is not as effective in producing runoff because a larger fraction of the moisture is used by vegetation. Two of the drought years, 1989 and 1991, had unusually heavy March rainfall amounts. But the wet Marches, although helpful, were not enough to offset the precipitation deficits of the three winter season months.

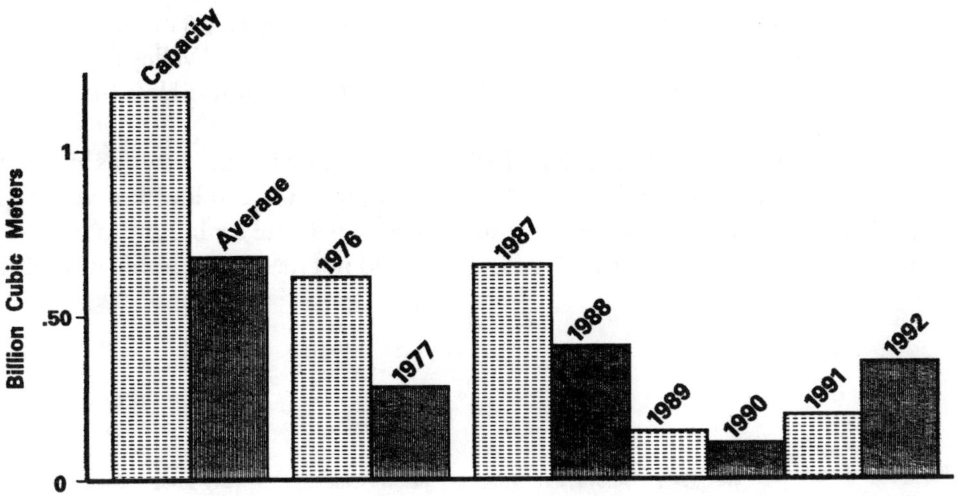

Figure 5. Storage in six major central coast reservoirs (October 1).

RESERVOIR STORAGE OF SURFACE WATER

California's reservoir storage proved its worth during this drought, especially during the first three years. However, by 1990, reserves were largely depleted and major curtailments in water delivery became necessary. Figure 4 presents October 1 storage in the 155 major reservoirs within California. Figure 5 shows the same data for six reservoirs in the Central Coast hydrologic region, roughly from Santa Cruz to Santa Barbara. The improvement from the relatively larger Southern California precipitation amounts in March of 1991 and in 1992, is evident in Central Coast storage.

TABLE 5. Reservoir storage on October 1

Year	1977	1986	1987	1988	1989	1990	1991	1992
				In billion cubic meters				
155 Major Res.	9.60	33.00	23.34	18.29	20.62	16.74	17.00	15.63
Major CVP	1.62	10.40	7.73	5.67	6.29	4.91	4.06	3.84
6 Major SWP	1.80	5.24	3.71	3.05	3.55	2.31	3.00	2.84
				In percent of historical average				
155 Major Res.	35	119	84	66	74	60	61	56
6 Major CVP	18	113	84	62	68	53	44	42
6 Major SWP	41	120	85	70	81	53	69	65

Statewide carryover storage in the 155 major in-state reservoirs on October 1 of 1992 at the end of the last year of the drought was about 5% less than a year before. This was the lowest of the recent 6-year drought, but still nearly 6 billion m³ more than the extremely low 1977 storage amount. Combined Central Valley Project and State Water Project carryover in late 1992 was about 0.4 billion m³ less than in the previous year.

Water storage levels in the CVP and SWP system could not be drawn as low as in 1977, because of federal and state endangered species laws, and low reservoir high temperature problems for salmon. Thus, probably half the six billion m³ of 1992 statewide storage in excess of 1977 amounts would not have been usable in 1993, if

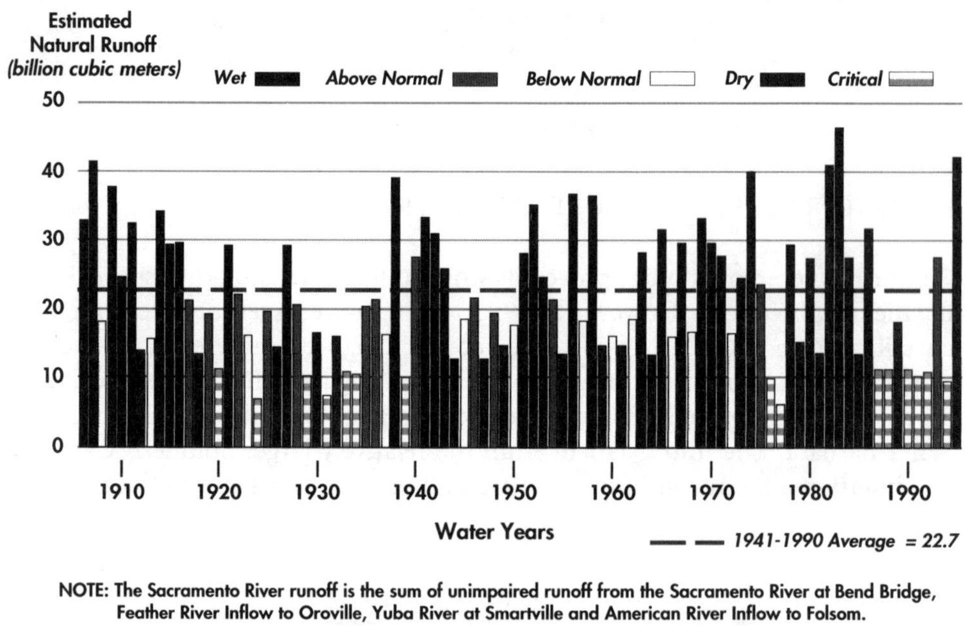

Figure 6. Sacramento River unimpaired runoff since 1906.

that year had been dry. Total net water use in California for irrigated land and urban purposes is normally about 42 billion m³. About 6 billion m³ more is needed, on the average, for required Sacramento-San Joaquin Delta outflow to meet water quality standards. Hence much of each year's supply must be generated from that season's runoff, with the smaller portion drawn from surface reservoir carryover storage.

TABLE 6. Drought frequency probabilities

Water year	*Length, years*	*Risk of Occurrence in Sacramento*	*Percent* San Joaquin*
1987	1	11	6.0
1987–1988	2	5	2.4
1987–1989	3	8	2.5
1987–1990	4	4	0.8
1987–1991	5	2.5	0.6
1987–1992	6	1.4	0.3

* Based on log-Pearson Type III Distribution .[2]

Comparison with Past Droughts in California

Average Sacramento River Index runoff during the 1987–1992 six-year period was about 12.3 billion m^3, or 54% of the average 22.7 billion m^3 of runoff. While unusual, this is not the driest of record. Runoff during the historical six-year critical dry period from 1929 through 1934 was 1% less at about 12.1 billion m^3. However, on the San Joaquin River system, the recent drought exceeded, by a large margin, the historical 1929–1934 runoff. Because 1932 was above average in the Southern Sierra, the earlier drought was eased somewhat in that region.

Extensive droughts in excess of three years seem to be rare in Northern California, except for the 1929–1934 and 1987–1992 periods, according to the historical runoff record (see Figure 6 for Sacramento River runoff since 1906) or the historical precipitation record. The latter goes back to 1850 for some stations and is reasonably complete since the early 1870s, when the major railroads were built. Long droughts in arid Southern California are more common.

The estimated recurrence frequency for a six-year drought period, like 1987–1992 for the Sacramento River region is 1.4%, or about once in 70 years, based on the 1906–1992 record. On the San Joaquin River, where the recent drought was more severe, the estimated recurrence frequency is only about 0.3%, approximately a 1-in-300-year event. These statistics represent both length (six years) and severity of drought. Estimated frequencies of drought periods within the 1987–1992 drought series in the two basins is presented in Table 6.

It is not wise to place much trust in the above statistics for extreme events, because the record is only about 90 years. Some long-term climate reconstructions show periods in the past which are different than the last 90 years. Drowned trees in the central Sierra Nevada indicate long dry periods on the order of a century around 700 and around 900 years ago (Stine, 1994).

Indirect indicators of rainfall and runoff are needed. The most promising tool for looking into the past year-by-year is by use of tree ring data. The thickness of annual growth rings of suitable trees indicates the wetness of the season. Tree ring widths are not a perfect match (for example they did not reproduce the 1976–1977 drought well) but have been useful to investigate how the measured runoff or precipitation record compares with a longer sweep of history.

TABLE 7. Sacramento River multi-year droughts constructed from tree-rings prior to year 1900

Period	*Length (years)*	*Average runoff Billion cubic meters*
1579–1582	4	15.3
1593–1595	3	11.4
1618–1620	3	16.3
1651–1655	5	15.2
1719–1724	6	15.5
1735–1737	3	15.0
1755–1760	6	16.4
1776–1778	3	15.0
1793–1795	3	13.2
1839–1841	3	15.8
1843–1846	4	15.2
1918–1920 (actual)	3	14.7
1929–1934 (actual)	6	12.1
1959–1962 (actual)	4	16.1
1987–1992 (actual)	6	12.3

A 420-year reconstruction of Sacramento River runoff from tree ring studies was made in 1986 for the Department of Water Resources by the Laboratory for Tree-Ring Research of the University of Arizona (Earle and Fritts, 1986; CA DWR, 1987). This study showed the 1928–1934 drought to be the worst in the reconstructed record, which began with year 1560 (Table 7). Periods are presented of consecutive years with runoff under 19.4 billion m^3, the historic median yearly runoff.

Since the tree-ring reconstruction doesn't always match the measured record for this century, the degree of precision of the tree ring derived information is not clear. Apparently, few droughts prior to 1900 exceeded three years and none lasted for over six years, although an eight-year period of less than average runoff occurred from 1839 through 1846.

John Bidwell, an early pioneer who arrived in California in 1841, confirmed that 1841, 1843 and 1844 were extremely dry years in the Sacramento area (Harding, 1965). He also talked about flooding in 1842 between San Jose and Sacramento. These historical notes seem to confirm that dry period in the tree ring reconstruction.

In conclusion, no predictable pattern of drought is apparent. Droughts exceeding four years in length seem to be quite rare. There are a multitude of factors which can influence California weather, some in ways which are not understood. El Niño, the warming of the ocean in the eastern tropical Pacific, does have influences around the world. However, no clear signal for Northern California is evident from the record:-some El Niño years are dry and some are wet.

(million cubic meters)

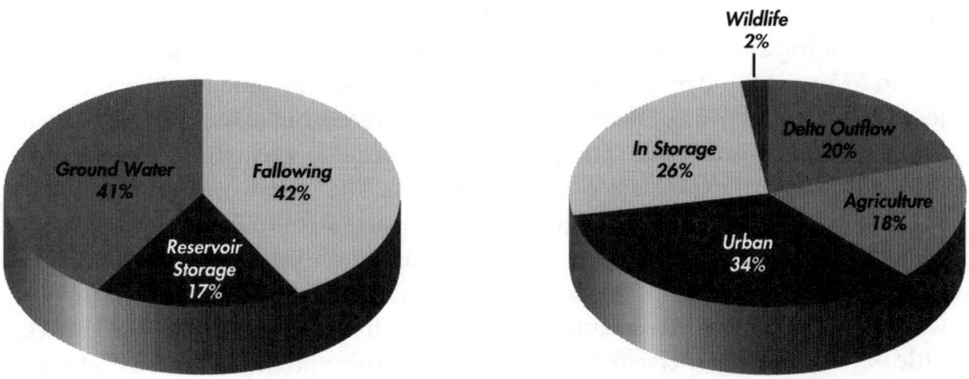

Sources of 1991 and 1992 Supplies

	1991	*1992*
Fallowing	518	0
Ground Water	318	199
Storage	175	39
TOTAL	*1011*	*238*

Allocation of 1991 and 1992 Supplies

	1991	*1992*
Agriculture	102	117
Urban	379	48
Fish & Wildlife	0	31
Delta Outflow	204	42
In Storage	327	0
TOTAL	*1011*	*238*

Figure 7. Water sources and allocations of the 1991 and 1992 State Drought Water Bank.

Drought Management Strategies

The first areas to feel the impact of the 1987–1992 drought were those which depended on local surface water supply or supplies imported by local agencies, particularly the San Francisco Bay region. The major storage reservoirs of the federal Central Valley Project and the State Water Project were sufficient to carry these projects through 1989. Initially it looked like there would be shortages in 1989, but heavy rains in March provided enough runoff to maintain deliveries that year. By the fourth year, 1990, major deficiencies in delivery of CVP and SWP supplies ensued.

Water year 1991 marked the nadir of the drought. The rainy season remained very dry until the end of February at which time projections were for one of the driest years on record with most usable reservoir storage gone. However, March rains were very heavy, about three times average statewide. The 'Miracle March' turned a desperate situation into a manageable one. A hastily formed water bank administered by the California Department of Water Resources bought water from those willing to sell at \$125 per acre-foot (\$101 per 1,000 m^3) and sold it to users with the most critical needs at \$175 per acre-foot (\$142 per 1,000 m^3) plus aqueduct pumping and transportation costs from the Delta which ranged from about \$25 to \$250 per 1,000 m^3 depending on how far along the Aqueduct the agency was located. A total of nearly 800 million m^3 was allocated, but about 40% remained in SWP storage for use in 1992. The difference, about 20%, consisted of losses—mostly in getting the water transferred across the Delta. The source of water bank supply and the uses for 1991 and 1992 are shown in Figure 7 (CA DWR, 1994, page 286). The 1991 water price was too expensive for most agricultural uses, except for those who needed water to save permanent crops and orchards. About 80% of the water went to urban users and around 20% to agriculture. The supply situation was a little better in 1992, especially in Southern California and the Central Coast region. That year the water bank was much smaller having a buying price of \$50 per acre-foot (\$41 per 1,000 m^3) and a selling price of \$72 per acre-foot (about \$58 per 1,000 m^3) at the Delta, while most water went to agriculture.

The sources of supply came from fallowing previously irrigated lands, pumping ground water in lieu of surface supplies, and some water from storage not needed by several local agencies. The basis of payment for fallowed land was the estimated consumptive use (evapotranspiration of applied water) of the annual crop which otherwise would have been grown. In the second year, 1992, no purchases from land fallowing were made. The need for water was less severe while there was concern for third party impacts in a loss of irrigated agriculture, that is, those who work on the land, sell supplies, or process the produce. Total water bank supplies in 1991, the worst drought year, were about 2% of normal water requirements. This does not seem much but it proved to be enough to take care of a number of critical situations. The California Aqueduct was the major means of transporting water bank water to users. The Central Valley Project also made some 'hardship' water available to some of its customers with the most serious needs.

Strategies used in coping with the drought included strong public education campaigns urging water conservation, rationing, shifting crops that use less water (to the extent markets can absorb the changes) and taking land out of production. In addition, thousands of wells were drilled, mostly in the Central Valley alluvial basins, and farmers pumped much additional ground water to try to survive. Restoration of the underground storage will probably take years in some areas, especially in the San Joaquin Valley. Underground storage is not likely to be restored in areas of continuing ground water overdraft and the lifetime of those sources has been shortened.

The 1987–1992 drought also severely affected fish, wildlife and forest resources. An estimated 42 million m^3 of merchantable timber were killed, enough to build 1.8 million homes (CA DWR, 1991, page 34). Some of the dead trees were harvested before decay set in, so not all was lost. Anadromous fish populations declined to record low levels, which negatively affected water supply as CVP and SWP operations, especially exports from the Delta, were further constrained to try saving fish and wildlife stocks.

Conclusions

The impact of drought can be eased through water storage, both surface and underground, carrying over supplies from the wet years to the dry years. Indeed, water deliveries were almost unaffected on the Colorado River system, which also suffered severe drought at about the same time. Total storage capacity on the Colorado River is 4 to 5 times average annual runoff. The California in-state surface reservoir capacity totals around 52 billion m^3, which is not much more than annual regulated water demands within the state. This reservoir storage helped, but is not adequate to deal with multi-year droughts. In the worst year, 1991, SWP Delta water export deliveries to urban customers were only 30% of the normal amount, while agricultural users received nothing. The CVP agricultural project water customers got only 25% that year. The water bank helped meet some of the most severe shortages, but was too expensive for most agricultural users. The volume of ground water storage (although much harder to fill) is much greater than surface storage and saved many water users from a worse fate. Additional ground water pumping in 1991 probably filled in about half of a shortfall estimated between 20 to 25% of total statewide water demand.

The State Water Project contractors have renegotiated, in the wake of the drought, how shortages are to be shared. These changes have been adopted, so future shortages in delivery will be shared equally by urban and agricultural users.

One new major surface storage reservoir, Eastside Reservoir, is in the beginning stages of construction in Southern California. It will add about 1 billion m^3 of carryover storage for dry years. Two other smaller dams are under construction with another 300 million m^3 of capacity. Except for these facilities, additional storage is not likely to be built, at least in the foreseeable future, because of cost and environmental concerns. Since the population and associated water needs continue to increase, many Californians can expect more frequent and more painful water shortages in future droughts as a result. Some changes may be forthcoming. In November 1996, state voters approved $995 million in bonds for water supply and ecosystem restoration, with a major share of improvements to be in the Delta and its tributary area.

California water supply system is decentralized. Local agencies and individuals furnish over half the water. There is no formal statewide criteria for drought (indeed there probably is too much variety in regional situations for any single measure to be successfully employed), but there is a very good system for sharing information. Water conditions are analyzed periodically, especially in the late winter and spring, by the Department of Water Resources to provide updated information on precipitation, snowpack, reservoir storage, and the runoff outlook. Basic hydrologic data is freely shared, becoming more and more available in computerized data banks. This wealth of information is used by all water and power system operators (private, local, state, and federal) to guide their actions as each season unfolds.

Notes

1. Net water demand is the amount of water needed in a water service area to meet all requirements. It is the sum of evapotranspiration of applied water in an area, the irrecoverable losses from the distribution system and the outflow leaving the service area. It does not include reuse of water within a service area, such as reuse of deep-percolated applied water or use of tail water
2. The log-Pearson Type III distribution is the recommended method in the United States of America for fitting annual flood peaks. The base 10 logarithms of the discharge (in this case, the annual or multi-year runoff) are plotted on probability paper at selected acccedence probability by the equation Log Q = x+ KS. Where Q is the runoff, x is the mean logarithim, S is the standard deviation of logarithms, and K is a factor which is a function of the skew coefficient and exceedance probability. The method is documented in U.S. Water Resources Council Buletin 17B, Revised September 1981, Guidelines for Determining Flood Flow Frequency, Washington, D.C.

References

CA Department of Water Resources (1987) *California Water: Looking to the Future.* Sacramento, CA.

CA DWR (1991) *California's Continuing Drought,* a Summary of Impacts and Conditions as of December 1, 1991. Sacramento, CA.

CA DWR (1994) *California Water Plan Update,* especially Chapters 1, 3, and 7 of Vol. 1. Sacramento, CA.

CA DWR, Annually, Bulletin 120. *Water Conditions in California,* especially the April 1 report. Sacramento, CA.

Earle, C.J. and H.C. Fritts (1986) *Reconstructing Riverflow in the Sacramento Basin Since 1560.* Laboratory of Tree-Ring Research, Tucson, AZ.

Harding, S.T. (1965) *Recent Variations in the Water Supply of the Western Great Basin.* Water Resources Center Archives, U.C. Berkeley.

Roos, M. (1995) *The Recent California Drought,* Proceedings of the 63rd Annual Western Snow Conference. Sparks, NV.

Stine, S. (1994) Extreme and persistent drought in California and Patagonia during mediaeval time. *Nature* 369:546-549.

ABOUT THE AUTHOR

Maurice Roos is currently Chief Hydrologist in the Division of Flood Management of the California Department of Water Resources in Sacramento, California, U.S.A. As a water engineer, he oversees work on flood forecasting, water supply and snowmelt forecasting, hydrology, and related matters. He also provides staff advice on drought, floods, climate change, and weather modification. Mr. Roos obtained a B.S. in Civil Engineering from San Jose State University in California in 1957. His Departmental career began with studies in the Sacramento San Joaquin River Delta. In 1965, he joined the Division of Planning and was engaged in statewide water planning, studies of reservoir system operation, evaluation of water requirements and potential water supplies. In 1979, he joined the Division of Flood Management.

Mr. Maurice Roos
Chief Hydrologist
California Department of Water Resources
P.O. Box 219000
Sacramento, California, 95821-9000
U.S.A.

Chapter 15

Cost-Benefit Analysis of Water Management Mega Projects in India and China

Hisham El-Bihbety and Harvey Lithwick

Introduction

Water mega projects—especially massive dams to provide hydro electricity, flood control and irrigation—are pivotal in many countries' economic development initiatives. Analyzing how the planning of such projects has dealt with environmental and related social issues therefore provides important insights into the value placed on environmental management within the development process.

This chapter examines the planning of two projects to construct the largest dams in the history of humankind: the Sardar Sarovar Dam in India's Narmada Valley, and the Three Gorges Dam in China. The former is in an advanced state of implementation, whereas the latter is still under consideration. It is therefore possible to assess the Sardar Sarovar Dam experience with the benefit of some hindsight, while we can study the more recent Three Gorges plan to see if any lessons have been learned from the Indian experience, especially with regard to accounting for the environment.

The key instrument of planning for both projects has been social Cost-Benefit Analysis (CBA). The problems that arose in implementing the Sardar Sarovar Dam project raised important questions about the appropriateness of CBA as a technique for appraising projects. In particular, it has been argued that the project's failure to address environmental and social issues is intrinsic to CBA, which should therefore disqualify it as a planning tool for managing economic development. This skepticism regarding CBA has had a negative impact on attitudes towards China's Three Gorges Dam project, frustrating its ability to secure funding from international capital markets.

The first section of this chapter reviews the Sardar Sarovar project's CBA which was done by the World Bank in 1985, as well as the independent review of the

H. J. Bruins and H. Lithwick (eds.), The Arid Frontier, 295–317.

project, which was funded by the World Bank in 1992. In the second section, the CBA for the Three Gorges Dam, done by the Canadian International Development Agency (CIDA) in 1988, is analyzed. This is followed by a discussion as to whether the shortcomings of the Sardar Sarovar project are intrinsic to CBA methods of project appraisal. We conclude with an assessment as to whether the troubled history of Sardar Sarovar might repeat itself in the Three Gorges project.

India's Sardar Sarovar Dam

BACKGROUND

After independence, dams became, in Nehru's words, India's 'secular temples'. The first proposal for damming the Narmada River and diverting its water was made by the British as early as 1863, but the first serious study to appraise the project was conducted in 1947. In 1959, after independence, the project was studied further and construction began in 1961. However, after two thousand villagers in the vicinity of Kevadia had been removed from their homes in 1960–1961, the project was suspended because of the disagreement between the three affected states regarding the distribution of the costs and benefits of the project. In 1969, the Narmada Water Disputes Tribunal (NWDT) was appointed to resolve this dispute. NWDT completed its mission in 1979.

In 1985, the World Bank (IBRD) conducted a formal CBA appraisal study of the project. The World Bank has been for several decades a major advocate of this technique in appraisals of the myriad of pubic projects it has helped finance. The study recommended investment in the project by an IBRD loan and International Development Agency (IDA) credit along with major Indian government and commercial co-financing. The government of India's share was to be US$ 1,374 million while the World Bank's share was US$ 450 million. The breakdown of the World Bank's share is as follows:

TABLE 1. Loan terms*, Sardar Sarovar Project

	US$ million	*Towards*
IBRD hard loan	200	Construction of the dam
IDA soft credit	100	Construction of the dam
IDA soft credit	150	Construction of the canal
	450	

* Soft and hard conditions refer to the term and conditions of loans. Hard loans are commercial loans at commercial interest rates, whereas soft loans tend to have some level of interest subsidy and/or loan repayment forgiveness under certain conditions.

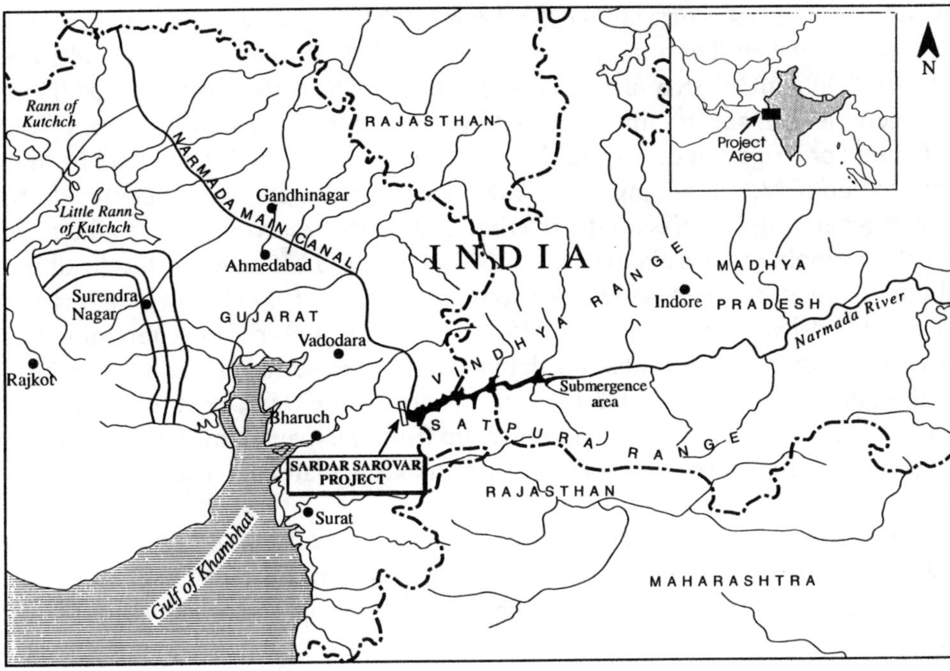

Figure 1. Site of the Sardar Sarovar Project.

Full scale construction started in 1987 (International Rivers Network, 1984:1). The feasibility study divided the project into two parts: the first part, 'Sardar Sarovar Dam and Power Project' (Report no. 5107–IN) included the dam and power units associated with it; the second part, 'Water Delivery and Drainage Project' (Report no. 5108–IN) included the canal and the irrigation network. The following is a summary of both these reports.

The Narmada River is the largest westward flowing river in India. The project is located on the Narmada River in Gujarat state, 170 kilometers upstream from where the river flows into the Khambhat in the Arabian Sea. The project consists of several components; a dam, river bed power house and transmission lines, main canal, canalhead powerhouse, and irrigation network. The project was planned to irrigate 1.9 million hectares of cultivable land and bring into production an additional 4–5 million hectares of agricultural lands which are prone to drought. Also, the project would provide 40 million people with drinking water. Water would also be used for industrial purposes. Furthermore, the project would generate at least 2,700 MW of electricity to Western India. The Chief Minister of Gujarat predicted that if the project is not implemented, the result will be a need to resettle hundreds of thousands of Gujarati citizens because of the drought (Morse & Berger, 1992:5–6).

The Sardar Sarovar is a gravity type dam, a 1210 m long wall of concrete across the valley rising 128.5 m above the river bed and 157.5 m above the lowest point of excavation. The reservoir would extend some 210 km upstream from the dam site in a narrow valley. The total reservoir capacity at Full Reservoir Level (FRL) of 138.7 is 9,500 km3 with a surface area of 370 km^2. The active storage is 5,800 km^3 between the FRL and minimum drawdown level (MDL) for power operations of 110.6 m. For irrigation diversions during severe drought, the MDL would be lowered to about 93.6 m adding about 1,640 km^3 of active storage. There would be power generating facilities on the riverbed and on the reservoir outlet to the main canal. The main canal is planned to be 460 km long and 250 m wide at its head and about 100 m at its tail. A network of secondary canals 75,000 km in length is also planned to deliver irrigation water to farmers. A large powerhouse containing turbines is scheduled to be built at the dam in addition to a smaller one near the head of the canal (World Bank, 1985:14–20). Vadgam Saddle Dam, a component of the Narmada Project, is located 800 m upstream from the main dam. The dam is 60m high and contains the channel power house intake regulator and controls the reservoir level.

It is assumed the construction of the Sardar Sarovar Dam and power complex will take 22 years to complete. All prices are expressed in constant 1984 rupees (Rs) and the discount rate is 12%. The exchange rate was taken to be Rs 9.28/US$ (World Bank, 1985:48).

PROJECT COSTS

The estimated construction cost for the ten components of the dam and power related project, over 110 months, was US$ 1,934 million (including land acquisition and rehabilitation), and the cost for the three water delivery and drainage components was US$ 521 million. The economic cost of the project was calculated based on the base costs plus contingencies and was then adjusted using a conversion factor for specific materials, such as cement, steel, skilled and unskilled labor. When a specific conversion factor could not be derived, a standard conversion factor was used, although its source was not explained in the study.

Annual operation and maintenance costs were found to amount to US$ 9.67 million a year for the dam and the power complex and US$ 71.4 million for the water delivery and drainage project. The cost of land acquisition in the reservoir area and for the infrastructure in the project command area was based on economic values of annual production foregone, estimated by taking the annual production per hectare in the future without the project. The total cost of land amounted to US$ 140.1 million.

Of particular interest is the treatment of environmental impacts. The study identified three categories of environmental effects: (1) 'fish and fisheries', as the dam would affect the migration of fish, quality of water, suitability for spawning, and fishery potential; (2) 'forest and wildlife', as the reservoir would submerge forests, agricultural land, cultural infrastructure, and would limit the wildlife habit; and (3) 'public health', because of the potential spread of three principal diseases; malaria, schistosomiasis, and filaria.

Although these effects were seen to be important, the World Bank study did not attempt to estimate their costs as is the standard procedure in CBA. It simply recommended several interventions and proposed studies to examine these issues (World Bank, 1985:28–19).

PROJECT BENEFITS

Agricultural benefits would mainly result from the provision of surface water from the Narmada River and the recycling of induced ground water . Crop intensity was expected to increase from 105% to 135–140%. At full development, annual production of food grains was estimated to increase in value by Rs 8,600 million. These agricultural benefits are deemed, by the study, to be conservative because they do not include the effect of future technological changes on agriculture production which would increase the productivity of the project itself. Import and export parity prices were estimated for major tradable crops such as cotton, rice, wheat and groundnut. These prices were based on IBRD projections of world market prices for 1995 (expressed in 1984 constant rupees at an exchange rate of Rs 11/US$). For non-traded crops, economic prices were based on real increases in economic prices of the tradable crops over the period 1984 through 1985.

The more intensive cultivation resulting from the completed project was estimated to provide the equivalent of approximately 0.7 million man-years of full time on-farm employment. Most of these employment opportunities would benefit landless laborers. The project would also provide 125,000 seasonal jobs annually during the construction period. Income would increase by Rs 10,400 million per year. This figure assumed that the ratio of the economic shadow wage to the market wage is in the range of 0.5 to 0.7. A farm labor conversion factor of 0.67 was used (World Bank, 1984:49, 53).

Villages and industries suffer from water that is scarce and of unsuitable quality. The calculation of project benefits included towns and villages within the project area but excluded those villages where water would continue to come from existing local water supply sources. The benefit calculations are based on the concept of opportunity cost, which in this case was the cost of alternative water supply schemes. The most extreme alternative considered was supply by tank truck. The estimated cost savings were Rs 10 per m3 (World Bank, 1984:50).

The benefits from power generation have been calculated by estimating the maximum willingness to pay for power in the industrial and agriculture sectors. Based on these estimates and on the average tariffs paid by consumers, the estimates were converted to an economic value by applying the assumed standard conversion factor (0.80) and the average efficiency of transmission and distribution from the power generator to consumers (0.82). The resulting economic prices amounted to Rs 0.66/ KWH.

Among other expected benefits, the project would introduce fisheries, but these benefits were not calculated. The study assumed that this benefit and the benefit resulting from the planned reforestation in the reservoir catchment area would help offset the economic costs from the submergence and changes in the downstream environment.

PROJECT ANALYSIS

The economic feasibility of the project, including its three functions of power generating, irrigation, and municipal and industrial (M&I) water supply, were calculated using a discount rate of 12%. The analysis gave a positive net present value of Rs 4,259 million (US$ 459 million). Also, the economic rate of return (ERR) was found to be 13%. This ERR was justified by arguing that worldwide experience demonstrated that large infrastructure projects tend to have low rates of return. The analysis also claimed that experiences from similar projects, for example in Malaysia, showed that the realized benefits of irrigation projects exceeded the calculated direct effects by 75%. Also, it was claimed that the analysis did not take into account any potential advances in agro-technology thereby understating benefits which might result.

A sensitivity analysis was undertaken. It calculated equivalent increases in different projects items which would result in a net present value equal to zero. Table 2 summarizes the results, which in general indicate rather wide margins for error:

TABLE 2. Sensitivity analysis, Sardar Sarovar Project

Costs	*Percentage Change*
Total cost	+15
Dam cost	+152
Main canal cost	+77
Total power cost	+75
Total irrigation cost	+24
Benefits	
Total Benefits	–13
Agriculture benefits	–22
Power Benefits	-38
Implementation Periods	
Dam	+22
Power facilities	+55

Since it is unlikely that all the assumptions would prove to be correct, the study conducted a risk analysis. The following factors were incorporated into the risk analysis; (1) costs of dam, powerhouses, transmission system, main canal, branches, distribution and drainage systems, and command area development; (2) implementation periods for the above operations; (3) errors in estimates of the economic conversion factors; and (4) deviation of agriculture gross returns, cost of cultivation, water demand, and cropping intensities. All the above factors were considered as random events following certain probability distributions. ERRs and net present values were then calculated for 500 different randomly chosen set of parameters. The risk analysis resulted in an expected ERR of 12% with a minimum value of 7.4% and a maximum value of 18.9%. Table 3 shows the probability of occurrence of the different ERRs.

TABLE 3. Risk analysis, Sardar Sarovar Project

ERR (%)	*Probability*	*Cumulative probability*
<8	2.2	2.2
8–10	13.4	15.6
10–12	34.4	50.0
12–14	34.4	84.4
14–16	13.4	97.8
>16	2.2	100

Based on these analyses, the study concluded that the project's return on investment would be satisfactory and the overall risk was reasonably low for a project of this magnitude (World Bank, 1984:58–62).

REVIEW OF THE SARDAR SAROVAR PROJECT

In 1991, an independent review of this project was commissioned by the World Bank. This was the first time that a lending or aid institution subjected its operations to an evaluation by an independent group. It was conducted by Bradford Morse, the formal Under-Secretary-General of the United Nations and head of the United Nations Development Program (UNDP) for ten years. The Terms of Reference included assessment of the resettlement and rehabilitation (R&R) of the population displaced and the amelioration of the environmental impacts of all aspects of the project including the main canal.

The reservoir was to submerge 37,000 hectares of land, hence 245 villages in which 100,000 people live would be submerged. Almost all the affected people in two of the three affected states are Tribal. In the third state, 40% of the affected people are Tribal. The submerged land is the most fertile land in India, although the project supporters had always argued that the affected lands are steep and rocky with only a marginal value (Morse and Berger, 1992:6). The main canal network requires 80,000 ha for construction which is more than twice as much as the reservoir. Taking into account resettlement associated with other projects related to the Dam, such as the canal and the construction of infrastructure at Kevadia, the number of relocated people could reach 300,000.

In 1980 the World Bank had decided not to support any project that does not have a clear assessment of the extent of displacement that might happen, and a detailed plan that shows how those who will be displaced are to be assisted to maintain or restore their standard of living. In the case of the Sardar Sarovar project, few if any of these requirements were met during the appraisal. No resettlement plan was prepared by the borrower and none was approved by the Bank. No adequate information was made available with respect to the submergence of villages, nor was there any consideration given to the impact on human populations of the canal and irrigation works. Moreover, the downstream effects were incompatible with the Bank's own resettlement and rehabilitation policy (Morse & Berger, 1992:41–43). The World Bank had been warned by its mission to India in 1984 that, in addition to the lack of infrmation about the size and the implications of the displacement, India's past record of reservoir-related relocation and its national policy regarding relocation did not meet the Bank's standards. Nevertheless, the Bank went ahead and funded the project which the review described as the most irresponsible decision ever taken by the Bank (Morse & Berger, 1992:44).

On the environmental level, the review concluded that measures to anticipate and mitigate environmental impacts were not properly considered in the design of the project because of a lack of basic data. Although the Bank insisted on an environmental work plan by the end 1985, nothing had been done at the time of the independent review in 1992. In addition, the review pointed to the following important elements that had not been considered: the impact associated with the backwater effect of sedimentation in the upper reaches of the reservoir; the downstream ecological implications of the dam's construction; the inability of the afforestation and catchment area treatment programs proposed upstream to meet the timetable of the project because of the lack of consultation with the villagers in the affected areas; an environmental assessment of the canal and water delivery system where salinity and waterlogging are potential problems; and the lack of precautions to prevent the spread of the serious diseases mentioned in the appraisal report, especially malaria (Morse & Berger, 1992:212–233).

The independent review concluded that much of what went wrong with the Sardar Sarovar project was a result of the failure of the Bank to insist that its own policies with respect to resettlement and rehabilitation and environmental matters be followed. It also pointed to the failure of the Indian government to consult the people in the affected areas which created hostility towards the projects and in turn threatened all aspect of implementation. The review could not make any recommendations for improving on implementation in these areas, as required by its Terms of

Reference, because of the absence of proper impact assessments and the lack of undisputed data.

China's Three Gorges Dam

BACKGROUND

The Chang Jiang or Yangtze, the largest river in China, carries nearly 40% of the country's annual runoff. From the grassy plains of Qinghai province, the river flows eastward, cutting through successive mountain ranges and ridges before meandering across Central China.

Figure 2. Site of the Three Gorges Project.

The Three Gorges Water Control Project was first suggested by Sun Yatsen in 1919 and the preliminary design was undertaken in the 1930s and 1940s. During the 1950s, the project was subject to intense political debates. In 1958, Mao Zedong made a decision in favor of the dam, but the Chinese economic crisis of the early 1960s and the rift with the Soviet Union led to a deferral of the project. In the 1970s, the issue re-emerged (Huenemann, undated:2–3). In early 1986, the Canadian International Development Agency (CIDA) and the Chinese Ministry of Water and Resources and Electric Power agreed to undertake a new feasibility study for the Three Gorges. Terms of Reference were prepared, and in June 1986 a consortium of consultants known as the CIPM Yangtze Joint Venture (CYJV) were retained to carry out the study. The purpose of the study was to evaluate the technical and economic feasibility of the Three Gorges project consistent with the criteria of international financial institutions, so as to help secure funding from them. Also, the report would provide impartial technical input to the Government of China in its decision-making process regarding the project. The study was completed in June 1988 (CIDA, 1988).

Chinese authorities claim that the major reason for the project is flood control, which would save hundreds of thousands of lives by protecting approximately 1 million km2 in the middle of the Yangtze valley. In addition, China's rapid economic growth had created a serious shortage of electricity supplies, making necessary new power sources. Growth of the industrial and agricultural sectors averaged 11% annually between 1978 and 1987 while output of electricity grew by only 8%. Shanghai's industrial sector has been operating at 30% below its capacity.

497.3 billion KWH of electric power is produced in China, with 80% from thermal plants and 20% from hydro plants. China has exploited only a small fraction of its potential hydro sites. Although renewable and non-polluting (in the air), delivery of hydro power is expensive due to the long distances between the most promising hydro sites in the deep south west and major consuming centers in other regions of the country.

ASSUMPTIONS OF THE FEASIBILITY STUDY

All costs and benefits in the report are calculated at the January 1987 price level. Net present values of annual cost and benefit streams are computed using a discount rate of 10% per annum (vs. 12% for Sardar Sarovar). The analysis period is fifty years from 1989 to 2039. The escalation rates used from 1987 to 1995 were based on World Bank information. These rates are 6.5% from 1987 to 1990 and 4.5% from 1991 to 1995. An additional 3% has been applied from 1996 to 2007. No escalation is applied after 2007.

The Reservoir Land Requisition Level is based on a 1 in 20-year flood level for permanent residences and a 1 in 100-year flood level for large factories. All displaced structures are to be relocated above elevation (El) 182, which corresponds with a 1 in 1000–year flood level at the dam. Resettlement planning was based on the criteria of international financial institutions which require improved living conditions of the affected population ('resettlement with development'). Environmental planning criteria were set to protect rare and endangered species and to provide improved facilities to mitigate other effects.

The project items examined in the feasibility study were: a concrete gravity dam 2.5 km long with crest level at El 185; a navigation structure made up of a single temporary lock and twin five-flight locks as permanent facilities; and two powerhouses downstream of the intake blocks.

The methodology used in the analysis followed the generally accepted techniques of the World Bank and other lending institutions to appraise projects. All costs and benefits of the project were analyzed together instead of using a separable-costs, separable-benefits methodology which is sometimes applied to a multi-purpose project. The nature of the project, where costs and benefits interact and affect each other, made it essential to apply this approach.

Economic pricing was a problem in a country where subsidies distort all prices. Therefore, financial costs, which are the costs of the project to the builder, had to be adjusted to reflect the 'real' costs of the project for the whole economy. Because of the complexity of the project with over 2,000 separate input items, economic prices could only be determined for the major project inputs such as; cement, steel, coal, petroleum products, timber, electricity for construction, labor, machinery, and thermal power plants. Shadow prices were also calculated for rail, water transportation, and agriculture crops.

The net present value criterion (present value of benefits minus present value of costs) was used to choose between alternative schemes. Both risk and sensitivity analysis were used to allow for uncertainties The main variables tested in the sensitivity analysis were: the discount rate, resettlement costs, the exchange rate, economic growth rates, coal prices, thermal and alternative hydro plant costs, projected electricity load, construction costs, and navigation traffic forecast. Because of the long construction period and the co-existence in China of a free market and state enterprises market, the sensitivity analysis had to take into consideration the ongoing process of price reform in China.

The project analysis period was 62 years where the first 12 years is for construction and the following 50 years is the operational lifetime of the project. The construction of support facilities is assumed to start in January 1989. It is assumed that the power will be available in year 12 (2000). Navigation benefits are also assumed to begin in 2000 while flood control benefits begin in 2004. Completion of the second powerhouse would be at the end of year 18 (2006).

The exchange rate is regulated in China and therefore equilibrium rates of 3.7 yuan/US dollar and 6 yuan/US dollar, based on a variety of economic indicators, were assumed in the sensitivity analysis.

PROJECT COSTS

The financial cost of construction of 16,113 million yuan was adjusted to an economic cost of 19,272 million yuan. These figures were based on a dam crest level of 185 m.

The project will result in the resettlement of many people along the Yangtze river. Between 500,000 to 1.2 million people are expected to be resettled depending on the flood control level (FCL) and the normal pool level (NPL), and based on a 1 in 20 year flood event. In addition to the number of people being moved, the estimated resettlement costs also include: rural job creation programs, the standards set for resettlement, infrastructure requirements, replacement of public buildings, *etc.*, and relocation of factories. The adjustment costs for the resettled, such as the need to learn new techniques of farming at higher altitudes with lower temperatures and on different soils were not considered, however. Based on four NPL/FCL alternatives, four alternative economic costs were estimated using conversion factors based on an assessment of the labor content of each activity and on the type of material required. The most significant adjustments were made to skilled labor which is known to be scarce in the reservoir area. All materials were assumed to be supplied from Chinese sources. The timing of the resettlement program was related to the construction program. The present value of the resettlement, at mid-1987, amounted to Yuan 4,000 million.

A major activity generating benefits in the gorges region has been tourism. The project will have a deleterious effect on this tourism. Flooding will remove a number of visually spectacular as well as historic sites and the consequent cost should be attributed to the project. Two percent of the total construction and resettlement costs were allocated for environmental studies and mitigation. 75% of this sum was to be incurred during project construction and 25% for ongoing monitoring and maintenance during the life of the project. The remaining costs are the capital and operating costs of the transmission system and the maintenance costs for the power plant and navigation facilities. The costs are calculated as the difference between the system transmission requirements 'with' and 'without' Three Gorges. Operation and maintenance of the power plants were taken to be 10 yuan per kilowatt installed. Operation and maintenance for navigation facilities were estimated to be 1% of the total cost of these facilities.

PROJECT BENEFITS

A detailed survey in the Jingjiang flood diversion area provided the basis of the estimates of existing and future flood damage potential. The damage potential includes agricultural and industrial losses, property damage, and the cost associated with the dislocated population during both the flood and subsequent rebuilding period. Damage to houses was estimated to be over 36% of the damage potential. Agriculture losses in shadow prices were estimated to be 23% and collective properties at 20% make up most of the remaining damage potential. Estimates of damage potential in other areas were linked to the estimates for the Jingjiang area with adjustments based on population and general statistics. Future economic development of the flood prone areas were taken into consideration. The Three Gorges project will be able to control 50% of the flooding that would otherwise occur on the middle of Yangtze between Yichang and Xilianghu with return periods of up to 1 in 200 years. (There is some evidence, however, that much of the Yangtze floodings were from below the dam site, and would not have been alleviated by the dam). The study had taken a conservative approach and did not include the benefit of saving lives and reducing the hardship and suffering because of floods. For example, the 1954 flood resulted in an estimated 30,000 lives being lost. Some 150,000 people died in the 1931 and 1935 floods on the Yangtze. The study also did not calculate the 'risk aversion' characteristic of the Chinese farmers living in the flooded areas. In other words, farmers in the flooded areas are willing to pay higher premiums for flood protection than would be required to cover the calculated economic costs of expected flood damages. The benefit from the project's flood control should be increased accordingly.

The Three Gorges project would create a completely new environment for navigation through the reservoir. The impoundment creates more consistent and less restrictive navigation conditions, increasing capacity and lowering costs for waterborne traffic on the river. The navigation benefits have been determined as the savings in transportation for future traffic. The problem faced by the analysts was to forecast the future traffic, so that navigation benefits were calculated for a range of future traffic levels. The medium scenario forecast was considered to be the most likely occurrence and was used as the basis for estimating the base project benefits. These forecasts had been developed using a 'bottom up' approach where each commodity now moving and likely to move in the future was analyzed separately. Then, a forecast of future river movements was developed by adding these individual commodity forecasts together. The study favored this approach over the use of surveys and statistical projection methodologies. The future capacity of the navigation facilities was based on a set of assumptions regarding the fleet mix, vessel size and lock cycle time and included an adjustment factor for random arrivals such as queuing and delays which can be kept at reasonable levels. Capacity can be expanded to some extent either by improving cycle time or fleet mix or by accepting longer delays. Ultimately, there is space available at the site to add another series of locks. The improved navigation environment will cause a reduction in operating costs which results from the following: larger tows can be brought into service, decline of the delay associated with one-way and winching sections, increase of daily operating hours, and reduction of fuel consumption due to the reduction of the channel velocity. The benefits of improved navigation were quantified by calculating the total annual costs of moving passengers and freight through the Upper Yangtze Corridor without and with the Three Gorges.

Power benefits represent a substantial portion of the benefits from the Three Gorges project reaching 70% of its total benefits. Depending on the NPL and FCL, the project has the potential to generate in the range of 13,000 to over 20,000 MW per year. The power benefits were determined as the avoided alternative cost of electric energy based on the assumption that China must in any event meet its increased energy demands. Hence, the total capital and operating cost for an electric generation system were calculated with and without the power from the Three Gorges project. Three scenarios were developed to generate the same amount of electricity without the Three Gorges. The first is an all thermal expansion which includes the presently committed hydro development. The second scenario includes some hydro which is not currently committed and some thermal developments. The third is a variation of the all thermal scenario but with a 3,000 MW interconnection between Central and East China. An added benefit of the Three Gorges project was the increase in energy production at Gezhouba which amounted to 1.5% of the average annual energy produced at Three Gorges. It should be noted that there is a potential contradiction between the flood control and power generation benefits that has not been fully explored. Simply stated, maximizing power production requires keeping the reservoir level at a maximum, which reduces significantly the potential to reduce flood damage.

INTEGRATED ANALYSIS

Because of the dependency of the different costs and benefits on the Flood Control Level (FCL), Normal Pool Level (NPL), and Power and Navigation Drawdown Level (PDL), and the large portions of the benefits that will be generated after many years, a sensitivity analysis had been done that also took into account the non-quantifiable factors. 13 NPL and FCL combinations were examined. In addition, these 13 schemes were analyzed using further sensitivity analysis on other factors. The impact of a 20% increase and decrease relative to the estimated value of the key costs and benefits was calculated. A positive net present value of the project resulted in all worse case scenarios (CIDA, 1988:6.1–6.9).

Significantly, many non-quantifiable elements were also considered in the analysis. Sedimentation is one of the problems which were not quantified in the study. It was argued that sedimentation near the head of the reservoir will be minimized if the combination of the FCL and NPL are kept below critical values. If the FCL is 145 m, the NPL should not exceed 165 m, or the perceived navigation benefits to Changqing may be significantly reduced by higher costs of channel maintenance. In addition there is the long term effect of sedimentation on the flood level. Sedimentation will accelerate the arrival of the 1 in 100 flood stage at Chongqing. These concerns made high NPLs and FCLs unfavorable solutions.

Social impacts of the resettlement were not included because of the difficulty in quantifying them. Since the differences among schemes are small, and because social impacts had not been considered, the study recommended that the scheme which affected fewer people be chosen. For example, one of the alternatives would generate more power and navigation benefits than a second, but the latter was deemed preferable because it has fewer impacts on people. By way of contrast, environmental impacts were given very little attention in the study (CIDA, 1988:6.15). Major environmental risks, such as earthquakes and landslides appear to have been ignored.

The study finally recommended the dam with NPL of 160m, FCL 140m, 22 units in the powerhouse, and 27 spillway blocks. The study also calculated the internal rate of return, the discount rate which equalizes the cost and the benefits, to be 15%. Sensitivity analysis yielded the following variances: variation of navigation traffic forecasts led to a change in net project benefits of between -4.8 and +10.8%. Economic growth alternatives had an impact on net benefits of +/- 19.5%. Power benefits represent 70% of the total project benefits and are dependent on coal prices, alternative thermal and hydro capital costs, and demand growth for power. The sensitivity analysis showed that a 20% decrease in the price of coal and in the cost of the thermal plants and other hydro sources would lead to a 14% decrease in the net benefits. When demand growth was held constant at 5% (instead of starting at 9% and decreasing to 5% by year 2010), the impact was a 16% decrease in the net benefits of the project. Exchange rate, discount rate, and cost assumptions, as well as the construction schedule of the project were also tested.

Risk analysis was undertaken for the recommended project to determine the combined effect of project-related uncertainties. Different risk scenarios were ex-

TABLE 4. Sensitivity analysis, Three Gorges Project

Sensitivity Factor	*Net Benefits*	*% Change*
Base Line Costs and Benefits	6,514	
Cost Sensitivities		
Construction Cost +10%	5,696	–12.6
Resettlement Cost +10%	6,109	–6.2
Transmission Cost +10%	6,453	–0.9
Economic Parameter Sensitivities		
Exchange Rate yuan 6/US$	8,607	+32.1
Discount Rate 12 %	2,644	–59.4
Construction Schedule		
1-year delay to 1st power	5,064	–22.5

amined based on the probability that a certain variable would take a specific value. The risk analysis incorporated the following major uncertainties: project construction cost, transmission and substation costs, resettlement cost, environmental mitigation costs, project operations, power benefits, flood control benefits, navigation benefits, and economic parameters. The calculated expected values of the benefits, the weighted average of all values in the distribution, were always higher than the base figures reflecting the conservative approach which was used in the analysis.

Financial analysis measures the overall project viability based on the financial rate of return (FRR), irrespective of the way in which the project may ultimately be financed. Financial values, not economic values, are used for costs and electric power revenues. FRR was found to be 8.95%. These calculations ignored the complexities of accounting methods, escalation and taxes. The study also implemented two sensitivity analyses for the type of financing: the concessional scenario, and the commercial scenario. Under both scenarios, the analyses found that there would be ample revenue to rapidly pay back all borrowing. The ability of China to finance the project had not been evaluated in the study.

The study also evaluated the project from a regional perspective rather than a national perspective. Positive impacts were found to outweigh negative impacts and the average income of the resettled population is expected to increase on average by 4% due to sectorial shifts and employment creation.

CRITIQUES OF THE THREE GORGES PROJECT

While the Canadian International Development Agency's (CIDA's) feasibility study considered the Three Gorges a viable project, opponents considered it a disaster, leading to major controversies, both internationally and within China. Concerns were raised over the immense investment, the long construction period, sedimentation problems and their potentially devastating effect on the inland port of Chongqing, resettlement difficulties, and doubts over China's technical capabilities to undertake such a massive project. Probe International reflected this opposition in its review of the CIDA feasibility study (Probe International, 1993).

With respect to resettlement, the review highlighted some important points. The CIDA analysis neglected the need to compensate the farmers who now live above the proposed reservoir level, but cultivate land below it. Also, the resettlement did not account for natural population increase and migration into the region which could increase the number of people who have to be resettled. With respect to the flood calculations, the review was concerned that CIDA flood probability analysis was too optimistic and was based only on historical data without taking into account the effect of deforestation during the last several decades which is expected to increase flood frequencies. Furthermore, the review predicted that the Chinese government will operate at higher reservoir levels than recommended by the study.

The reason has to do with the greater weight the government, for economic and political reasons, is likely to attach to maximizing power production.

Sedimentation problems represented a large portion of the criticism of the Three Gorges feasibility study. The review criticized the quality and the quantity of the basic field data that CIDA used in the sedimentation analysis, although CIDA's sedimentation calculations were not analyzed.

Comparative Analysis

The first two sections presented the feasibility studies and the critical reviews for the Sardar Sarovar and the Three Gorges projects. Some of the critiques of the Sardar Sarovar project were based on *ex post* results, which can be used to help us investigate in this section the likelihood that similar problems might be encountered during the construction of the Three Gorges Dam. We also address the criticism that appraisal based on CBA was the cause of many of Sardar Sarovar's problems, and therefore the results of the Three Gorges feasibility study, which used a similar appraisal technique, should not be accepted as the basis for decision making.

The similarities between the two projects makes a comparative analysis appropriate. Both are mega-projects with large capital investments. The projects are in developing countries which are more sensitive because of the uncertainties that surround such large projects. Both projects involve large scale resettlements and unknown environmental impacts. However, the objectives of the two projects were quite different. The Sardar Sarovar project is primarily an irrigation project with the potential to promote development in areas adjacent to the project. Drinking water for 40 million people will be also provided by the project. In contrast, flood control is claimed by the Chinese authorities to be the major reason for the construction of the Three Gorges Dam. The human rights record of the Chinese government has led opponents of the Three Gorges Dam to question this dedication to saving lives. They see the true objective as power generation for China's development program. Nevertheless, regardless of the intention of the Chinese government , flood protection will save lives and these constitute real benefits from the Three Gorges Dam.

To be conservative, the CIDA feasibility study deliberately omitted an estimate of the benefits resulting from lives saved (Huenemann, undated, 11). The failure to include this benefit not only understated the net present value of the project, but it contradicted the argument that flood control is the main reason for the dam. The appropriate conservative approach would be to assign a modest value to this benefit, rather than entirely omitting it. The reason may reflect an unwillingness to address the ongoing controversy regarding the value of human life.

We have prepared the following table to give an idea of the value of these benefits omitted from the CIDA study. The table calculates the net present value of the benefits gained from protection from one flood. The absence of information on the probability of floods and the expected number of lives lost per flood led us to produce a contingency table. Different estimates are given regarding the date at which a flood might occur after the end of construction. Also, various estimates are provided for alternative numbers of lives saved. The computation of the value of lives saved began with a widely used Canadian estimate, in which the human consequences of the loss of a life was valued at CDN$ 6.1 million (Ontario Safety Research Branch, 1994:10). This value was then adjusted to represent the Chinese value of life using the GNP per capita ratio of the two countries. The World Bank projects the future growth of China's GNP per capita to be higher than that of Canada. At the time when the CIDA study was undertaken (1988), GNP per capita in China was US$ 400 and in Canada was US$ 17,690. At the end of the 12 year construction period, the Chinese GNP per capita should reach US$ 684. For simplicity, the growth of the GNP per capita in both China and Canada is assumed to be equal during the analysis period (50 years after construction), although in fact the former is projected to grow much faster. Based on the above assumptions, the cost of a Chinese life, at the end of the construction period, amounts to CDN$ 236,000 which is equivalent to US$ 168,500. The values in the table are in billions of Yuan that have been discounted to mid-1987.

These estimates are only for the value of lives saved as a result of the project. The benefit from reducing the hardship and suffering of nearly one million people who

TABLE 5. Discounted net benefits from preventing one flood (billions of Yuan discounted to mid-1987)

Years after Construction	*20,000*	*Lives 30,000*	*Saved 40,000*	*50,000*
0	3.98	5.96	7.95	9.94
10	1.53	2.3	3.07	3.83
20	.591	.887	1.18	1.48
30	.228	0.342	0.456	0.570
40	0.079	0.132	0.178	0.220
50	.034	0.051	0.068	0.085

would likely become homeless due to flooding was not measured. If we take into account the facts that:

(1) CIDA's estimates of the net present value of the project amounted to Yuan 6.269 billion,

(2) the assumptions on which the calculations in the table are based on are very conservative, and

(3) that between 1931 and 1935, 150,000 people died in floods.

It can be reasonably concluded that had the benefits of saving lives been accounted for, the project would have appeared to be a much more attractive investment.

The calculation of power generation benefits of the Three Gorges was criticized as demonstrating a persistent error of CBA. This calculation was based on the cost of the alternative of thermal power generation. This method of calculation, which is really just a cost effectiveness measure (Huenemann, undated:11–14) omits the greater benefits of using environmentally more friendly hydro power. The study's omission of environmental benefits is probably for the same reason that it omitted the benefits from saving lives. Environmental benefits are controversial and difficult to calculate especially in developing countries where prices are distorted. The same omission in calculating benefits occurred in dealing with the Municipal and Industrial Water Supply in the Sardar Sarovar Project. The World Bank study calculated this benefit based on the cost of getting this water through the second best alternative which was moving the water by tank trucks. The benefit side, once again, was ignored, which in this case would have included the environmental benefits (associated with not having to use trucks) less the employment opportunities which could have been created in the trucking industry. In both cases, this error in calculation likely underestimates project benefits.

The method of dealing with resettlement is the major problem highlighted in the independent review of the Sardar Sarovar Project. The project appraisal of the World Bank totally ignored the cost associated with this anticipated resettlement due to the dam, the reservoir area, and the main canal. It can be argued that the Indian government did not provide the necessary information to estimate the cost of resettlement. The declared policy of the World Bank at the beginning of the 1980s was that no funds should be provided unless a clear resettlement plan is provided by the borrowing country. The Bank's deliberate disregard of its own policy leads one to wonder whether the critics of Multilateral Development Banks are correct, namely that they have a strong bias towards lending large amounts of money (Mistry, 1995:83). This bias reflects the institutional culture in which officers tend to be promoted more on the size than on the quality of their loans.

However, the Sardar Sarovar appraisal's treatment of resettlement was not repeated in the Three Gorges study. The CIDA feasibility study accounted for virtually all the anticipated costs associated with resettlement. Also, the resettlement costs were included in both the sensitivity and the risk analyses. Probe International, in criticizing the feasibility study, argued that the Chinese government would operate the Dam at higher Normal Pool Level and Flood Control Level to maximize the power production, and this would cause higher resettlement costs. Probe also suggests that the Chinese government would not in fact pay the resettlement costs that were identified in the study. This argument, regarding the intention of the Chinese government, cannot be tested or analyzed. From a purely technical perspective, a 20% increase in the resettlement costs still gives a positive net present value of Yuan 5,300 million to the project and that is what counts.

Both appraisals suffered from a lack of information with regard to environmental consequences. The exact impact of dams on the environment remains very difficult to ascertain and measure, although both projects have been exposed to intensive attacks on this issue. For example, damage to estuarine and seacoast fisheries, such as occurred with the Aswan Dam in Egypt, was not estimated. Benefits from reservoir fisheries, on the other hand, were also not counted. Similarly, more exotic but potentially significant environmental effects, such as greenhouse gas emissions (methane, *e.g.*) from decomposition in flooded reservoirs were also omitted. It is extremely unlikely that important socio cultural costs, such as the destruction of sites of religious or familial significance to relocated residents, were considered. Similarly neglected were health threats created by new catchment areas (*e.g.* schistosomiasis)

Unfortunately, the absence of data on these issues makes it impossible for us at this distance to shed any more light on this issue. Because of this problem, both feasibility studies recommended programs of research and efforts to monitor environmental impacts. However, the Sardar Sarovar appraisal did not assign any costs for such studies, while CIDA's appraisal allocated 3% of the total construction and resettlement costs for environmental research and monitoring.

A specific problem of sedimentation was overlooked in the World Bank appraisal of the Sardar Sarovar project. The Bank did not even mention the issue, although evidence shows the persistence of sedimentation problems in dam projects. In contrast, CIDA's feasibility study had undertaken an intensive examination of the issue based on earlier Canadian experience with sedimentation in the James Bay Dam in Quebec. Several recommendations were made regarding the operational NPL and FCL which would minimize the sedimentation problem. However, the feasibility study only examined the sedimentation effect on the reservoir capacity. The reduction of land fertility down stream from the dam, which is an expected consequence of the sedimentation problem, was ignored. One way to measure this additional cost would have been to measure the cost of increased use of fertilizers that would be required as an offset.

The World Bank appraisal of Sardar Sarovar assumed a 50 years operational lifetime for the project, but it included the construction period within the 50 years frame. The total construction period of the Sardar Sarovar was assumed to be 22 years, with some benefits such as power benefits to be realized after 9 years. The study should have at least started the 50 year operational period *after* the 9 year construction period. In failing to do so, the analysis underestimated the project benefits in the latter years of the project. This mistake was not made in CIDA's feasibility study where the study assumed that the operational lifetime of the project, also 50 years, would start after completing the first phase of the project (12 years).

The problems raised in this section, together with the sensitivity and the risk analyses of both projects, demonstrates that the World Bank appraisal for Sardar Sarovar omitted several crucial issues which affected the end result of the feasibility study, particularly underestimating costs. The failure to raise these issues even if they could not be quantified allowed the implementation authorities to ignore them, while leaving no opening for the Bank subsequently to pursue the issues with the Indian authorities. Furthermore, the risk analysis showed that with all these omissions the probability of yielding a negative net present value was 50%. Funding the Sardar Sarovar project based on its own flawed appraisal was an irresponsible decision by the World Bank. One suspects that the Bank had already decided to fund the project and the appraisal study was used to legitimize this decision. It is therefore inappropriate to condemn the CBA technique for the shortcomings of the Sardar Sarovar project. Most problems were properly addressed in CIDA's appraisal, even if, in some cases, there could be no quantification.

Almost all the omissions in CIDA's feasibility study were on the benefits side and are due to the conservative approach that was taken. The risk analysis of the Three Gorges Study showed that the probability that the project would realize a negative net present value is less than 10%. A major benefit, saving lives, should have been included in the analysis.. Although the appraisal yielded overall acceptable results, if these omissions had been included, the project would have been far more attractive, helping the Chinese authorities overcome their current difficulties in raising the required funds for the project in the international capital market.

Conclusion

The debate over large dams has been ongoing for years. Environmentalists and human rights activists have been criticizing these mega projects because of their negative impacts on the environment and also because of the population costs involved. Sardar Sarovar Dam is a case which has been used and will continue to be used by opponents of dam construction as a demonstration of the false promise of dam projects. The Cost-Benefit Analysis technique itself has come under attack but our investigation indicates that the flaws were not in the technique, but in the failure of the World Bank appraisal to use the technique in a manner that it itself endorses. The Three Gorges Study demonstrated that the technique could be sensibly utilized. However, although the CBA of the latter project endorsed its viability, China has had to struggle to raise funds in international financial markets. China's human rights record has not helped, but it appears that the Sardar Sarovar project has played a key role in turning opinion against the Three Gorges project.

Our aim is not to provide unqualified support for the Three Gorges project. But the CBA that was done provides an opportunity to move the debate from emotion to fact, allowing appraisers and evaluators to deal with the technical questions of appropriately measuring costs and benefits, and to allow for the insertion of mechanisms during project implementation that could increase the project's likelihood of success.

Cost-Benefit Analysis is an indispensable tool for appraising projects. It provides a framework for comprehensive and consistent analysis of very complex matters. Without it, mega-projects will be decided on the basis of biased or even dishonest criteria. Badly applied, as in the case of Sardar Sarovar, the same perverse result is likely. The correct lesson, therefore, is not to throw out a valuable tool, but to insist that it be applied properly, and with professional integrity.

Acknowledgement

Pieter Louppen, cartographer of the Department of Geography and Environmental Development at Ben-Gurion University of the Negev prepared the two excellent maps for this chapter.

List of Abbreviations

CBA	–	Cost Benefit Analysis
CIDA	–	Canadian International Development Agency
CYJV	–	CIPM—Yangtze Joint Venture
El	–	Elevation
ERR	–	Economic Rate of Return
FCL	–	Flood Control Level (allows for flood control capacity)
FRL	–	Full Reservoir Level
IBRD	–	International Bank for Reconstruction and Development (World Bank)
IDA	–	International Development Agency
MDL	–	Minimum Drawdown Level (allows for least amount of withdrawals)
MWL	–	Maximum Water Level
NPL	–	Normal Pool Level (assumes no flood control allowances or other contingencies)
NWDT	–	Narmada Water Disputes Tribunal
PDL	–	Power and Navigation Drawdown Level (level required to enable these uses)
PMF	–	Probable Maximum Level
R&R	–	Resettlement and Rehabilitation
TOR	–	Term of References

References

Canadian International Development Agency, CIDA (1988) Three Gorges Water Control Project Feasibility Study. Canadian International Development Agency, Ottawa.

Huenemann, R.W. (undated) *A Persistent Error in Cost-Benefit Analysis of Power Projects: The Case of the Three Gorges Dam in China.* University of Victoria mimeo, Victoria, British Columbia.

International Rivers Network (1994) Sardar Sarovar Project: An Overview.

Mistry, P. (1995) Multilateral Development Banks. FONDAD.

Morse, B., and T. Berger (1992) Sardar Sarovar: Report of the Independent Review. Resource Futures International.

Ontario Safety Research Branch (March 1994) The Social Cost of Motor Vehicle Crashes in Ontario. Toronto.

Probe International (1993) *Damming the Three Gorges; What Dam Builders Don't Want You to Know.* Probe International, Ottawa.

World Bank (1985) Narmada River Development—Staff Appraisal Report. World Bank Publications, No. 5107–IN , Washington.

ABOUT THE AUTHORS

Hisham El-Bihbety has a Bachelor of Science in Engineering from Cairo University and a Masters of Development Administration from Carleton University. At present, Mr. El-Bihbety is a Systems Analyst with a leading consulting company in Toronto. He has extensive experience managing information systems through designing and implementing applications using different development environments. This experience includes analysis of client needs and requirements, inter-system compatability, and re-engineering of working processes. His work as a design engineer for four years with an innternational consulting office, has provided him with the opportunity of working on a diverse range of projects in many countries.

Mr. Hisham El-Bihbety
School of Public Administration
Carleton University
Ottawa, Ontario
Canada

Harvey Lithwick is Professor of Economics at the School of Public Administration, Carleton University, Ottawa, Canada, and Senior Fellow at the Negev Center for Regional Development at Ben-Gurion University of the Negev, Beer-Sheva, Israel. He has published widely on the subjects of economic growth, project evaluations, urban economic development, and regional economic policy. He has served as an Assistant Secretary to the Ministry of Urban Affairs of the Government of Canada, and economic adviser to the Prime Minister of Canada. Among his most recent publications is *Frontiers in Regional Development*, co-edited with Prof. Yehuda Gradus, and published in 1996 by Rowman & Littlefield.

Prof. Harvey Lithwick
Negev Center for Regional Development
Ben-Gurion University of the Negev, Beer-Sheva, 84105
Israel (fax: 972-7-6278991; e-mail: lithwick@bgumail.bgu.ac.il)

Part Six

Renewable Energy Potential for Development of Arid Lands

Chapter 16

Solar Energy in Arid Frontiers: Designing a Photovoltaic Power Plant for Kibbutz Samar, Israel

David Faiman

Introduction

Kibbutz Samar is an agricultural community in the Arava valley, about 30 km north of the Red Sea port of Eilat (Figure 1). This area is located in the hottest and driest part of Israel, having a P/ETP index of 0.01, *i.e.* being hyperarid.[1] The principal source of income for Kibbutz Samar is based on specialized production of fruit and vegetables which are grown through irrigation, but without the use of chemical fertilizers. In fact the community takes such pride in their environmental awareness that they recently took a collective decision to try and generate all of their electricity requirements (about 200 kW) using solar energy. To this end, they contacted our Center, in order to inquire how they might realize this plan.

Two-hundred kW is an awkward size for a solar energy power plant. On the one hand, it is far too small to interest any of the commercial companies that have helped put Israel on the world map as regards solar power production. Here one has in mind the Ormat Corporation which developed the 5 MW solar pond generator at the Dead Sea (Tabor & Doron, 1990) and Luz Corporation (now Solel Solar Systems, Ltd.), whose concentrating-mirror systems have been used in California to generate hundreds of MW of electricity (Jaffe *et al.*, 1988). Both of these so-called 'solar-thermal' kinds of technology use heat, generated by solar energy, to drive turbines. Their financial viability are heavily governed by economies-of-scale and 200 kW is simply too small.

Photovoltaics, on the other hand, is a technology that is more modular and, consequently, less sensitive to system size. Unlike the solar-thermal technologies photovoltaic cells do not produce heat as an intermediate stage of the electricity production process. Instead, by virtue of the electronic structure of the materials from which they are made, the cells convert light directly to electric current. With

H. J. Bruins and H. Lithwick (eds.), The Arid Frontier, 321–336.

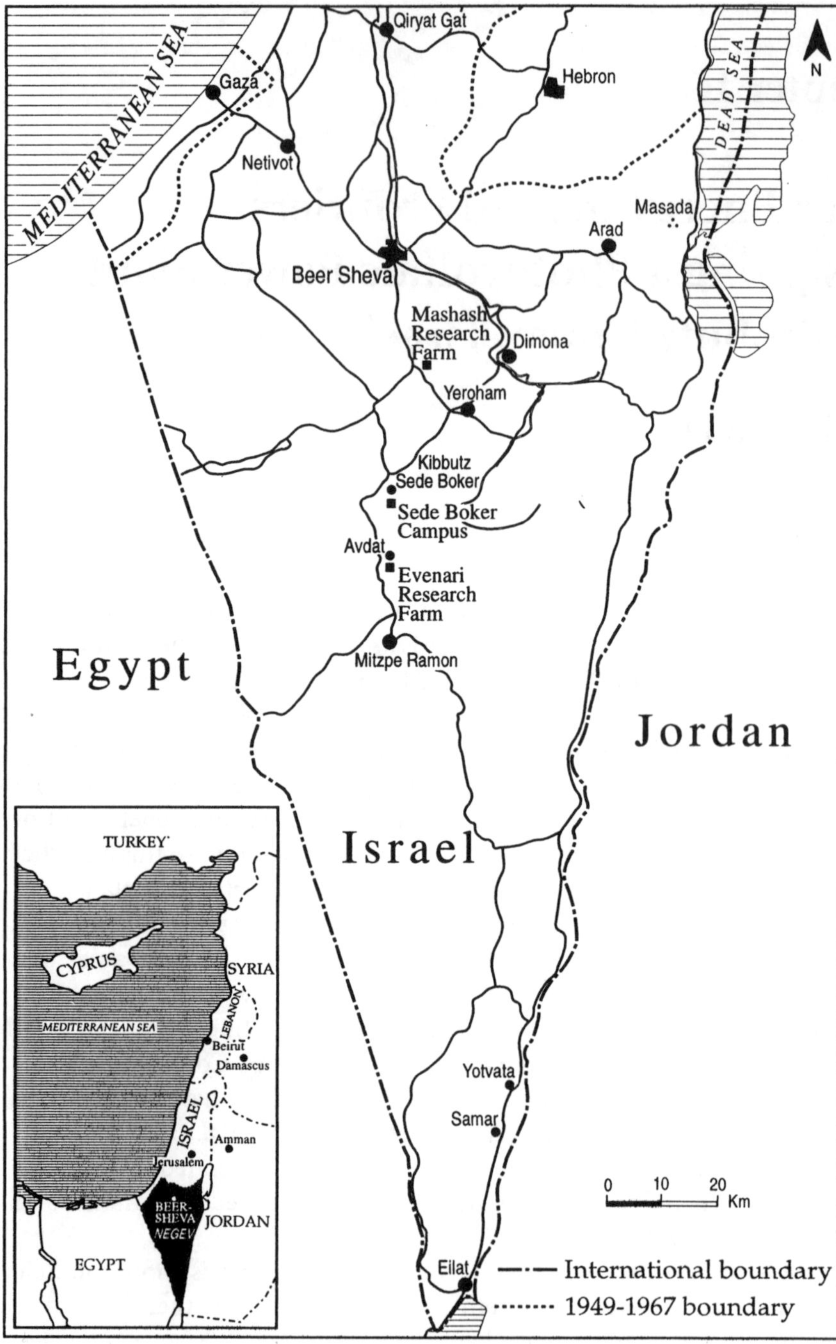

Figure 1. Location of Kibbutz Samar in the Arava Rift Valley (Southern Negev); the Sede Boker Campus is situated in the Central Negev Highlands.

the consequent absence of pumps and heat transfer fluids photovoltaic cells can be used in systems as small as wrist watches or as large as power stations. The only drawback with this kind of technology, as compared to the solar-thermal variety is that, at the present time, it is substantially more costly (Faiman, 1990).

Even though the economics of such a system seemed dubious at first, the project interested our Center for a number of reasons. First, a survey of solar radiation in the Negev desert, with which we have been involved for several years, indicates that Samar is located in the optimal part of the Negev for a photovoltaic system (Faiman et al., 1990). Secondly, no photovoltaic systems of anywhere near this size have ever been built in Israel. The two largest, respectively 4 kW and 10 kW, happen to be located at our Center and have been the subjects of extensive study. The time was therefore ripe to think about larger systems, particularly in view of everything that had been learned from the smaller ones. Thirdly, there was the prospect of effecting considerable cost-savings by virtue of the kibbutz members' readiness to perform much of the construction work themselves. Finally, scientists at our Institute had, some years previously, developed a computer simulation program for use in the design of photovoltaic power stations (Gordon & Wenger, 1991). It would be a fascinating exercise to run this program using the latest component cost figures and solar radiation data from the Samar vicinity, in order to determine the degree to which a photovoltaic power plant might be cost effective at this best of all possible sites in Israel.

The Solar Resource

The first question that needs a quantitative answer, when considering the use of any solar energy conversion device, is: How much solar energy is there? In the case of Kibbutz Samar the answer is readily available since less than 10 km away, at Yotvata, we have one of the Negev Radiation Data Base stations. The amount of available solar radiation at any given site depends upon the manner in which the solar collecting panels are oriented. If the panels are mounted on computer-controlled, two-axis trackers, they will always point directly at the sun and, thereby, collect the maximum amount of available solar energy. Such a set-up is shown in Figure 2. Alternatively, one could mount the panels on a horizontal axis that rotates them from east to west, in step with the sun's daily passage across the sky. Such a setup, depicted in Figure 3, employs fewer moving parts and intercepts almost as much summer radiation as a full 2-axis tracking system because, in summer, the sun passes relatively high overhead. In winter, however, such horizontal axis systems pay a severe energy penalty because the sun is too low in the sky to illuminate the panels in an efficient manner. If we want to do away with having to use moving parts altogether, the panels could be spread out, horizontally, on the ground so as to maximize the amount of summer radiation that is collected or, better still,

Figure 2. The 2–axis sun-tracker at the Sede Boker Campus: 75 modules constitute the photovoltaic system discussed in the text. The remaining 3 modules are parts of other experiments.

Figure 3. The NS horizontal-axis tracker at the Sede Boker Campus, showing some of the 60 modules comprising the system. Background: Mirror-enhanced modules in a different study.

TABLE 1. Monthly daily average global solar radiation on 1 m^2 of a variety of fixed and sun-tracking plane surfaces for a typical meteorological year at Yotvata (close to Kibbutz Samar)

Month	*Horizontal* *($kWh\ m^{-2}\ day^{-1}$)*	*30° fixed tilt* *($kWh\ m^{-2}\ day^{-1}$)*	*NS horiz. axis* *($kWh\ m^{-2}\ day^{-1}$)*	*Full tracking* *($kWh\ m^{-2}\ day^{-1}$)*
Jan	3.38	4.52	4.25	5.41
Feb	4.39	5.45	5.59	6.55
Mar	5.98	6.65	7.34	7.91
Apr	6.48	6.52	7.89	8.06
May	7.39	6.90	9.23	9.31
Jun	8.01	7.11	10.25	10.34
Jul	7.87	7.13	10.10	10.19
Aug	7.23	7.08	9.39	9.57
Sep	6.18	6.73	7.79	8.27
Oct	4.81	5.79	6.13	6.99
Nov	3.75	4.94	4.77	5.94
Dec	3.04	4.14	3.79	4.95
Year ($kWh\ m^{-2}$)	2087	2220	2635	2846

mounted on equator-facing stands, tilted so that the angle they make with the horizontal is precisely equal to the geographic latitude angle of the site. This is the fixed angle that maximizes the amount of annual collectable energy.

Table 1 shows typical monthly amounts of solar radiation that would be intercepted by solar collectors located at this site, arranged in each of the above four types of configuration (Faiman et al., 1995).

From Table 1 one sees that a horizontal solar collector of area 1 m^2 intercepts about 2,000 kWh of solar energy per year. Tilting the collector southward at an angle of 30° to the horizontal results in a 6% increase. Rotating the collector from east to west about a horizontal axis, so as to follow the daily movement of the sun, results in a 26% increase compared to a horizontal collector. Having the collector track the sun so that the latter is always at normal incidence results in a 36% increase over the horizontal situation. It must also be realized that these figures can fluctuate from year to year, by about ±5%, as a result of varying amounts of cloud cover and dust in the air.

There is also a more subtle aspect of the solar resource in that the voltage output of a photovoltaic cell is quite sensitive to the temperature at which it operates: As the temperature rises the voltage falls. When designing a photovoltaic power station one therefore needs to take into account the ambient temperature and wind

TABLE 2. Temperature and wind parameters from the Typical Meteorological Year data base at Yotvata

Month	T_{mean} (°C)	T_{max} (°C)	T_{min} (°C)	*Wind speed* ($m\ s^{-1}$)	*Prevailing direction*	*Prevailing occurrence (%)*
Jan	12.7	22.8	3.6	2.8	N	47
Feb	13.5	26.3	3.2	2.7	N	34
Mar	17.2	32.9	0.9	3.1	N	38
Apr	22.6	38.4	6.7	3.3	N	42
May	25.3	39.5	14.3	2.2	N	43
Jun	30.2	44.5	18.7	3.4	N	45
Jul	30.9	41.1	20.0	3.3	N	42
Aug	31.8	44.5	18.5	3.4	N	52
Sep	27.9	41.0	16.7	2.9	N	43
Oct	26.4	36.3	17.2	3.1	N	40
Nov	19.2	31.9	6.7	3.0	N	57
Dec	16.1	25.6	7.2	3.0	N	61

speed at various times of the day and year. Table 2 gives a monthly summary of these parameters for the Yotvata station (Faiman et al., 1995).

In Table 2, T_{mean} refers to the average hourly temperature during the actual month of data that was employed for the Typical Meteorological Year data base. It is *not* a long-term average. For photovoltaic system design purposes the correlation between solar radiation and ambient temperature and wind conditions is most important. Therefore, independent averages must not be used. T_{max} and T_{min} refer, respectively, to maximum and minimum temperature averages measured hourly during the month. Wind speed is the monthly average of hourly averages. Prevailing wind direction and occurrence indicate, respectively, the direction which occurs the largest number of times among the hourly average wind direction data and the percentage of data that actually point in this direction.

Tables 1 and 2 give average monthly figures to enable the reader to compare, at a glance, this site with others that might be more familiar. As will be explained below, however, it is the individual hourly data—not the monthly averages—that will be used for computational purposes.

Designing a Photovoltaic System

A typical photovoltaic panel or 'module', to use the accepted terminology, of the kind currently in mass production, generates a DC current of about 3 amperes under conditions of maximum solar illumination (about 1,000 W m^{-2}), is capable of giving about 50 W of electric power, depending on the temperature at which the module operates. However, the corresponding module voltage of about 16 V is too small for standard domestic appliances and, of course, its DC current output is similarly unsuitable. For a 240 V AC appliance one must therefore think in terms of strings of, say, 15 modules connected in series followed by a DC→AC inverter. In this manner one receives about 750 W of power. Some 4,000 of these photovoltaic modules would consequently be needed to produce the 200 kW of power that Kibbutz Samar wish to generate, but only under conditions of optimal solar illumination. (Manufacturers of photovoltaic modules employ a 'peak watt' notation, denoted W_P, to indicate the amount of power a module would produce under so-called 'standard test conditions', specifically, a solar illumination level of 1,000 W m^{-2} and a module temperature of 25°C). As will be seen below, however, a 200 kW_P system would probably never give 200 kW, even under 'peak' illumination conditions, passing clouds would result in even less power and at nighttime there would be none at all.

There are a number of options that may be employed to smooth out the availability of power from a photovoltaic system. One could, for example, employ storage batteries which would be charged by the varying solar output from the photovoltaic modules. The batteries themselves would then provide a constant power output. Provided one does not withdraw power at a greater rate than the average rate at which the batteries are being charged the frequency of power failures can be very small. For such systems, however, the batteries require attention and eventual replacement. Furthermore, the large number of storage batteries required would constitute a significant part of the system cost.

In the case of Kibbutz Samar there is a simpler option. The Kibbutz is connected to the national power grid and the Israel Electric Corporation is required, by law, to purchase electricity from anyone who has some to sell. Samar could, accordingly, connect their photovoltaic power station to the grid lines, sell surplus electricity during the day and purchase some at night. If income balances expenditure the photovoltaic system would thus, effectively, supply all of their annual electricity needs. The kibbutz decided to opt for the grid-connection alternative rather than to employ storage batteries.

Having thus decided that the photovoltaic system is to be grid-connected the next important decision for the Kibbutz Samar project relates to the array geometry. That is to say, not only must one decide on the series-parallel strategy to be employed for interconnecting the individual photovoltaic modules (many possibilities exist for 4,000 modules) but one must also decide whether the module arrays should be stationary or sun-tracking. For this purpose it is convenient to consult Table 1.

As we have already seen, Table 1 indicates that 2-axis tracking can be expected to produce the highest annual electricity output per given number of (expensive) modules purchased. But there are two hidden potential snags: Relatively large land requirements and many moving parts to maintain. Specifically, it is not feasible to place 4,000 modules on a single tracker. Rather, one would need to think in terms of some 60–70 trackers, if they were of the kind we have at the Sede Boker Campus (Figure 2). Such trackers need to be spaced relatively far apart from one another in order to prevent mutual shading. Shading, it must be emphasized, is a major concern in the design of photovoltaic systems since it converts active, power-producing, solar cells into passive, power-consuming resistors. In order to minimize this effect blocking diodes are employed but even then a shaded tracker, or even part of a tracker, can result in a serious loss of power. For two-axis trackers the land/module area ratio of large systems (Arnault et al., 1984) is required to be typically 8:1 in order to minimize shading losses. Furthermore, each tracker requires two computer-controlled motors in order to follow the sun's motion. This represents a large number of moving parts to maintain and service for a system that is located in a relatively remote area.

Although its energy output is somewhat lower, as can be seen in Table 1, a north-south horizontal axis tracker would be preferable for two reasons. First, the rows can be packed together more closely. A 10 kW_p system, with this geometry, at the Sede Boker Campus has a land/module area ratio of about 5:1. Secondly, the entire array is moved by a single computer-controlled motor. If one thinks of a 200 kW_p system as being something like the Sede Boker system (Figure 2) replicated 20 times, then only 20 motors would be necessary compared with over 100 motors for a comparable 2-axis tracking system.

Finally a non-tracking system, although poorer still in annual energy output, as indicated in Table 1, has no moving parts to worry about and occupies the minimum amount of land. Table 1 indicates that tilting the modules southward at an angle of 30° to the horizontal involves only a 6% annual energy increase compared to horizontal panels but such a fixed system actually exceeds the performance of a NS-axis tracker during the winter months of November–January. For truly horizontal panels the land/module area ratio would be 1:1. By tilting the modules one needs either to place them all in a single line (which would be 1–2 km in length) or provide sufficient space between the rows in order to minimize shading. Suppose, for example, one wanted to ensure that there would be no shading between 9 am and 3 pm at any time of the year. Then the land/module area for 30°-tilted photovoltaic modules would need to be approximately 2.5:1 at the latitude of Kibbutz Samar (and most other parts of Israel).

The choice between the last two geometries is interesting. NS-axis tracking requires only twice the amount of land area than a fixed array, and 20 tracking motors are probably a manageably small number when one considers the relatively large energy increase that may be expected. This choice should be made based on the relative economics of the two systems, as will be discussed below.

Finally, whatever array geometry is chosen, a decision must be taken regarding the size and type of DC→AC inverter and power conditioning equipment. For 2-axis trackers, each individual tracker would normally feed its own power-conditioning system of a few kW capacity. In effect, each tracker is a self-contained mini power station. For larger arrays, such as the NS-tracking or stationary types, economies-of-scale may play an important role via the use of large inverters. But the calculations are tricky. World experience indicates that the power-conditioning part tends to be the *Achilles heel* of photovoltaic power stations (Thomas & Stevens, 1994). Some types of small inverter (*i.e.* in the 10 kW range) have been found to operate quite reliably but there is not very much experience with large inverters (*i.e.* in the 100 kW range). If a small inverter fails, the loss of revenue, until the inverter is repaired, is not very large. This is not the case if a large inverter fails. If, as could well be the case for Kibbutz Samar, the inverter manufacturer is overseas, many small inverters may be more practical in the long run although a small number of larger units would probably cost less initially.

The Sede Boker Experience

An important aspect of system design decision making for Kibbutz Samar is the experience gained from the mini photovoltaic power stations at the Ben-Gurion National Energy Center, (Sede Boker Campus). At this laboratory we have had the opportunity to test a large variety of photovoltaic modules, and several systems, over a time period of about a decade.

For photovoltaic modules, the overwhelming lesson learned is that manufacturer's specifications always over-rate the power output that may be obtained. This is not deliberate deception. Rather, it results from the fact that mass-produced modules are tested in the factory (for quality control purposes) using a set of lamps designed to simulate, as well as possible, natural sunlight. But photovoltaic material is acutely sensitive to the spectral content of the illumination. Indeed, the spectrum of natural sunlight changes from noon to sunset, even on a clear day, and module output (suitably corrected for changing illumination levels and cell temperatures) is observed to vary by some 5% over this period of time. Clearly a set of electric lamps can not be expected to correctly simulate sunlight at all times (if, indeed, at all). At the Sede Boker Campus we test modules at noontime on clear days (Berman et al., 1995) when their power output is highest. We find that these figures are typically 10%-20% lower than manufacturer's specifications after the latter have been adjusted to the operating conditions under which we perform our tests. This is a most significant result for large system design: It warns us that the Kibbutz Samar system will either cost 10%–20% more than expected, or will produce 10%–20% less annual energy.

On the positive side, it is also our experience that modules made of single-crystal or polycrystalline silicon cells have extremely stable properties for many years. So if a reliable power-conditioning system can be installed the system may be expected to provide many years of service.

At the system level it is our experience too that most problems involve the so-called 'balance of system' components rather than the photovoltaic modules themselves.

Table 3 illustrates the observed annual performance of our 2-axis tracking system at the Sede Boker Campus (Figure 1). This system contains 75 photovoltaic modules fabricated from single-crystal Si solar cells and, being a complete sun-tracker, the only variations in solar intensity it experiences are those associated with passing clouds and time-of-day. Using manufacturer's ratings for the modules the system would be rated as 3.98 kW_p but using our measured module performance figures the true rating should be 3.53 kW_p. Real systems, however, do not operate at Standard Test Conditions: the solar radiation may vary from zero to more than 1,000 W m^{-2} as a result of time of day, passing clouds, etc. and module temperatures may reach 50°C, or more, in summer.

From Table 3 one sees that in spite of a measured efficiency of 11.9% for the individual modules (measured under the standard test conditions described above) the final AC system efficiency was only 8.7%. The DC→AC inverter for this system proved to be quite reliable and of relatively high efficiency (85% on average). One t

TABLE 3. The measured annual system efficiency of a 2-axis tracking photovoltaic system at the Sede Boker Campus (STC = 'standard test conditions')

Manufacturer's advertised module efficiency (at STC)	13.4%
Measured mean module efficiency at Sede Boker (at STC)	11.9%
Annual observed 2–axis system DC efficiency	10.2%
Annual observed 2–axis system AC efficiency	8.7%

hus sees that the advertised 53 W_p photovoltaic modules, employed for this system, need to be down-rated to 34–35 W_p in order to indicate true system performance for this kind of system. It should be mentioned that a weakness that manifested itself with this system was a problem with one of the two tracking motors. This resulted in undesirably long periods of system down-time (which have been corrected for in Table 3 but which, nevertheless, resulted in loss of revenue for this grid-connected system).

One of our NS-axis systems (Figure 2), on the other hand, employs 60 photovoltaic modules fabricated from polycrystalline Si solar cells. This system exhibited some problems caused by accidental self-shading at certain times of the year. Furthermore, system self-shading regularly caused small power losses in the early morn-

TABLE 4: Measured annual system efficiency of a NS-axis tracking photovoltaic system at Sede Boker (STC = 'standard test conditions')

Manufacturer's advertised module efficiency (at STC)	9.8%
Measured mean module efficiency at Sede Boker (at STC)	9.0%
Annual observed NS-axis system DC efficiency	7.0%
Annual observed NS-axis system AC efficiency	5.4%

ing and late afternoon. Table 4 summarizes the measured annual output from this NS-axis tracking system. The system is rated at 2.88 kW_p using the manufacturer's rating but its true peak rating, based on measurements under Negev sunshine, should be 2.64 kW_p.

Referring to Table 4 we may start with the manufacturer's advertised efficiency of nearly 10% for the type of photovoltaic module employed in this system. This figure is measured at the standard test conditions (STC) referred to above. Our own tests, however, under real Negev sunshine, indicated a module efficiency of only 9% when reduced to STC conditions. However, when varying radiation levels, module temperatures and solar incidence angles are included, the mean annual system efficiency is even lower: 7%, according to our observations. Lastly the observed average DC→AC inverter efficiency was less than 80% for the specific inverter employed by this system. This reduced the total annual solar energy → AC electrical energy efficiency to a mere 5.4%. Thus for a 'back of the envelope' estimate of how many modules would be needed for a photovoltaic system of any desired power, the nominally 48 W_p modules of Table 3 would need to be de-rated to an average contribution of only 26–27 W_p from each in order to take into account all of the various system losses that must be included for the present system. Here it should be pointed out that the tracking motor gave no problems. The DC→AC inverter of this system did, however, switch off on many occasions. This down-time has been corrected for in Table 4 but, as in the case of motor failures, it too results in loss of revenue.

Simulating Expected System Performance

As illustrated by the two examples of photovoltaic systems at the Sede Boker Campus, by using only solar radiation data and manufacturer's module ratings one cannot obtain a reliable estimate of the amount of energy that will be generated by the system. One must, instead, perform calculations of a more detailed kind in order to be able to study the potential economic value of building a system. For such purposes a computer simulation is convenient because it enables one to study the effect of various changes in system design on the overall economics. The basic ingredients of a computer simulation are as follows.

First, one gives the computer a file of hourly meteorological data for an entire year. These data enable the computer to calculate the intensity and angle of incidence of the sun's radiation and the corresponding module temperature.

Second, one specifies the array geometry: *i.e.* series/parallel connections among the modules, their tilt angle (if fixed), the orientation and spacing between rows. This enables the computer to identify the degree of shading that may occur for certain modules at various times of the day or year.

Third, one gives the computer technical information about the type of photovoltaic modules to be used. This information is of a kind that enables the computer to calculate the module's electrical output power for any given level of illumination and module temperature. It also enables the computer to take into account any partial shading of the module that may occur.

Fourth, one specifies certain technical information about the DC→AC inverter so that the computer may calculate the amount of AC energy produced during each hour.

The computer program naturally contains all of the necessary formulae, electrical, thermal and geometrical to enable it to perform the calculations. The results may be displayed in any convenient format, *e.g.*, graphical, spread sheet, etc.

For the purposes of the Kibbutz Samar project we have used PVISRAEL, a simulation program developed a few years ago by colleagues at our Institute (Gordon & Wenger, 1991). This permits us to obtain, for each month of the year, the number

TABLE 5. Hourly simulated output at Kibbutz Samar (in kWh, AC) of a 1 kW_p photovoltaic system with fixed south-facing array, tilted at 30° to the horizontal

Hour	*Jan*	*Feb*	*Mar*	*Apr*	*May*	*Jun*	*Jul*	*Aug*	*Sep*	*Oct*	*Nov*	*Dec*
6				0.09	0.77	1.07	0.62	0.18				
7		0.15	0.81	2.62	3.18	3.18	2.61	2.25	2.01	1.04	0.15	2.51
8	2.79	3.47	4.95	7.30	7.77	7.25	6.96	7.01	6.87	6.76	4.25	7.82
9	8.67	9.32	11.01	13.06	13.24	12.75	12.63	13.14	13.09	13.43	10.30	12.44
10	14.93	15.01	16.00	18.58	17.62	17.07	17.20	17.94	17.96	18.78	15.89	16.41
11	19.09	17.14	18.43	20.91	19.86	19.70	19.99	21.06	20.94	21.67	18.82	18.14
12	20.35	18.84	19.24	21.54	20.81	20.77	21.17	22.28	21.98	22.42	19.70	16.88
13	18.79	18.30	19.82	20.94	20.41	20.33	20.87	21.92	21.42	21.22	18.58	14.81
14	15.91	16.35	18.11	18.64	18.15	18.51	19.04	19.94	19.06	18.27	15.25	10.41
15	12.10	12.26	13.97	14.86	14.79	15.09	15.56	16.32	15.00	13.62	11.07	4.77
16	6.23	7.71	8.70	9.75	10.14	10.09	10.55	10.89	9.17	7.20	5.22	0.16
17	1.35	2.86	3.49	4.25	4.45	4.57	4.88	4.78	3.24	1.30	0.28	
18			0.27	0.58	0.86	0.97	1.11	0.86	0.13			

of kWh of electrical energy that would be generated each hour of the day. This monthly-hourly output format is convenient because the Kibbutz intends selling the electricity to the utility company. The Israel Electric Corporation employs different tariffs depending on the time of day, the day of the week and the season of the year. One may, accordingly, convert the energy output given by the computer simulation directly into an estimate of annual income.

Table 5 presents a typical result of such a simulation for Kibbutz Samar. The specific module and inverter details that were employed for the simulation summarized in Table 5 need not concern us as they are in no sense final. However, a table such as this, normalized to 1 kW_P (*i.e.* manufacturer's rating), of photovoltaic modules is a powerful aid to sizing the system and computing the corresponding economics. The total amount of AC annual energy generated by this 1 kW_P system sums to 1,665 kWh which represents a conversion efficiency of about 8.7%. Folding in the varying Israel Electric Corporation tariffs the annual income would amount to approximately $123, *i.e.* the average value of the electricity is about $0.074/kWh. Suppose the system were to be purchased with a 20 year loan at 8% annual interest. It would then have to cost no more than $1.23/$W_P$ in order to be cost competitive with conventional electricity in this region. More favorable financing, *e.g.* with government support, could help raise this latter figure and render it more commensurate with current photovoltaic module prices of around $4/$W_P$.

Similar matrices to Table 5 were computed for NS-axis and 2-axis tracking systems and, of course, the predicted amounts of annual electricity generated by such systems were found to be correspondingly larger. However, an essential part of maintaining low system costs is the fact that for stationary photovoltaic modules the kibbutz members can perform all of the construction work themselves. There is also no additional financial cost for motors and guidance systems or for their maintenance. The practicalities of being in a relatively remote area therefore strongly suggest that a stationary system will prove to be the most reliable in the long run and hence the most cost-effective.

Conclusions

In summing up the importance of this project it is convenient to draw a distinction between the scientific, technical and strategic aims. At the scientific level there are a number of questions associated with solar simulation programs that require further research and the Kibbutz Samar project may be expected to go a long way towards answering them. Without dwelling too much on details, one has in mind subtle problems such as the varying sensitivity of photovoltaic module output power to changing spectral content of the incoming light—a 5% effect that is usually neglected by simulation programs for lack of data. There are also more prosaic problems such as the fact that most simulation programs use as input the manufacturer's efficiency ratings for their modules rather than the, now-known, lower, true ratings. Thus the actual performance of the Kibbutz Samar system will help to make improved simulation programs more accurate than at present.

At the technical level this system, being the first of its size range in Israel, will provide much valuable information concerning the matching of photovoltaic systems to the local grid. Such information exists, of course, in several other countries but every region has its own power consumption time-of-day patterns and relevant information is not easily obtained from overseas systems. Specifically, the Kibbutz Samar project will enable the local utility company to study the appropriate tariffs for such systems. For example, one method of charging for electricity is to allow the kibbutz electricity meter to run forwards, as usual, when Samar is drawing electricity from the grid, in particular at night, and backwards when it is generating power. But should such consumers be encouraged to 'buy cheap and sell dear'? This is clearly beneficial for both parties provided that the utility company needs extra power during the day. But, presumably, were enough photovoltaic systems to be built, there would come a point when the Israel Electric Corporation would have a large amount of wasted generating capacity the sole role of which would be to provide photovoltaically-powered consumers with electricity at night. Such a situation will not, however, arise in the near future since Israel has about 6,000 MW of generating capacity compared to the 0.2 MW_p that Kibbutz Samar plans to build. One may thus look forward to a successful project spawning many duplicates—perhaps 100 or more, with their concomitant benefit to the environment and the power company.

At the strategic level there are a number of important conclusions. The Samar system will operate under the most arid of desert conditions that photovoltaic systems are likely to encounter. The life-time and O&M costs of this system and its components will therefore play an important role in decision-making both for Israel and for many other desert areas that do not yet have adequate electricity supplies. An important aspect of these decisions will no-doubt be influenced by the fact that the Kibbutz Samar system will have been built and maintained by farmers with little technical training.

Secondly, this 200 kW_p system will generate over 300,000 kWh of electrical energy annually. This will save some 500 tons of carbon dioxide—the undesirable by-product of burning fossil fuel in order to generate this much electricity by conventional means—from being released into the atmosphere. But it is important to learn the monetary cost of employing photovoltaics to effect such environmental cleansing. Kibbutz Samar—being located in an ideal climatic position for benefiting from photovoltaics and by virtue of its members willingness to significantly lower the balance-of-system costs by their own labors—will have a lowest-possible cost system, given the present status of hardware costs. This will provide a new direction for environmental investigations. For, instead of trying to estimate what the true cost of electricity would be were the utilities to be forced to reduce CO_2 emissions to zero, one will be able to turn the question around. Given the present (*i.e.* Kibbutz Samar) cost of photovoltaic electricity one could ask what kind of technology (if any) could be employed together with fossil-fueled power stations in order to achieve zero greenhouse gas emission for this price.

Acknowledgements

The initiative for this project came from Brian Medwed of Kibbutz Samar. The quantitative results tabulated in this chapter involved the labors of my colleagues, David Berman, Daniel Feuermann, Peter Ibbetson, Howard Wenger and Amos Zemel. The Yotvata climatic data, employed for simulating the Kibbutz Samar system, were provided by the Israel Meteorological Service. The research was funded by the Israel Ministry of Energy and Infrastructure.

Notes

1. Hyperarid according to the classification system by UNESCO (1979), based on the P/ETP index: average annual precipitation, divided by the average annual potential evapotranspiration.

References

Arnault, R.J., E. Berman, C.F. Gay, R.E.L. Tolbert and J.W. Yerkes (1984) The ASI one-megawatt photovoltaic power plant. *In* S.V. Szokolay (ed.) *Solar World Congress.* Pergamon Press, Oxford, p. 1624.

Berman, D., S. Biryukov and D. Faiman (1995) EVA laminate browning after more than 5 years in a grid-connected, mirror-assisted, photovoltaic system in the Negev desert: Effect on module efficiency. *Solar Energy Materials and Solar Cells* 36:421.

Faiman, D. (1993) The present costs of solar electricity. *Perspectives in Energy* 2:127.

Faiman, D., D. Feuermann, P. Ibbetson and A. Zemel (1990, 1992, 1993) Data processing for the Negev radiation survey. *Israel Ministry of Energy and Infrastructure final reports*: RD-14-90; RD 03-92; RD-06-93; and RD-07-93.

Faiman, D., D. Feuermann, P. Ibbetson and A. Zemel (1993, 1994) Data processing for the Negev radiation Survey: Data for 1993 and 1994. (Work performed under contract No. 94-11-038 for the Israel Ministry of Energy and Infrastructure, First interim report, March 1995.)

Gordon, J.M., and H.J. Wenger (1991) Central-station solar photovoltaic systems: Field layout, tracker, and array geometry sensitivity studies. *Solar Energy* 46:211.

Jaffe, D., S. Friedlander and D. Kearney (1988) The Luz solar electric generating systems in California. In W.H. Bloss and F. Pfisterer (eds.) *Advances in Solar Energy Technology.* Pergamon Press, Oxford, p. 519.

Tabor, H.Z., and B. Doron (1990) The Beith Ha'Arava 5 MW(e) solar pond power plant (SPPP) Progress report. *Solar Energy* 45:247.

Thomas, M.G. and J.W. Stevens (1994) Highlights of recent balance of system research and evaluation. In L. Mrig (ed.) *Proc. Photovoltaic Performance and Reliability Workshop,* Sept. 21–23, 1994, Lakewood, Colorado, (NREL/CP-411-7414 DE 95000251), p. 572.

UNESCO (1979) Map of the world distribution of arid regions. UNESCO, Paris, Man and the Biosphere (MAB) Technical Notes 7.

ABOUT THE AUTHOR

David Faiman, born in England in 1944, obtained his Ph.D. in 1969 at the University of Illinois, U.S.A. He joined the Ben-Gurion University of the Negev in 1976, holding joint appointments with the Department of Physics, Beer-Sheva and the Jacob Blaustein Institute for Desert Research, Sede Boker Campus (which latter he helped establish). He founded and headed the Applied Solar Calculations Unit (later re-named the Center for Energy and Environmental Physics) in 1977. He was appointed Director of the Ben-Gurion National Solar Energy Center in 1991. Current status: Full Professor with Tenure. Author or co-author of approximately 100 articles on: the physics of elementary particles; solar-thermal technology; solar radiation studies; the economics of solar power; photovoltaic devices and systems; fullerenes. Current major areas of research: photovoltaics and solar radiation studies.

Prof. David Faiman, Director
Ben-Gurion National Solar Energy Center
Jacob Blaustein Institute for Desert Research
Ben-Gurion University of the Negev
Sede Boker Campus, 84990
Israel (e-mail: faiman@bgumail.bgu.ac.il)

Chapter 17

Rural Electrification with Renewable Energy: Geothermal Power in Arid Regions

Tsvi Meidav

Introduction

Abundant sources of alternative energy are available in various parts of the world. 'Big is Beautiful' was the unstated motto of Western power planners, in designing power additions in the Northern hemisphere. Economies of scale favored large central plants. Costly high voltage, low loss transmission lines paid for themselves because they carried large loads and operated at high capacity factors. Furthermore, emissions to the atmosphere were considered a secondary factor, not taken into account until about a decade ago. 'The solution to pollution is dilution' was another, now discredited motto, *i.e.* tall enough smoke stacks will distribute the pollution over a large enough area, making the concentration of pollutants small enough that its effects could be ignored. Hence, development of cleaner alternative energy resources, which are amenable to smaller, distributed networks, lagged behind those that were large and relied on fossil fuels.

Nearly half of the world's population does not enjoy the benefits of electrical power, due to distance from a grid, small local demand which does not justify interconnection, or locations which cannot be connected readily (islands or high mountain regions). Some of those regions, such as the High Andes or the Afar Desert of Ethiopia, are blessed with geothermal resources which could provide electricity at reasonable prices without interconnection to a national grid.

After defining the characteristics of several renewable power technologies, we attempt to provide some guidelines to the selection and suitability of these techniques in remote, rural areas. Particular attention is paid to the utilization of geothermal energy by itself or in combination with other renewable technologies. The value of renewable energy is briefly discussed from the point of view of environmental quality. The benefit to society from the utilization of renewable energy resources, such as geothermal energy, lies in the fact, that besides conserving non-renewable fossil resources such as oil and gas, renewable resources do not generally

H. J. Bruins and H. Lithwick (eds.), The Arid Frontier, 337–350.

cause emission of carbon dioxide or other gases which contribute to global warming. Various governmental and research institutions have estimated the environmental costs of electricity, when that electricity is generated by the utilization of fossil fuels (Ottinger *et al.*, 1990). These environmental costs, often referred to as 'externality costs' are hidden costs of utilizing polluting sources of electricity. The cost of remedial activities such as the costs of planting trees which would absorb carbon dioxide of an equivalent amount to that emitted by carbon dioxide emitting power plants is called the externality cost.

This chapter deals specifically with the applicability of geothermal energy in remote, arid regions where geothermal energy may occur. We discuss the feasibility of integrating geothermal sources, which are particularly suitable for continuous use, with other sources of energy which can be economically run intermittently, as peaking power.

Environment, Health and the Choice of Energy

The 1992 Earth Summit called attention to the fact that indiscriminate use of fossil energy can cause irreparable harm to the global environment. This harm would be expressed primarily in global warming, as much as 4.5°C by the end of the 21st century. Global warming could cause major changes in the agricultural productivity of vast regions of the earth, a substantial rise in the level of oceans which would inundate island and coastal areas, and other catastrophic events.

Countries which were signatories to the Earth Summit pledged to significantly reduce carbon dioxide emission to the atmosphere, such that the total global emission by the year 2000 would be no greater than that in 1990. Emission of carbon dioxide to the atmosphere amounts to about 1.2 kg per kWh produced from coal burning; the CO_2 emission drops to about 0.9 kg/kWh if oil is utilized, and to about 0.4 kg/kWh if natural gas is employed. Although the earth is rich in fossil fuel resources, it is evident that addition of electricity supplies in the future must reduce the amount of electricity needed per unit of product manufactured in industry or the amount of lighting needed at home or the office, as well as increase the utilization of alternative energy resources over those utilized in the past, if the increase in global warming is to be arrested (Gore, 1993; Romm & Lovins, 1992). The alternative resources include geothermal, wind, biomass and solar, if CO_2 emission is to be lowered.

The externality cost of fossil fuels (*i.e.* the cost of mitigating the damage caused by a given fuel), is not taken into account in the conventional analysis of electricity production cost per unit of energy produced. That cost depends upon the fuel. Thus, the cost of global warming or preservation of the human habitat, should be a factor of major importance in evaluating alternative energy paths. This paper assumes *a priori* that alternative energy paths are always preferable to those based upon fossil fuel energy in power production, because alternative energy sources such as wind, solar or geothermal do not contribute to global warming.

The Public Service Commission of Nevada (1991) estimated the externality costs associated with different fuels is as shown in Table 1. These externality costs include the various pollution components associated with each of the fuels considered, including CO_2, carbon monoxide, methane, nitrogen oxides, sulfur oxides, reactive organic gases and particulate matter.

TABLE 1. Externality costs of different fuels, Nevada

Fuel Type	*\$/kWh*
Coal	0.050
Oil	0.031
Natural Gas	0.023

These externality costs deal only with direct environmental impacts of the specific fuels and do not include the external military or political costs of assuring the security of some fossil fuel costs (Romm & Lovins, 1992). Those latter costs could greatly increase the true national costs of the fossil fuels, but are paid out of other pots of money, such as the defense department, and are therefore not included in the true cost of fossil energy.

In the absence of clear regulatory requirements, electric utility planners generally ignore the externality costs associated with the specific fuels that are selected for the next plant to be constructed. Ignoring externality costs of fuels would tend to aggravate the perilous state of the global environment. This undesirable approach to energy supply could be reversed if governments would impose some uniform taxes on energy supplies, which reflect the environmental costs of the pollutants. Such a tax would encourage the abatement of pollutants or their displacement by alternative energy sources, and by demand side management. In 1991, the European Community proposed the imposition of a uniform tax of \$100 per ton of carbon emitted into the atmosphere. That proposal has not been adopted to date, because of the lack of support for it by the U.S. Congress.

An example of the external cost of energy utilization is provided by the World Health Organization which asserts that respiratory infections are now the leading killer of children under five, accounting for more than 3.5 million deaths annually (Lenssen, 1993). Likewise, it is estimated that acid rain in China, a heavy coal user, now affects more than 14% of the country's population, such as an increase in various respiratory and other diseases associated with coal-burning power plants.

Internalizing the external costs of energy supply would lead to a better allocation of economic resources (Friedrich & Voss, 1993) and would accelerate the development of demand-side management and alternative energy resources. Reduction of energy losses associated with long transmission lines in rural regions may become a significant contributor to increasing the overall efficiency of power generation. Development of small, alternative energy resources, when locally available, may become an important factor in reducing the amount of pollution associated with increase in the power supply additions as well in the unit cost of electricity provided (Sinha & Kandpal, 1991; Johansson *et al.*, 1993).

DECENTRALIZED VERSUS GRID CONNECTED ELECTRICITY

Remote regions may find that a local supply of electricity, despite its inherent inefficiencies, may be more cost effective than the extension of long, costly transmission lines to the region. In some isolated areas and islands, local supply, although more expensive than grid-connected electricity at short distances from the source, is the only viable solution to electrification. Experience in India has shown that decentralized energy technologies based on local resources can be a viable alternative for rural electrification (Sinha & Kandpal, 1991).

The cost of electricity is partially dependent upon the percentage of the time that the power plants is actually generating electricity (*i.e.* its load factor). Low load factors and long transmission and distribution lines, coupled with high transmission and distribution losses, make many rural electrification programs economically unattractive. Average load factors in parts of rural India, for example, may be as low as 6% to a high of 16%. According to Sinha and Kandpal, the losses of power between a central power plant and the finally distributed power to the consumer in rural India are often as much as 40% of the power produced at the power plant.

Because of these factors, local supply of electricity may be the most effective in remote areas, even though the apparent unit cost of electricity may be substantially higher than that paid by customers in a densely populated region which is well served by central power plants. Renewable energy sources such as biomass, geothermal, wind, and solar may be the appropriate sources of power in the smaller, less dense or remote areas which cannot be effectively served by the central grid.

Renewable Sources of Electricity

Abundant sources of alternative energy are available in various parts of the world. These include hydro, biomass, geothermal, wind and solar. Energy substitution, or demand-side management offers an important supplement to supply additions, especially in the case of geothermal.

The following describes the different sources of energy, with a discussion of advantages and shortcomings of each.

BIOMASS

Biomass (plant matter) accounts for about 15% of world energy use and 38% of energy use in developing countries (Hall *et al.*, 1992). Biomass, primarily in the form of agricultural and industrial waste, burned to fuel conventional steam turbines to produce electricity, has been in use for many years. Biomass is often characterized as 'non-polluting' despite the large emission of carbon dioxide and other pollutants, because the amount of CO_2 emitted in the combustion process equals the amount absorbed from the atmosphere during the growth of plants and photosynthesis of atmospheric CO_2. This is, however, only correct for young trees or annually harvested biomass, such as the bagasse remaining from the extraction of juice from sugar cane.

In developing countries, growing biomass for energy in an environmentally sensitive manner, can provide livelihood for farmers and pay for restoration of land. Large scale biomass production as a crop will be limited to regions of high rainfall or abundant resources of fresh water. A common example of biomass conversion into electricity is the burning of bagasse, the left-over residue of sugar cane, in a conventional steam boiler plant. Likewise, the residues of growing of corn, wheat, rice and other field products provide a large mass of residue suitable for electricity production.

Approximately 9,000 MW of biomass-based electric power is installed in the U.S., as a result of incentives provided by the government to private cogenerators who are considered Qualifying Facilities. Biomass plants have hitherto tended to be small (because of the dispersed nature of the feedstock), low pressure (because of decreased unit costs of boilers at low pressures) and low efficiencies, in the range of 14–18% in California, compared to about 40% for a modern coal plant. As a result, biomass units in the U.S. have to rely on low-, zero- or negative cost of the biomass fuel to be operated economically (Williams & Larson, 1993). It is expected that with the advent of biomass-integrated gasifier/gas turbines (BIG/GT), unit costs of electricity production will dramatically decline in the near term, even for smaller power plants, bringing it down to about 5 cents per kWh for the total cost of generation, in 1993 dollars. The actual cost will be higher when various other external costs are taken into account, such as financing costs, interest during construction and profit (if constructed by private investors). Typically, a privately owned facility must charge between six and seven cents/kWh (in 1993 dollars) to be minimally profitable.

Biomass-based power generation enjoys the advantage of being capable of load-following, with power output that can be increased or decreased with demand. Wherever available, biomass offers substantial advantages because of its flexibility of supply. A large quantity of biomass matter may be available in agricultural communities during the relatively short season following the harvest, such as after the processing of sugar cane, but become scarce during the growing season. Adequate planning is needed for assurance of a continuous supply of the biomass, to ensure a steady output from the power plant.

WIND POWER

The great advantage of wind power is that it can be constructed in small modules, has a relatively low environmental impact, and involves a fundamentally simple technology. The chief shortcoming of wind power is that it produces power only when the wind blows. Wind power supply typically reaches its peak in the afternoon or night time in many areas. The typical availability factor of wind turbines has been in the range of 15–20%, due to the vagaries of wind. With the advent of more flexible wind turbines which can operate at a wider range of wind speeds, it is anticipated that wind power availability may increase to 25% in good wind areas. Likewise, the cost of wind-based power plants is continually declining, and is now already in the range of $800–1,000 per kW of installed capacity. 'Installed capacity' is the ability of a plant to deliver a given amount of power. For example, an installed capacity of one megawatt means that the plant is capable of delivering one megawatt of electricity, if provided with the fuel required to operate the plant. Since the wind in good wind areas blows only about 25% of the time at the required velocity to operate the wind turbine, the cost of the electricity is greater than that of a power plant of similar cost which operates an greater percentage of the time.

The major problem with wind power is its irregularity. It can therefore be a valuable addition to a grid only if the region already has a reliable baseload source of power, and where the wind contributes only a small fraction of the total power needed. Alternatively, battery storage can augment the availability of wind power in remote areas. The cost of battery storage can exceed the cost of generation.

SOLAR POWER

Solar power enjoys the advantage of tracking the increase in electric demand during the day (assuming cloudless weather). Like wind power, it suffers from low availability, which is about 25% of installed capacity. Solar-thermal electricity could be produced in 1992 for about 9.3 cents/kWh in large installations (Johansson *et al.*, 1993). It is anticipated that with further research, the cost of solar-electric power would ultimately decline to about 6 cents/kWh (in 1992 dollars). The cost of solar power may double, if the cost of battery storage is added to the cost of generation (Entingh *et al.*, 1994).

Photovoltaic power generation enjoys the advantage of simplicity and the possibility of having small scale units installed, but suffers from the high cost of electricity that is generated with it at present, which is estimated at 25–35 cents/kWh in 1992 (Johannson *et al.*, 1993). Continuous supply of solar-based power electricity is possible, if battery storage is utilized. However, the addition of such an installation may increase the cost of solar power to more than twice the above cited figure (Entingh *et al.*, 1994). Despite its high cost, photo-voltaic power generation can offer a significant contribution in remote areas which are far from an interconnected grid. Furthermore, being available during peak demand, solar power may justify the high costs associated with it even in some grid-connected systems. Keeping in mind the very high losses that are often associated with grid connected electricity in remote areas, the cost of photo-voltaic power may not be exorbitant.

GEOTHERMAL ENERGY

Anywhere on the earth's surface, temperatures of rocks increase at a rate of 25–35°C per km of depth. Temperature gradients may increase much more rapidly, as much as 500°C per km, along geologic plate boundaries and their adjacent areas (see Figure 1), where volcanic activity has occurred in the last million years. In areas of low or moderate temperature gradients (less than 60°C per km), such as Europe and Eurasia, geothermal water are already being pumped for direct heat applications such as space heating and greenhouse operations on a major scale. It is estimated that over 11,000 Megawatt of thermal were in use by the end of 1989 on a worldwide basis, and that this rate would double by the year 2000 (Palmerini, 1993).

Arid regions characterized by the availability of geothermal energy of commercial quality (*i.e.* high enough temperature to be economically usable) include large parts of the Ethiopian Rift Valley, substantial portion of the High Andes of South America and others. There are rural areas in non-arid regions which are remote from other sources of renewable energy and could benefit from available high quality geothermal energy. These include many of the islands of Indonesia and the Philippines, parts of Central America (Nicaragua, Guatemala, El Salvador and Honduras).

Electricity production from geothermal energy has been in use along the world's geothermal belts (Fig. 1) for the past 80 years. At present, about 6,000 MW of geothermal power are in operation, with power plants varying in size from 300 kW to 135 MW each. The temperature of the geothermal reservoirs tapped for power production vary from as little as 90°C to nearly 400°C.

Efficiency and cost of geothermal energy plants used for power production vary greatly with reservoir temperatures. At temperatures higher than about 170°C, the simpler flash technology can be utilized. Flash technology involves tapping the high temperature liquid resource at some depth below surface, letting it flash into a mixture of steam and boiling water in the borehole, and then separating the steam from the boiling water at the surface. The steam, under pressure, is conducted into a turbine to generate electricity (Figure 2). The separated boiling water is usually reinjected into the subsurface. When the temperature of the geothermal fluid is

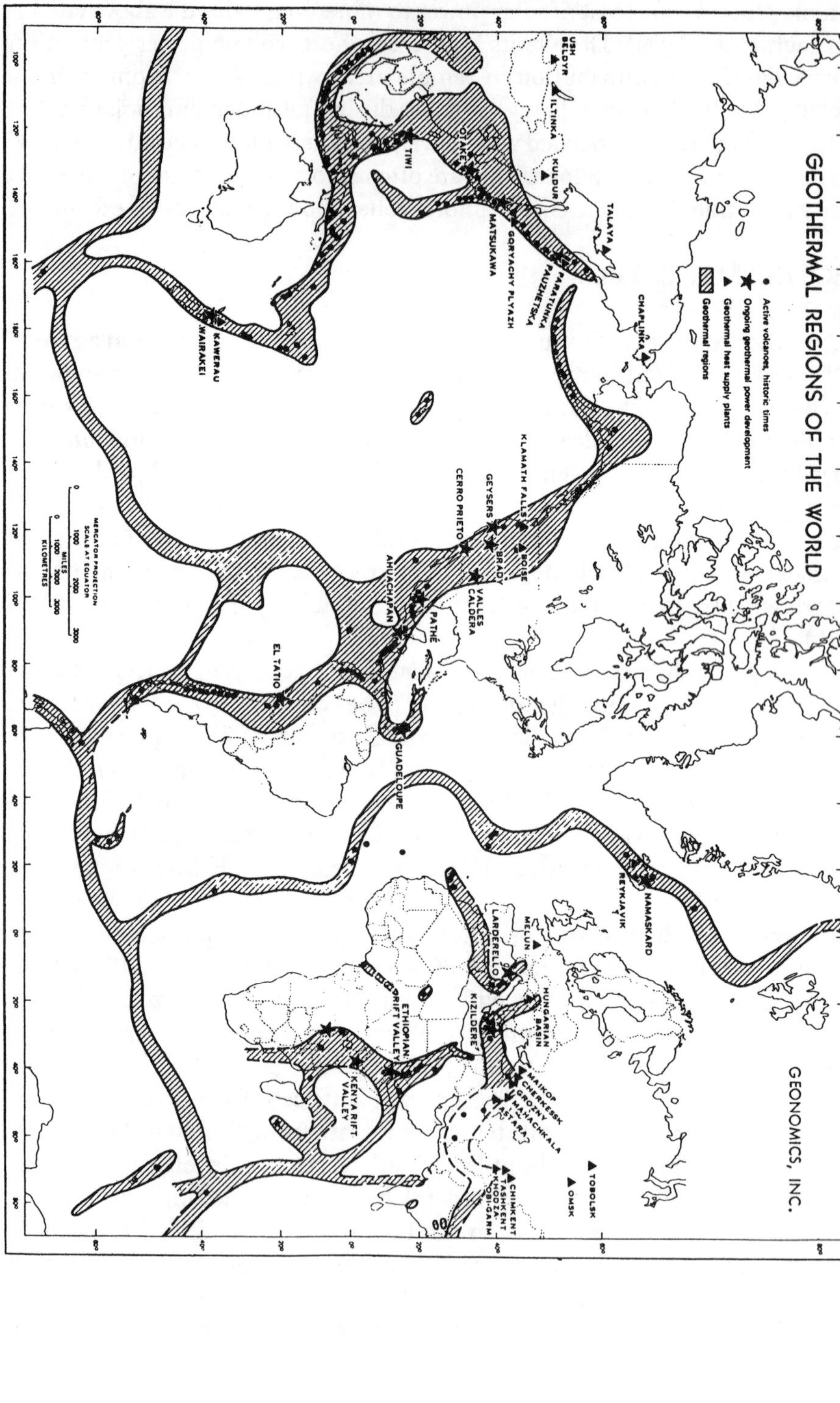

Figure 1. Geothermal Regions of the Wrold.

lower than about 175°C, not enough steam is produced when the well is allowed to flash. In that case, the binary technology is utilized.

Binary systems are typically utilized where reservoir temperature is less than about 180°C. In the binary power plant (Figure 3), a pump is inserted in the well at depth. The hot water is pumped in a liquid state into a heat exchanger on the surface. A low boiling point fluid, such as hydrocarbon, is utilized as the heat transfer medium. The working fluid is made to boil, generating high pressure gas which makes the turbine turn. The hydrocarbon is then compressed again into a liquid, and the cycle is repeated. Because of the utilization of a lower boiling point fluid, the binary system permits the extraction of a greater amount of heat per unit of geothermal fluid at lower reservoir temperatures, thereby allowing a more effective utilization of the geofluid for electrical generation. However, because of the greater complexity of the binary system, the capital cost of binary power plants tends to be approximately 1.5 to 2 times that of flash systems.

The overall capital cost of geothermal plants, like those of hydro projects, tends to be greater than that of a fossil fuel plant of comparable size. This is because in the case of a geothermal system it is necessary to drill wells and install gathering lines, in addition to construction of the power plant itself. However, on a levelized lifetime costs, geothermal power projects are price-competitive with the most inexpensive fossil plants, even without regard to the externality costs that should be imputed to the latter. This is because with geothermal plants, as with hydro, there is no need to buy fuel from an outside source at a cost which is likely to escalate with inflation. The operation and maintenance costs of a geothermal project typically increase over time at a rate of one third to one half that of the inflation rate.

Geothermal wells will usually decline with time at a very steady, predictable rate. As a result, geothermal plants have usually proven to be highly reliable baseload plants, with very little change in output from summer to winter. This makes geothermal power highly valuable as a steady-output baseload source of electricity, which is both a blessing and a problem.

In the U.S., geothermal plants have availability factors which range from 88% to 98%. If liquid cooling is feasible, the differences in output between day and night, or summer and winter, are negligible. Air cooling is practiced when water cooling cannot be used. If air cooling is deployed, a substantial difference in output can take place between day and night and between summer and winter. This difference in output could amount to as much as one third. The output is higher when the air temperature is lower, and lower when the air temperature is higher.

In industrial countries and in large, centrally-connected grids, the variation in demand of electricity between day and night is accommodated in different ways. The grid manager will bring additional capacity on line, or order excess capacity to be shut down, as changes in demand take place. This is also the way in which he deals with the sudden increase in electricity availability from variable output renewable resources. This applies to wind-power in particular. However, since wind-

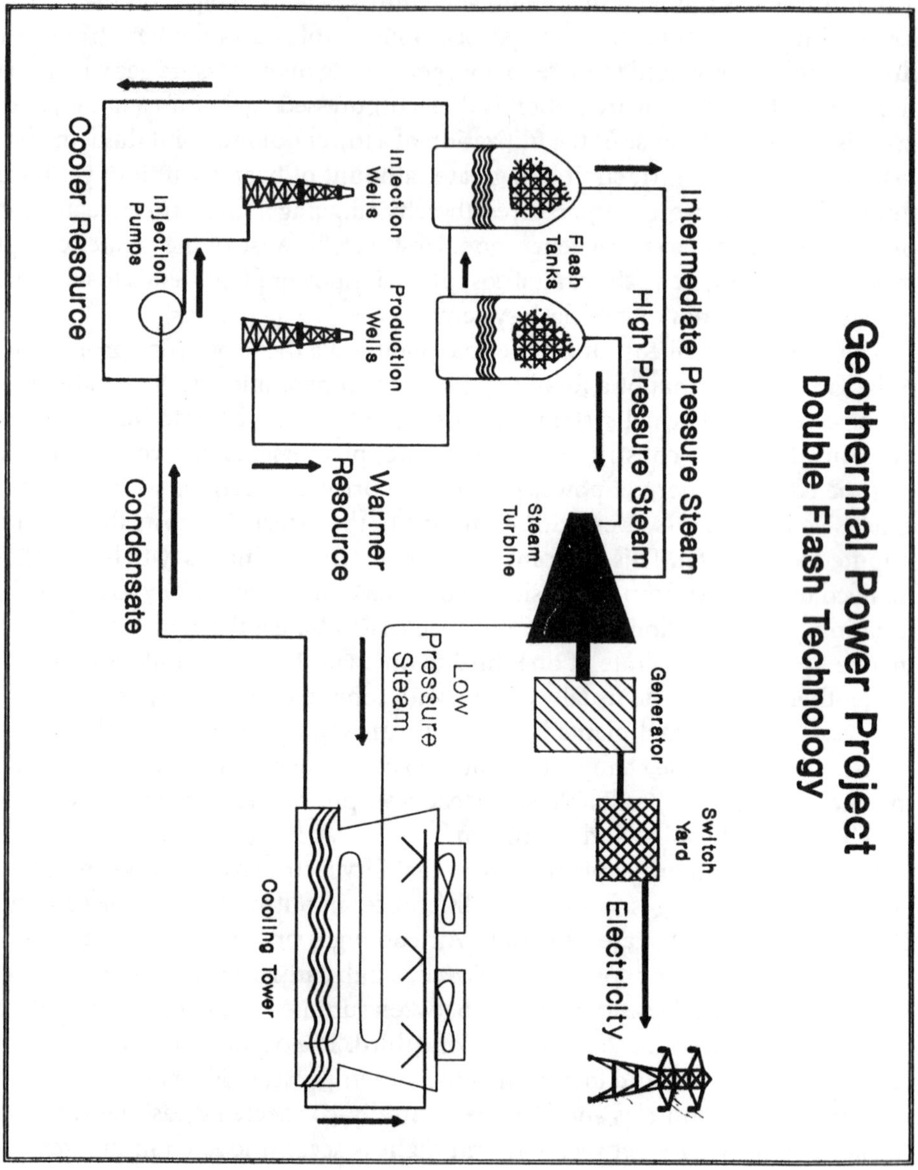

Figure 2. Geothermal power project: Double flash technology.

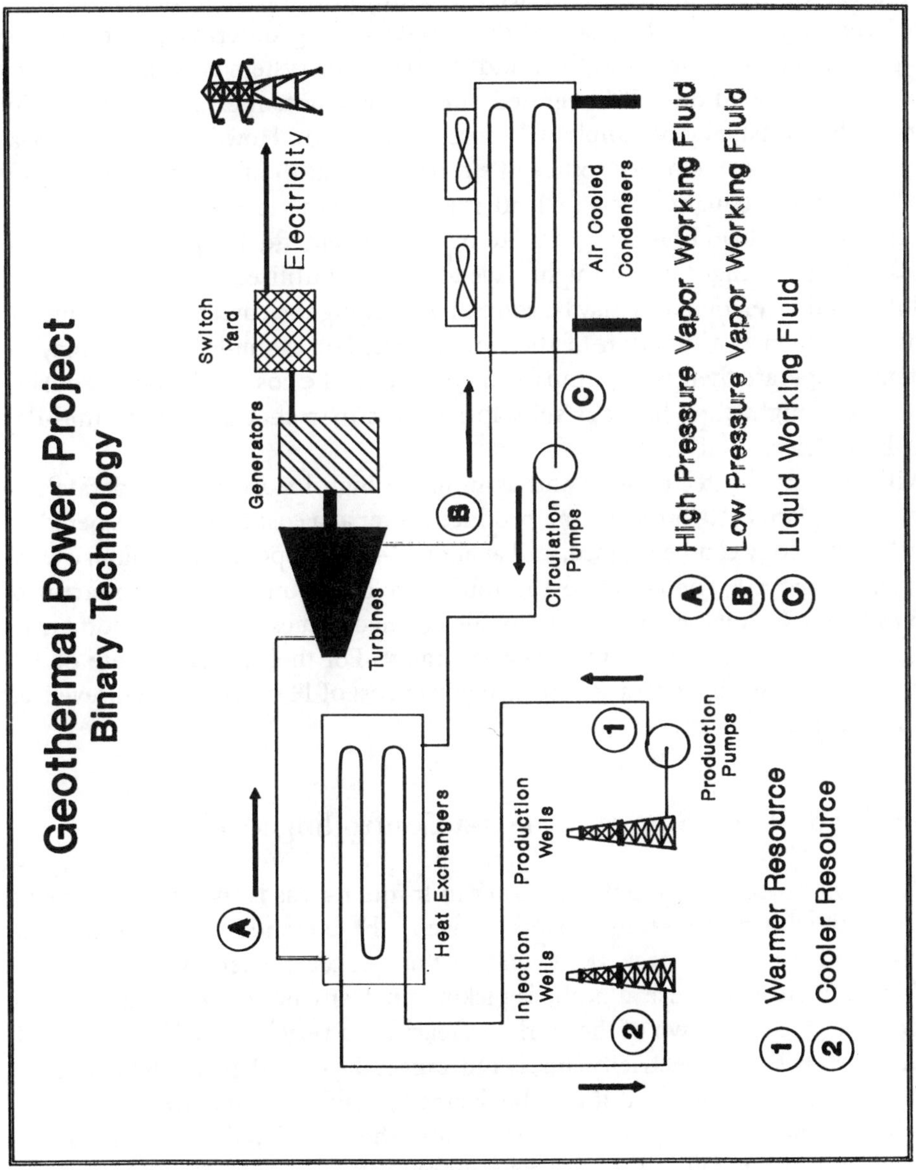

Figure 3. Geothermal power project: Binary technology.

based power in any large grid provides only a minuscule fraction of the power in the total system, there is no need for storage. Likewise, the increase in power output at night from an air-cooled geothermal plant is readily absorbed in a large power network by reducing the output from fossil-fuel-based power plants.

In rural areas which are not connected to a central grid and where demand greatly varies from day to night, the near-steady output from a geothermal plant may not provide the necessary power at the midday peak, if the system is designed to provide an average load only. If integrated into a larger system of diverse electricity sources, this constancy of supply will pose no problem. However, if geothermal power is to serve as the major source of electricity, what should its design capacity be? If geared to the night time base load, it will not serve the needs of the community during peak load hours. If geared to peak load demand, the plant would have excess capacity at night which might have to be left unutilized. In some cases, this problem of uneven demand may be solved by charging the customer with a fee for electricity which is a function of the time of use. For example, farmers may be content to operate their well pumps at night only, if the cost of the power during the off-peak night time hours is substantially less than that during the mid-day peak-hour price.

With present day technology and assuming a high load factor, moderate size geothermal plants (10–50 MW) can produce power at a cost of 5–7 cents per kWh if flash technology can be utilized, and at about 6–9 cents per kWh, if binary technology is deployed. The above cost numbers are based on a utilization factor of 80% or better. Entingh *et al.* (1994) provide detailed cost estimates for geothermal plants in the 100 kilowatt to one megawatt range. For their modal system of 300 kilowatts they come up with a geothermal power cost of 11 cents per kWh, levelized in constant 1993 dollars.

Diversity of Power Sources as a Cost Controlling Mechanism

The difficult problem in supplying power in frontier areas is the uneven demand for electricity during a 24–hour period. Typically. the demand for power starts near sunrise, and reaches its peak at the end of the workday, when factories are still working, but electricity use at home is picking up. Demand for electricity starts to decline after the shutdown of the workday, reaching a very low level between midnight and sunrise. Yet, electricity supply must meet the peak demand, which occurs in the late afternoon hours, if it is to be deemed reliable. In rural areas, the difference in demand between peak daytime use and the low of night time use can exceed a factor of ten. What is the electric supply entity to do? If it creates a single supply system that would meet the peak demand, that supply system may sit idle most of the time, because the average use of power (the so-called 'load factor') may not exceed more than a small fraction of its total capacity to produce electricity.

For example, if the peak load in a given area is 5 megawatts during the daytime, but drops to 200 kilowatts after midnight (for street lighting, refrigeration and the like), the cost of amortizing the 5 megawatt unit becomes a very high component of the cost of electricity that is actually utilized. A better strategy would entail having two sources of electricity, each providing part of the total (peak) load required, but because of proper design of these two smaller units, the average costs of the power produced is substantially lower than that of having one large machine which sits idle most of the time.

In frontier areas which have a geothermal potential but where connection into a grid is not feasible, the development agency may make use of the availability of geothermal energy with some other source of electricity by constructing a hybrid plant. The hybrid plant may consist of a combination of geothermal, as a base load supplier, and a biomass plant as a peak load source. Alternatively, if any hydro potential exists in the area, utilize a small hydro plant as the source for peaking power. Alternatively, the geothermal plant may provide the power for a pumped storage operation, if local conditions permit it, thereby obviating the need for other sources of energy. Finally, but least preferably, use a fossil fuel plant as a standby peaking power supply source.

In our hypothetical case of a 5 megawatt peak load demand in a given area, a geothermal plant may provide a constant supply of, say, four megawatts continuously, but utilize the excess capacity at night to pump water into a reservoir at a higher elevation. The hydro head created by the pumped storage could be utilized for providing peaking power when needed

Because of the fact that a geothermal plant operates most efficiently if utilized continuously, the economic utilization of geothermal energy would require that the local users would adopt practices that permit a high load factor utilization of that energy. For example, pumped irrigation at night could increase the load factor of the plant, thereby decreasing the costs of the electricity per unit.

The World Bank, the United Nations Development Program, the U.S. Department of Energy, and a number of other bodies are now beginning to examine the possibility of utilization of small scale geothermal installations in isolated rural areas, where geothermal resources have already been investigated, but where no development has taken place as yet.

References

Entingh, D.J., E. Easwaran, L. McLarty (1994) Small geothermal electric systems for remote powering. Presented at the U.S. Department of Energy Geothermal Program Review, San Francisco, April 28, 1994.

Gore, A. (1993) *Earth in the Balance.* Penguin Books, U.S.A.

Hall, O.H., F. Rosillo-Calle, R. H. Williams, and J. Woods (1992) Biomass for energy: supply prospects. *In* T.B. Johansson, H. Kelly, A.K.N. Reddy, and R.H. Williams (eds.) *Renewable Energy.* Island Press, Washington, D.C.

Huang J.P. (1993) Electric consumption and economic growth. *Energy Policy Journal* June 1993.

Johansson T. B., H. Kelly, A.K.N. Reddy, and R.H. Williams (1993) Renewable fuels and electricity for a growing world economy. *In* T.B. Johansson, H. Kelly, A.K.N. Reddy, and R.H. Williams (eds.) *Renewable Energy.* Island Press, Washington, D.C.

Lenssen, N. (1993) Providing energy in developing countries. *In: State of The World.* A Worldwatch Institute Report.

Ottinger, R.L., D.R. Woooley, N.A Robinson, D.R. Hodas, and S.B. Babb (1990) *Environmental Costs of Electricity.* Pace University Center for Environmental and Legal Studies, Oceana Publications, Inc.

Palmerini, C.G. (1993) Geothermal energy. *In* T.B. Johansson, H. Kelly, A.K.N. Reddy, and R.H. Williams (eds.) *Renewable Energy.* Island Press, Washington, D.C.

Public Service Commission of Nevada (1991) In Re Rulemaking Regarding Resources Planning Changes Pursuant to SB 497, Docket No. 89–752.

Romm, J.B., and A.B. Lovins (1992) Fueling a competitive economy. *Foreign Affairs,* Winter 1992/1993.

Sinha C.S. and T.C. Kandpal (1991) Decentralized v grid electricity for rural India. *Energy Policy,* June.

Williams, R.H., and E.D. Larson (1993) Advanced gasification-based biomass power generation. *In* T.B. Johansson, H. Kelly, A.K.N. Reddy, and R.H. Williams (eds.) *Renewable Energy.* Island Press, Washington, D.C.

ABOUT THE AUTHOR

Tsvi Meidav received his Ph.D. from Washington University, St. Louis, Missouri in 1961. He has 35 years of experience in natural resources exploration and management. Of these, 24 years have been devoted almost exclusively to geothermal. From 1967 to 1970, Dr. Meidav served as Associate Professor of Geophysics, University of California at Riverside. There he conducted research to determine the location and extent of the various geothermal fields in the Imperial Valley of California. He co-discovered all of the major geothermal fields now under development in the Imperial Valley.

As technical manager of geothermal energy development for the United Nations (1970–1973), Dr. Meidav was responsible for carrying out technical pre-feasibility studies of geothermal energy projects in various developing countries. As president of Trans-Pacific Geothermal Corporation, Dr. Meidav was responsible for the implementation of various geothermal projects. This included resource exploration, management of feasibility studies, contract negotiation with utilities, and negotiation with banks and other funding bodies for financing of each project. He has published several dozens of papers related to geophysical exploration and geothermal energy resources development.

Dr. Meidav is the former President of the National Geothermal Association (U.S.A.) and a co-founder and former Director of the International Geothermal Association. He currently serves as a Director of both the Geothermal Energy Association and of the Geothermal Resources Council (U.S.A.).

Dr. Tsvi Meidav
Trans-Pacific Geothermal Corporation
1901 Harrison Street, #1590
Oakland, California 94612-3501, U.S.A.

Chapter 18

The Fuelwood Crisis in Arid Zones: Runoff Agriculture for Renewable Energy Production

Beth Sauerhaft, Pedro R. Berliner and Thomas L. Thurow

Introduction

Fuelwood is any type of wood harvested from either living or dead trees and shrubs, in order to burn it as a source of energy for cooking foods, boiling water or space heating. Fuelwood is an important renewable energy resource in developing countries with limited means to purchase petroleum products or other alternative fuel sources. As human populations have grown, an increasing number of trees have been cut without replacing them with new trees, while forested land has not been protected to ensure continued production in many parts of the world (Kummer, 1992; Pimentel *et al.*, 1986; Bowonder *et al.*, 1987; Cleaver & Donovan, 1995; FAO, 1994).

World fuelwood consumption has steadily increased in the last thirty years. The FAO (1994) estimates that three billion people depend on wood as their prime source of household energy, consuming more than one half the wood produced. For example, in dry, mountainous regions of western Pakistan, northwestern India and northern Nepal, fuelwood supplies respectively 100%, 65% and 85% of the domestic fuel needs (FAO, 1981). Fuelwood provides more than 90% of total energy consumption in Sub-Saharan countries such as Ethiopia and Mali, and at least 75% in 13 of the 16 least developed nations of this region (Montalembert and Clement, 1983). A survey of 45 cities in developing countries documented that 23% of the average household energy consumption is based on wood, which represents 14% of household fuel expenditures (Barnes *et al.*, 1994).

Large quantities of fuelwood are required to sustain domestic needs: approximately 6–10 Giga Joules (GJ) or 0.5–1 m^3 of fuelwood per person per year, varying with location and efficiency of cooking equipment. In cold mountainous regions, the yearly per capita needs can be as high as 25–30 GJ or 3 m^3 of fuelwood. In

H. J. Bruins and H. Lithwick (eds.), The Arid Frontier, 351–363.

Africa, fuelwood consumption is estimated at 0.6 m^3 per capita for urban households and 0.8 m^3 per capita for rural households (Anderson & Fishwick, 1984).

Concern about a fuelwood shortage in the developing regions of the world has increased during the last 10–20 years, as human populations continue their rapid growth while exploiting existing forests. Recent trends indicate that present wood reserves will not be able to support the fuelwood needs of existing and projected world populations adequately (Eckholm, 1975; Lanly, 1982; Montalembert & Clement, 1983; Eckholm *et al.*, 1984; Pimentel *et al.*, 1986; Bowonder, 1987; Harrison, 1987; Postel & Heise, 1988; Gregersen *et al.*, 1989; FAO, 1981, 1994, 1995; World Resources Institute, 1995). This shortage, as will be discussed in greater detail later, is primarily a localized problem, particularly in arid, semi-arid, and high mountainous regions of the developing world. Harvesting fuelwood in an unsustainable way has detrimental environmental impacts such as accelerated erosion.

Its production can be significantly enhanced in arid regions of the world through implementation of runoff agriculture systems. In these systems the runoff from rainfall is harvested and conveyed to storage basins (Evenari, Shanon & Tadmor, 1982). The main limiting factor for tree production in dry areas is thus removed in fields which receive runoff water. The long roots of trees are able to access water that percolated deep into the soil of the storage basins. Thus many arid-zone tree species are able to survive in dry seasons or drought years.

The Fuelwood Shortage

Uncontrolled wood harvesting and deforestation to expand agronomic cultivation and grazing lands are primary causes of the fuelwood shortage (World Energy Outlook, 1995; Cleaver & Donovan, 1995; Schelhas & Greenberg, 1996). Once cut, it is difficult for trees to re-establish themselves in dry environments, because the forest micro-environment necessary for greater regeneration has been altered, usually resulting in lower infiltration rates, more wind and more extreme temperatures (Weber, 1986). Removal of the canopy exposes the soil surface to the beating action of the rain drops. Soils in arid and semiarid zones are typically unstable: aggrevates break down and a crust is formed on the surface of the soil which decreases its infiltrability. Only small amounts of water are absorbed by the soil as a result of this crust formation, and tree or shrub regrowth is severely hampered.

The Southern African Development Coordination Conference perceived the impending fuelwood shortage to be sufficiently critical to warrant the commissioning of a study to determine how best to assure future supplies (Prior & Cutler, 1992). At least 16 of the 45 sub-Saharan African countries faced fuelwood deficits in 1985 in part or all of their territory (FAO, 1985). The authors of the FAO report concluded that to meet current needs the population of these countries would have to cut the trees faster than they can grow! Another 18 countries also suffer an acute scarcity which cannot be solved by overcutting (FAO, 1985). Inhabitants of arid regions in Middle Eastern countries faced deficits ranging from 7,000 m^3 in Jordan to 16 million m^3 in Turkey (Montalembert & Clement, 1983). The arid and semi-arid zones of Latin America had an overall fuelwood deficit of 45 million m^3 in 1980 (Montalembert & Clement, 1983).

Fuelwood shortages frequently result in a change of preparation techniques for food and decreased hygiene, such as less frequent boiling of water to kill parasites. This may lead to inadequate nutrition and poorer health. If firewood is unavailable, the first alternative fuels to be exploited are generally agricultural residues and livestock manure. The ecological consequences of using these fuelwood substitutes are decreased fertility and soil structure degradation which result in lower crop productivity, as nutrients from residues or dung are not incorporated into the soil.

CHANGE IN NUTRITIONAL STATUS

Impacts of fuelwood scarcity on nutritional status are difficult to analyze due to various associated variables whose effects are difficult to isolate. Nevertheless, fuelwood shortages are associated with changing nutritional situation in rural households (Brouwer *et al.*, 1989; FAO, 1994). The search for fuelwood becomes more time consuming for women who already carry heavy bundles of wood and other goods for kilometers. They are forced to walk further and burn more calories from an already insufficient diet (Harrison, 1987; Cleaver & Donovan, 1995). They spend less time cooking and often prepare less nutritious foods whose preparation time is shorter and therefore demand less fuel. This can result in a decline in caloric intake and/or quality for the family. Consumption of snacks and sweets has increased in many countries because it is cheaper than cooking more nutritional food, due to the high price of wood (*cf.*, Dewees, 1989). As women are forced to spend more time collecting fuelwood, they have less time for their other tasks while children may be forced to collect wood which effects their opportunity to receive schooling.

COMPROMISES IN HEALTH CARE

Women also conserve wood by cutting back on hygiene and space heating. The calories needed to maintain body temperature increase as home heating decreases. The incidence of intestinal infections also rises as a result of drinking unboiled water and eating with dirty hands from insufficiently cleaned plates. Thus lack of fuelwood puts the physical condition of family members in jeopardy (Brouwer *et al.*, 1989).

The use of wet or inferior wood and other substitutes can provide problems to the health of women and other members of the family by causing them to inhale more smoke. A study carried out on women in India, indicated that exposure to benzo-a-pyrene, a common by-product of wood combustion, was equivalent to smoking 400 cigarettes a day (Batliwala, cited by Brouwer, 1989; Barnes *et al.*, 1994). Smoke inhalation also poses risk to fetal development. Children exposed to this smoke have higher risks of bronchitis.

AGRICULTURE, SOIL QUALITY AND BIODIVERSITY

Soil structure deteriorates as deforestation exposes soil to direct raindrop impact. Structural deterioration is further exacerbated when farmers use dung and crop residue as a substitute for scarce fuelwood. This deterioration of the soil leads to reduced infiltration rates, especially if soil crusting occurs (Sumner & Stewart, 1992). This increases the rate of runoff which leads to more severe downstream flooding and accelerates soil and nutrient loss (FAO, 1994; Pimentel *et al.*, 1987). Nutrient cycling is altered with deforestation as leaf litter is removed from the forests.

Deforestation can lead to a decline in biodiversity as both plant and wildlife habitat is destroyed (Schelhas & Greenberg, 1996). Planting or preserving forests may mitigate the threat to biodiversity, as certain animal and plant species will find protection in such a habitat. Forests also provide corridors between habitats for migrating species. Indigenous tree species, planted in clusters within runoff agriculture systems could therefore provide plant and wildlife habitat in addition to renewable wood resources for humans (Schelhas & Greenberg, 1996).

In a critique of existing interventions in southern Africa, O'Keefe & Munslow (1989) believe that natural resource development projects have failed to address the fuelwood problem correctly. Such project evaluations have overestimated the size of the problem on national levels. However, the fuelwood supply on a national level is not a problem in southern and eastern African nations with the exception of Lesotho. Shortages occur, rather, in certain locations where the problem is critical. Moreover the above authors are not satisfied with existing data, which do not place fuelwood in the category of either plantations or natural woodlands. The amount of wood in both categories is well-documented by UNEP and FAO papers, but some of the fuelwood may not have been included. The writers of these UN documents do acknowledge that part of the feul wood harvest goes undocumented, since it belongs to none of the measured categories. Also, with the unreliability of various demographic data, the size and direction of the problem are unknown. If feulwood shortages are a local, rather than a national, problem, it must be addressed differently in planning and management terms.

Possible Solutions

There are two possible approaches to mitigate fuelwood shortages:
1) Improve the energy efficiency of fuelwood combustion.
2) Improve fuelwood production systems.

IMPROVED TECHNOLOGY

Improving the burning and transmission of heat in stoves can substantially reduce the amount of wood consumed per meal. Well designed wood stoves burn wood at 18% efficiency as opposed to traditional stoves which burn at about 6% efficiency (Barnes *et al.*, 1994). A barrier to adoption of new stove technology by rural households is the required investment to purchase such a stove. However, due to the higher fuelwood costs, city dwellers often find that shifting to more efficient stoves may be economical (Kammen, 1995).

IMPROVED FUELWOOD PRODUCTION SYSTEMS: RUNOFF AGRICULTURE

The World Bank, in its 1995 report on Agriculture, Poverty, and Policy Reform in Sub-Saharan Africa (SSA), set as one of its goals for solving the fuelwood problem, the "systematic management of forest cover...and plantations of fuelwood species" (Cleaver & Donovan, 1995).

Runoff agriculture can be implemented as a sustainable technique for managing forest cover as well as growing plantations of fuelwood species. When runoff agriculture is used in conjunction with agroforestry, these two techniques have the potential to provide fuelwood, reduce erosion (Barrow, 1982; Reij *et al.*, 1988), improve soil quality by aiding in nutrient cycling (Tilander, 1993; Reij *et al.*, 1988; Cook & Grut, 1989), and provide fodder for livestock (Hazra, 1994; Lövenstein *et al.*, 1991; Cook & Grut, 1989) in addition to building materials and windbreaks (Cook & Grut, 1989; Weber, 1986).

In many arid and semi-arid locations, tree growth on any significant scale would not be possible without the use of stored water in the soil profile. In those areas large fractions of the rainwater are lost to evaporation or surface runoff and, therefore, little rainwater reaches the root zone (Reij *et al.*, 1988).

The runoff water can be harvested by channeling the surface runoff to fields where it can be stored in the soil for access by deep-rooted tree species. En route to the area of infiltration, the water also transports nutrients from the soil surface of the catchment area. The nutrient enriched water is channeled into catchment basins or limans (fields with bunds to hold the water) where it infiltrates into the soil and becomes available for both shallow and deep root uptake over an extended period of time (Reij *et al.*, 1988). The result is an environment more suitable for tree growth than without runoff. As the trees become established, they in turn alter the microclimate and soil environment within the liman (Vandenbeldt, 1990). These alterations, such as less extreme temperatures, increased shade and, if nitrogen fixing tree species are used, improved nutrient content and cycling (Cook & Grut, 1989; Young, 1989), make an environment more conducive to continued tree growth and understory crop production. The important feature of this system is that trees within the basins can be harvested as the supply of water for their regrowth comes from an external source and is not affected by activities within the basin.

Rainwater Harvesting for Runoff Agriculture

Early evidence of runoff agriculture can be found in Israel in the Negev Desert from the time of the ancient Israelite period (ca. 1200–587 BCE), the Nabatean period (300 BCE–100 CE), and the Roman-Byzantine (100–600 CE) periods (Evenari *et al.*, 1982; Bruins, 1986; Bruins *et al.*, 1986).

Runoff agriculture at a reconstructed Nabatean farm in the Negev has yielded 1.4 tons $ha^{-1}yr^{-1}$ of almonds and pistachios, and up to 10 tons $ha^{-1}yr^{-1}$ of olives (Evenari *et al.*, 1982, Evenari, 1989) in addition to figs, pomegranates, grapes, carobs, peaches, apricots and loganberries (Evenari *et al.*, 1958). Other runoff systems used today in various parts of the world successfully increase wood production. Near Teheran, Iran, terraces on contour lines were covered with asphalt to increase runoff. Subsequent increases in growth height, crown cover and stem cross section were respectively 61.5%, 60.9% and 53.0% for *Robinia pseudacacia* L.; 14.6%, 15.4%, and 31.6% for *Cupressus arizonica* G.; and 29.4%, 79.6% and 23.9% for *Fraxinus rotundifolia* Mill.(Kowsar *et al.*, 1978), as compared to similar species which did not receive runoff. In the Thar Desert, Pakistan, runoff agriculture experiments in microcatchments for seedling establishment showed a 77% survival rate for trees grown in sloping catchments versus 64% for those planted on flat ground without the runoff catchment (Sheikh *et al.*, 1984).

Rainwater harvesting has also been used in Africa to increase agronomic crop yields (Bruins *et al.*, 1986; Reij *et al.*, 1988), but detailed system analyses over many years, including data on maintenance and social aspects are generally lacking. In Tigray, Ethiopia, the use of runoff has increased yield of a variety of crops such as sorghum, up to three-fold during a dry period which had 20% less rain than normal (Bennett, 1985). In Sub-Saharan Africa, the Haussa in Niger maintain planting pits to harvest runoff and produce an average of 522 kg/ha of millet as compared to 0 kg on land not using planting pits (van Dijk & Reij, 1994). Sorghum yields on plots receiving runoff were between 2.3 and 3.4 times greater than on control plots, while cowpea yields were between 3.5 and 7.7 times greater (MoALD, 1984, Reij *et al.*, 1988) in Kenya. Farmers in Burkina Faso constructed stone bunds to capture runoff and yields on these plots were 12% to 90% higher than on control plots (Wright, 1985, Reij *et al.*, 1988). The increased grain yields indicate the potential in these regions for increased biomass production with the addition of runoff through rainwater-harvesting systems.

Fuelwood Production through Runoff Agrogorestry

Fuelwood production using runoff agroforestry has been carried out successfully with a variety of tree species at the Mashash Experimental Farm (Lövenstein *et al.*, 1991).

Research carried out at the Jacob Blaustein Institute for Desert Research in the Negev Desert between 1991 and 1994 showed the potential of runoff agriculture to produce fuelwood (Sauerhaft, Ph.D., in progress). *Acacia saligna* (Labill.) H. Wendl. trees were grown in a system artificially flooded to simulate rainwater harvesting with an amount of water comparable to an average annual runoff harvest in the Central Highlands Negev—700 mm (Evenari, 1989). In this region, Evenari (1989) estimated that a mean rainfall of 100 mm which produces 20–30 mm of runoff, could in theory produce 500–750 mm of runoff for a field whose catchment area is 25 times larger in size. Mean annual rainfall (1971–1991) is 96 mm, most of which falls between October and April. The *Acacia saligna* was chosen in this instance because of its tolerance to drought and salinity (NAS, 1980), and its high production potential of quality fuelwood. It also fixes nitrogen and rapidly regrows when cut. Both are important factors in arid regions of the world.

The trees were grown at three different planting densities in rows equidistant from one another. After flooding, the trees were pollarded (*i.e* all the branches were cut back) to a radius of 0.5 m and a height of 2 m. Wood yields are listed in Table 1.

TABLE 1. *Acacia saligna* wood yield, above 2 m height, pollarded material only.

Planting Density (m)	*Average Wood Production*		
	kg/ha	m^3	Calorific value* (million Kcal)
1	2525	5.2	11.83
2	3104	6.3	14.54
3	2611	5.3	12.23

**Acacia saligna* density and calorific values taken from Gough *et al.*, 1989.

Basal diameters at 10 cm above the ground were measured following flooding, repeated periodically throughout each season, and at the end of each growing season. Relative diameters, by density, are shown in Figure 2.

Trees were planted in a runoff agroforestry system at three densities, 4 m spacing between all tree rows: 3 m (840 trees/ha), 2 m (1,330 trees/ha), and 1 m (2,520 trees/ha) between trees. Yields are in kg/ha^{-1} of oven dry matter, based on averages of two years of data.

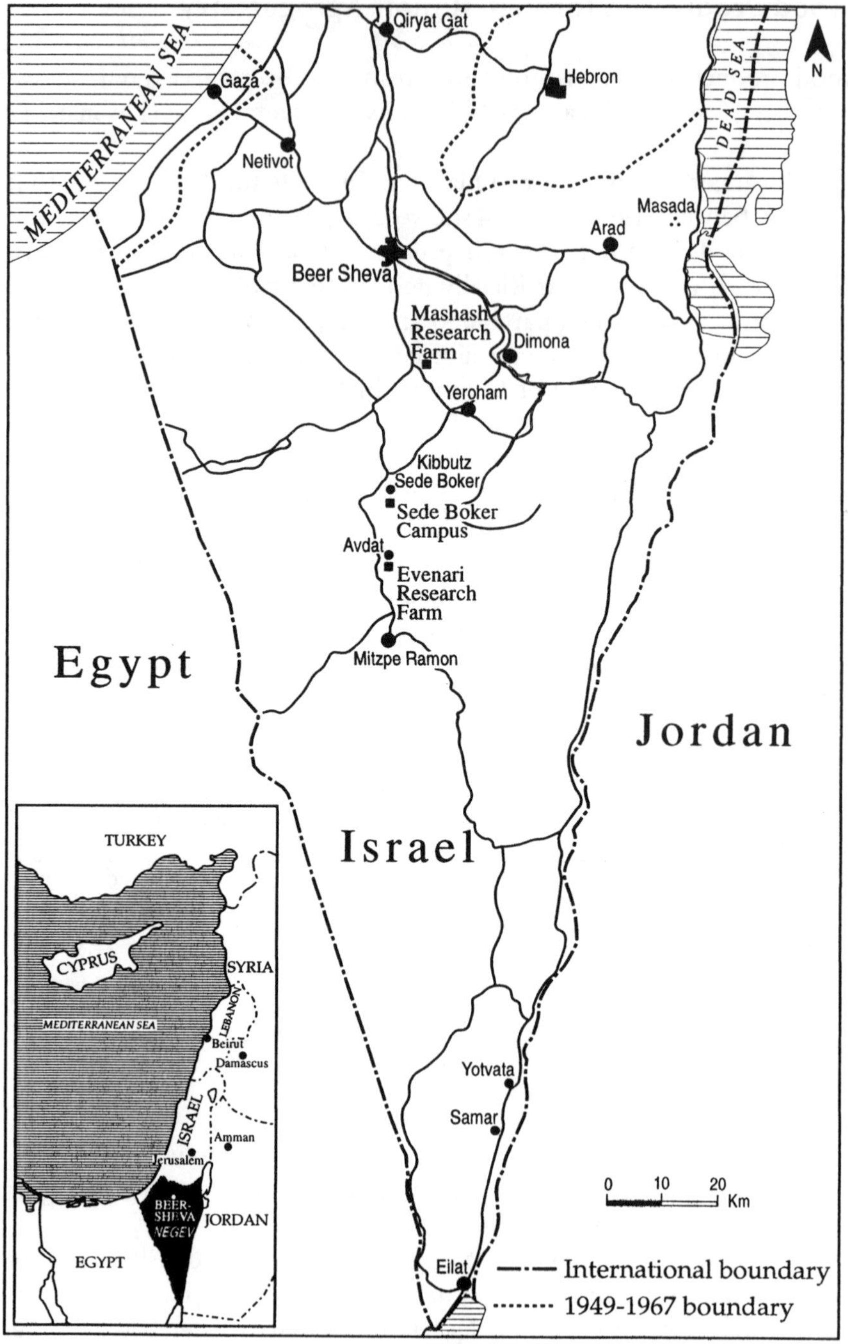

Figure 1. Location of the Sede Boker Campus, the Evenari and Mashash Experimental Runoff Farms in the Negev desert.

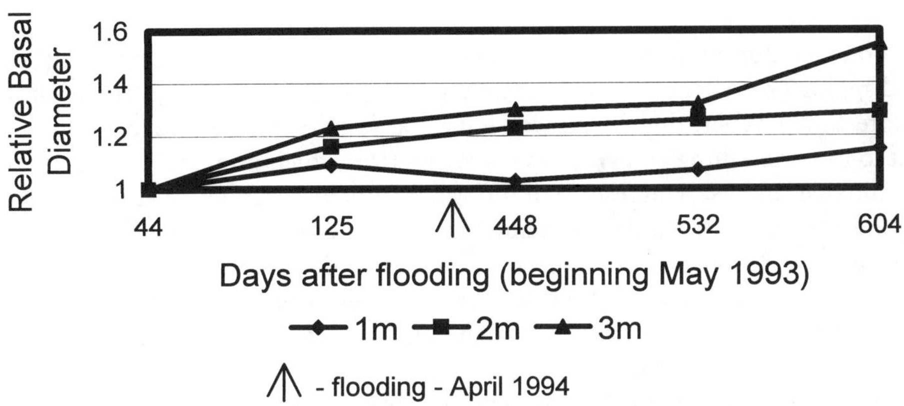

Figure 2. Development of relative mean basal diameters, as influenced by planting density. Points are plotted according to days after flooding where the first flooding is day zero and subsequent measurement dates follow from this day onwards for two years.

The results indicate that planting density influences the type of wood product, not the total amount of biomass produced. There was no significant difference in production between planting densities. At the higher density, tree basal diameters were significantly smaller than at the lowest density. Therefore, demand for fuelwood size and tools available for cutting wood would determine planting density.

Based on these results, there seems to be a potential for runoff agroforestry production systems. One hectare of land under this particular system can produce the fuelwood needs of a 5–6 person family based on average yearly fuel requirements mentioned earlier. Previous research carried out in the Negev Desert supports these findings (Zohar *et al.*, 1988; Lövenstein *et al.*, 1991). On two, three and four year rotations, respectively, 29.4, 40.8 and 39.8 kg/tree of *Eucalyptus occidentalis* wood and 15.3, 26.1 and 33.2 kg/tree of *Acacia salicina* wood were produced in the dry arid climate of the Central Negev Desert of Israel, having a P/ETP index of about 0.05. The rotation is the length of time between harvests. Firewood from the trees was not harvested every year, but after two, three or four years, respectively. The importance of the results, besides the success to produce the firewood in such a dry climate, is that a longer rotation does not give a significantly larger harvest. Cutting branches for firewood every year may, therefore, give the highest production. If this technique is viable under the extreme conditions under which it was tested, it will probably be effective under similar dry conditions in other parts of the world.

Summary

Wood is the dominant source of household energy in most of the arid and semi-arid regions of Africa, Asia and Latin America. Its use is leading to widespread deforestation. Where there is a critical fuelwood scarcity, people often have no choice but to use dried dung and crop residues, practices which undercut the prospect of sustainable land use. Harvesting runoff water from local rainfall in arid regions can significantly increase plant production in sustainable systems. Some of the fuelwood tree species adapted to arid climates are well-suited to runoff agriculture systems because they can survive the long dry periods. Their root systems can utilize runoff water coming from a contributing catchment area, stored deeply in the subsoil of the runoff collecting field.

Combining agroforestry systems with rainwater harvesting has the potential to improve yields of fuelwood in arid regions. Available species, growing conditions and socio-economic factors will determine the tree or intercrop species that may be used. Further research and extension efforts are needed to enhance understanding of what appears to be a promising practice of using agroforestry and runoff collection systems to increase fuelwood production in arid and semi-arid land.

References

Anderson, D., and R. Fishwick (1984) Fuelwood consumption and deforestation in African countries. World Bank Staff Working Papers No. 704. World Bank, Washington, D.C.

Barnes, D.F, K. Openshaw, K.R. Smith, and R. van der Plas (1994) What makes people cook with improved biomass stoves? A comparative international review of stove programs. World Bank Technical Paper Number 242, Energy Series. The World Bank, Washington, D.C.

Barrow, E.G.G. (1982) Use of micro-catchment for tree planting in soil conservation in semi-arid areas. *In* D.B. Thomas and W.M. Senga (eds.) *Soil and Water Conservation in Kenya.* Proceedings of a 2nd National Workshop. Nairobi, March 10–13, 1982. Institute for Development Studies and Faculty of Agriculture University of Nairobi. Occasional Paper No. 42. Pp. 324–332.

Bennett, J. (1985) Tigrayan test-bed. *World Water* May 26–27.

Bowonder, B, S.S.R. Prasad, and N.V.M. Unni (1987) Deforestation around urban centres in India. *Environmental Conservation* 14(1):23–28.

Brouwer, I.D., L.M. Nederveen, A.P. den Hartog. and A.H.C. Vlasveld (1989) Nutritional impacts of an increasing fuelwood shortage in rural households in developing countries. *Progress in Food and Nutrition Science* 13:349–361.

Bruins, H.J. (1986) *Desert Environment and Agriculture in the Central Negev and Kadesh-Barnea during Historical Times.* Published Ph.D. Thesis, Agricultural University of Wageningen, The Netherlands; Midbar Foundation, Nijkerk.

Bruins, H.J., M. Evenari and U. Nessler (1986) Rainwater-harvesting agriculture for food production in arid zones: The challenge of the African famine. *Applied Geography* 6:13-32.

Cleaver, K.M., and W.G. Donovan (1995) Agriculture, Poverty, and Policy Reform in Sub-Saharan Africa. World Bank Discussion Papers. Africa Technical Department Series #280.

Cook, C.C., and M. Grut (1989) Agroforestry in Sub-Saharan Africa: A farmer's perspective. World Bank Technical Paper No. 112. The World Bank. Washington, D.C.

Dewees, P.A. (1989) The woodfuel crisis reconsidered: Observations on the dynamics of abundance and scarcity. *World Development* 17(8):1159–1172.

Dijk, J., van and C. Reij (1994) Indigenous water harvesting techniques in Sub-Saharan Africa: Examples from Sudan and the West African Sahel. *Water Harvesting for Improved Agricultural Production.* Proceedings of the FAO Expert Consultation. Cairo, Egypt, 21–25 November, 1993. FAO Water Reports No. 3. FAO, Rome. Pp. 101–112.

Eckholm, E.P. (1975) The other energy crisis: Firewood. Worldwatch Paper 1.

Eckholm, E., G. Foley, G. Barnard, and L. Timberlake. (1984) *Fuelwood: The Energy Crisis that Won't Go Away.* An Earthscan Paperback.

Evenari, M. (1989) *The Awakening Desert: The Autobiography of an Israeli Scien*tist. Springer-Verlag, Berlin.

Evenari, M., Y. Aharoni, L. Shanan, and N.H. Tadmor (1958. The ancient desert agriculture of the Negev. III. Early beginnings. *Israel Exploration Journal* 8(4):231–268.

Evenari, M., L. Shanan and N. Tadmor (1982) *The Negev: The Challenge of a Desert* (2nd ed.). Harvard University Press, Cambridge, MA.

FAO/UNEP (1981) Tropical forest resources assessment project: Forest resources of tropical Asia. UN 32/6.1303–78–04 Technical Report 3. Rome.

FAO (1985) Crisis *of Sustainability: Africa's Land Resource Base.* AGD/801/1. Rome: FAO.

FAO (1994) *The State of Food and Agriculture.* FAO Agriculture Series #27. 7.

FAO (1993) *Yearbook. Forest Products 1982–1993.* 1 FAO Forestry Series #28. Rome: FAO.

Gough, D.K., R.E. Bell, P.A. Ryan, and C.T. Bragg (1989) Drying and burning properties of the wood of some Australian tree species. *In* D.J. Boland (ed.) T*rees for the Tropics: Growing Australian Multipurpose Trees and Shrubs in Developing Countries.* Australian Centre for International Agricultural Research, Canberra, pp. 177–186.

Gregersen, H., S. Draper and D. Elz (1989) *People and Trees: The Role of Social Forestry in Sustainable Development.* The World Bank, Washington, D.C.

Harrison, P. (1987) *The Greening of Africa.* International Institute for Environment and Development, Great Britain.

Hazra, C.R. (1994) Forages from rainfed agroforestry in Bundelkhand region of India and its influence on soil fertility. *In* P. Singh, P.S. Pathak and M.M. Roy (eds.) *Agroforestry Systems for Degraded Lands. Papers presented at the International Conference on Sustainable Development of Degraded Lands Through Agroforestry in Asia and the Pacific.* New Delhi, November 25–30, 1994. Science Publishers, Inc., 1:140–147

Kammen, D.M. (1995) Cookstoves for the developing world. *Scientific American* July 64–67.

Kowsar, A., P. Mehdizadeh, E. Vaziri, and L. Boersma (1978) Water harvesting for afforestation: II. Survival and growth of trees. *Soil Sci. Soc. of Am.* J. 42(4):650–656.

Kummer, D.M. (1992) Upland agriculture, the land frontier and forest decline in the Philippines. *Agroforestry Systems* 18:31–46.

Lanly, J.P. (1982) Tropical forest resources. FAO Forestry Paper #30. Rome.

Lövenstein, H.M., P.R. Berliner and H. van Keulen (1991) Runoff agroforestry in arid lands. *Forest Ecology and Management* 45:59–70.

MoALD (1984) Runoff harvesting for crop, range and tree production in the BPSAAP-area. BPSAAP Interim Report, ch. 12:78–113. As cited in C. Reij, P. Mulder, and L. Begemann (1988) Water harvesting for Plant Production. World Bank Technical Paper No. 91. The World Bank, Washington, D.C.

Montalembert, M.R. de and J. Clement. (1983) Fuelwood supplies in the developing countries. FAO Forestry Paper #42. Rome.

National Academy of Sciences, Panel on Firewood Crops (1980) Firewood crops—Shrub and tree species for energy production. National Academy of Sciences, Washington, D.C.

Nessler, U. (1980) Ancient techniques aid modern arid zone agriculture. *Kidma* 5(4):3–7.

Pimentel, D., W. Dazhong, S. Eigenbrode, H. Lang, D. Emerson, and M. Karasik. (1986) Deforestation: Interdependency of fuelwood and agriculture. *Oikos* 46:404–412.

Pimentel, D., J. Allen, A. Beers, L. Guinand, R. Linder, P. McLaughlin, B., Meer, D. Musonda D. Perdue, S. Poisson, S. Siebert, K. Stoner, R. Salazar, and A Hawkins (1987) World agriculture and soil erosion. *Bioscience* 37:277–283.

Postel, S. and L. Heise (1988) Reforesting the earth. Worldwatch Paper 83.

Prior, J. and D. Cutler (1992) Trees to fuel Africa's fires. *New Scientist* August 29:35–39.

Reij, C., P. Mulder and L. Begemann (1988) Water harvesting for plant production. World Bank Technical Paper No. 91. The World Bank, Washington, D.C.

Schelhas, J. and R. Greenberg (1996) Introduction: The value of forest patches. *In* J. Schelhas and R. Greenberg (eds.) *Forest Patches in Tropical Landscapes.* Island Press, Washington, D.C.

Sheikh, M.I., B.H. Shah and A. Alee. (1984) Effect of rainwater harvesting methods on the establishment of tree species. *Forest Ecology and Management* 8:257–263.

Sumner, M.E. and B.A. Stewart (1992) *Soil Crusting: Chemical and Physical Processes.* International Symposium on Soil Crusting: Chemical and Physical Processes. Lewis Publishers. Boca Raton.

Tilander, Y. (1993) Effects of mulching with *Azadirachta indica* and *Albizia lebbeck* leaves on the yield of sorghum under semi-arid conditions in Burkina Faso. *Agroforestry Systems* 24:277–293.

Vandenbeldt, R.J. (1990) Agroforestry in the semiarid tropics. *In* K.G. MacDicken and N.T. Vergara (eds.) *Agroforestry: Classification and Management.* John Wiley and Sons, New York.

Weber, F.R. (1986) *Reforestation in Arid Lands.* Volunteers in Technical Assistance. Virginia, U.S.A.

World Resources (1994–1995) The World Resources Institute, UNEP, UNDP. Oxford University Press, Oxford.

World Energy Outlook (1995) International Energy Agency, Frxance.

Wright, P. (1985) La conservation des eaux et des sols par les paysans. *In* W. Ohm and J. Nagy (eds.). *Technologies appropriees pout les paysans des zones semi-arides de l'Afrique de l'Ouest: 60–67.* Purdue University, West Lafayette, U.S.A. As cited *In* C. Reij, P. Mulder and L. Begemann. (1988. *Water Harvesting for Plant Production.* World Bank Technical Paper No. 91. The World Bank, Washington, D.C.

Young, A. (1989) *Agroforestry for Soil Conservation.* ICRAF. Nairobi.

Zohar, Y, J.A. Aronson and H. Lövenstein. (1988) Cultivation of multipurpose trees in rain water harvesting systems in the arid zone of Israel. *Commonw. For. Rev.* 67(4):339–349.

ABOUT THE AUTHORS

Beth Sauerhaft received her Masters in Environmental Management in Forest Ecology from the School of the Environment at Duke University. She is a doctoral candidate in the Department of Rangeland Ecology and Management at Texas A&M University and carried out her research at the Jacob Blaustein

Institute for Desert Research in the Negev Desert of Israel on the use of agroforestry in fields set up to simulate a rainwater-harvesting system. She has investigated the effects of within row tree density, and an intercrop on the harvested biomass, soil moisture and nutrient quantities in an alley-cropping system. She has successfully completed consulting jobs in watershed management, which consisted of natural, cultural and recreational resource inventories and design of management plans for watersheds in the United States and Canada.

Beth Sauerhaft
45 Deerfield Rd.
Chappaqua, NY 10514
USA

Pedro R. Berliner holds a Ph. D. in agricultural meteorology and soil physics from The Hebrew University of Jerusalem. His main interest is the movement of water in the soil-plant-atmosphere continuum. Dr. Berliner is the head of the Runoff Agriculture Unit at the Jacob Blaustein Institute for Desert Research (Ben-Gurion University of the Negev). The Unit maintains two experimental runoff farms: The Evenari Farm near Avdat and the Mashash Farm south of Beer-Sheva. In addition, fields at the Sede Boker Campus are used for simulated rainwater-harvesting investigations. Basic research related to runoff agroforestry systems has been carried out. An additional sphere of his present research is the quantification of radiation, water and carbon dioxide fluxes in runoff agroforestry systems.

Dr. Pedro R. Berliner
Wyler Laboratory for Arid Lands
A. Katz Center for Desert Agrobiology
Jacob Blaustein Institute for Desert Research
Ben Gurion University of the Negev
Sede Boker Campus, 84990
Israel

Dr. Thomas L. Thurow is an Associate Professor in the Department of Rangeland Ecology and Management, Texas A&M University, College Station, Texas, U.S.A.. The focus of research concentrates on how rangeland management influences the quality and quantity of water yield, particularly at the landscape level of resolution. His agroforestry research concentrates on how trees can be incorporated into agricultural or pastoral production systems to enhance environmental and economic stability, especially with regard to soil and water management.

Prof. Thomas L. Thurow
Department of Rangeland Ecology and Management
Texas A&M University
Animal Industries Building, Rm 225
MS 2126
College Station, TX 77843-2126
USA

Chapter 19

Lightweight Housing in Arid Zones: Thermal Comfort and Energy Use

David Pearlmutter and Isaac A. Meir

Introduction: Housing in the Desert

Throughout the arid lands in which man has settled, traditional solutions to housing have most often relied on readily accessible building materials, and on techniques which were adapted both to the characteristics of these materials and to the constraints of local climate and geography. Such vernacular housing is exemplified by the cliff dwellings and adobe pueblos of the American Southwest, as well as by the vaulted stone dwellings built by Nabateans and Byzantines in the Negev. In these examples, it may be noted that the construction techniques developed by local inhabitants employed locally available heavy materials with a high thermal inertia, capable of moderating the sharp fluctuations of heat and cold which characterize the desert climate.

Over the course of generations, the historical advance of trade and industrialization has increasingly widened the scope of potential building solutions in any given region, ultimately giving rise to the 'internationalization' of design and construction. Prominent expressions of this trend include: 1) the far-reaching influence of foreign and universalist schools of thought in planning and design; 2) the importation of building materials, manufactured components and construction systems; and 3) the increased use of mechanical air conditioning for interior climate control. Due to this latter development, the burden of providing thermal comfort has been largely shifted from the building itself (through its architectural design) to extraneous means, relying on a seemingly endless supply of cheap energy.

In most desert regions, with sparse populations distanced from centers of growth,the technological limitations which dictated the nature of vernacular housing persisted longer than in more climatically amenable, and rapidly developing, environs. As modern housing solutions were eventually implemented in such frontier areas, they were quite often the product of a foreign context—developed to meet the needs of different populations under vastly dissimilar environmental con-

H. J. Bruins and H. Lithwick (eds.), The Arid Frontier, 365–381.

ditions. The consequences of such a process are illustrated in the following case study, which examines the policy of providing lightweight housing in Israel during the recent wave of large-scale immigration.

The Context for Lightweight Immigrant Housing in Israel

In response to the massive wave of immigration to Israel in the early 1990s, government policies were formulated which led to the addition of a large number of lightweight buildings to the country's housing stock—which was hitherto almost exclusively based on conventional heavyweight construction, largely relying on locally available materials. The distribution of such lightweight housing spans the various geographical regions of the country, from the heavily populated urban center on the Mediterranean coastal plain to the sparsely settled, arid Negev region in the south.

Among the many questions raised by the widespread introduction of this housing type is that of climatic appropriateness, both in terms of thermal comfort and of energy consumption. The following discussion focuses, firstly, on the considerations and decision-making process which led to such a policy, and secondly, on the implications of this policy for inhabitants in an arid frontier region.

CONSIDERATIONS OF CENTRAL AUTHORITIES

Political changes during the late 1980s in the former Soviet Union and elsewhere precipitated an influx of immigrants to Israel, numbering by 1996 over 700,000, *i.e.* an addition of about 15% to the existing population (Central Bureau of Statistics, 1996). While under 30,000 housing units per year were built in Israel during the previous decade, a need was perceived in 1991 for 500,000 additional units—largely for the accommodation of over one million anticipated new immigrants—over the coming five years. The increased demand for housing, however, was not accompanied by a sufficient increase in capital investment, a fact which placed great pressure on the building industry and led to an increase in the cost of housing (Tal-Yosef, 1991).

As a result the government Ministry of Construction and Housing became directly involved in all aspects of the planning and provision of housing, reversing the trend toward privatization which had occurred over the previous 30 years. Public housing, which in 1989 comprised only 18% of all new residential construction, had by late 1991 risen to over 75% (Ministry of Housing, 1991; Fialkoff, 1992). Government housing starts thus increased from 3,000 to 60,000 per year (Eldor & Evans, 1992).

The government stated a number of positions which it held as basic values for national development and the provision of housing, including *a)* a dispersion of the population from the central coastal plain to peripheral areas which would be developed (similar to the population dispersion policy employed in the previous massive immigration wave of the 1950s), and *b)* an emphasis on single-storey housing solutions which would allow economical expansion in the future and increase the attractiveness of particular areas for continued settlement (Abelson, 1992).

A major part of the strategy to house the thousands of new immigrants was the use of pre-fabrication, as a means to control costs and regulate production. For such solutions the authorities turned in many cases to foreign suppliers, and to locally unconventional building methods such as lightweight construction. Examples include housing units of lightweight wood construction which are assembled from imported components, and as an 'emergency' measure, the installation of thousands of factory-built lightweight mobile homes, or 'caravans' (Milard, 1991; Faitelson *et al.*, 1992).

While stopping short of requiring industrialized building systems, the Ministry of Housing offered incentives for rapid completion of housing units, which encouraged the use of such technologies. The Ministry also provided advice on the use of specific lightweight building components and assemblies, which ultimately encouraged their use in new housing. In addition, the Ministry called for the establishment of ten local manufacturing plants, intended to produce 10,000 lightweight housing units per year (Collins, 1991).

Two general types of lightweight housing were included in plans for coping with the wave of immigration. The first type, labeled 'advanced building,' includes various lightweight systems, many imported, which are intended to last 30–50 years. The second type is the previously mentioned temporary housing, consisting of small manufactured units generally known as 'caravans,' which are intended to provide shelter for a period of up to five years. Over 20,000 of these temporary units were planned, about 90% to be imported. The decision to import the vast majority of these units was based on the rationale that efficient production systems were already in place overseas, including assembly lines and access to raw materials, making foreign supply less expensive than local production, even with transportation costs included. Insignificant in its numbers before the start of the immigration wave, lightweight construction by mid-1991 already accounted for 7% of all housing starts, with industrialized building in general comprising over one-third (Sigora, 1991).

Approximately 30% of the new housing which accompanied the immigration wave was concentrated in the Negev (Ministry of Housing, 1991). This sparsely populated dry region, constituting over 60% of the country's land area and less than 7% of its population, represents a peripheral frontier area in Israel's highly centralized political and economic system (Gradus, 1984). Nevertheless, the availability of land at low prices and the relative lack of regional planning constraints made the Negev the ideal region for the location of large immigrant housing projects of all types, including those which employed lightweight construction. According to statistics from the end of 1991, the proportion of all temporary lightweight housing which had been located in the Negev was in excess of the proportion of overall construction in the region, indicating that *this technology was by no means deemed inappropriate for the southern area of the country* (Ministry of Housing, 1991).

Critique of Housing Policy

Several questions are raised by the approach taken to provide immigrant housing, in particular its application to the Negev region. The government's policy has been most notably challenged on the following counts:

1) A common apprehension among residents and local authorities has arisen that the low durability of lightweight housing (whether intended as temporary or permanent shelter) would lead to the formation of slums, and result in the deterioration of land values (Kolka, 1991). In addition, it has been suggested that the general assurance of quality in the 'emergency' housing was compromised by the government's decision to forego the normal approval process, and to allow imported building systems to follow either Israeli regulations or those prevailing in the country of export—which, of course, do not necessarily provide measures which are appropriate to meet local conditions.

2) Because of local building restrictions and the nature of the structural system, the lightweight construction in most cases was limited to single-story applications at relatively low density (Kidma, 1990). The implied inefficiency of land use in undeveloped regions, where land costs are comparatively low, is seen as removing from these areas a large source of potential investment which could otherwise come from future development.

3) In most cases, the major obstacle to successful absorption of immigrants has been the lack of employment, not housing (Kidma, 1990). As with other industrialized approaches to housing, the majority of lightweight building techniques which have been introduced are capital-intensive rather than labor-intensive systems (Becker & Lux, 1984). The small amount of manpower required is mostly in the form of specially trained assembly crews, which must be imported from overseas along with the buildings themselves, and the amount of employment generated locally is minimal when compared with that provided by conventional construction (Sigora, 1991). The implications are especially damaging for towns in the Negev, which suffer from unemployment rates which are among the highest in the country. From considerations such as these, the charge has been made that the establishment of emergency lightweight housing could help turn areas such as the Negev into 'dumping grounds,' adversely affecting the local markets for land, housing and labor (Kidma, 1990).

One factor which has not been adequately addressed, however, is the *lack of responsiveness to environmental conditions in this approach to housing.* While the low

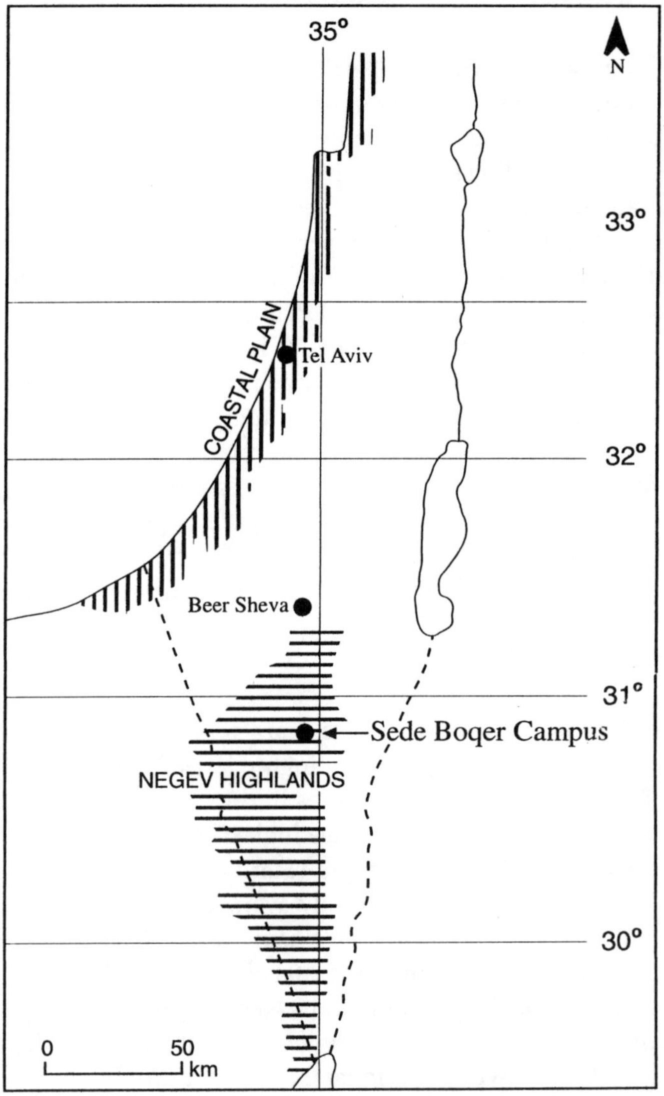

Figure 1. Map of Israel showing the location of sites in the Negev Highlands and Mediterranean Coastal Plain regions.

thermal inertia of lightweight housing was cited by the National Building Institute as a factor detrimental to its use in Israel (Becker & Rector, 1982), such analysis was applied *without geographical distinction between regions of varying climatic conditions.* Hence the present analysis focuses on the thermal behavior of lightweight housing, and its potential liabilities when compared to conventional heavy construction, when sited in two climatically, as well as demographically and politically, dissimilar environments: the peripheral Negev and the central Coastal Plain.

Climatic Analysis

In order to examine and illustrate the particular climatic implications of lightweight construction for inhabitants of the desert periphery, the thermal behavior of a typical lightweight housing unit was compared with that of a conventional heavyweight unit, the latter built according to common local construction techniques. This comparison was made initially in the arid Negev Highlands, which has a P/ETP index of 0.05, *i.e.* an arid region close to the hyper-arid zone boundary of 0.03. An additional comparison was made under the Mediterranean (semi-arid to sub-humid) climatic conditions of the central Coastal Plain, (P/ETP index about 0.50) where much of Israel's population is concentrated (Fig. 1).

These comparative investigations were made through the use of computerized thermal simulations, which predict the daily temperature pattern and/or energy requirements of the given building during the summer and winter seasons. Physical monitoring of actual light and heavy structures was also conducted during the summer and winter, using buildings located in the Negev Highlands only. These on-site measurements provided additional data for validation of the simulation results, and for further analysis of the buildings' comparative behavior.

REGIONAL CONDITIONS

Summer conditions in the coastal area are characterized by average daily temperatures in July ranging from 22–29°C and a high relative humidity of 65–75% at all hours (Bitan & Rubin, 1991). Temperatures in the Negev Highlands for the same period range more widely from an average high of 32°C in the daytime, with humidity under 30%, to under 18°C at night (Stern *et al.*, 1986). Global solar radiation in both locations is high, with a daily average of over 6 kilowatt-hours per square meter on the coast, and nearly 8 kWh/m^2 in the Negev.

In January, daily temperatures in the coastal plain range on average from 9–17°C with a high occurrence of cloudy skies. In the Negev Highlands temperatures again fluctuate widely, as the daily maximum ranges from 10–20°C and the nightly minimum averages under 4°C, with occasional freezing conditions. The desert region is characterized by high levels of solar radiation in the winter as well as the summer, with global insolation in January averaging over 3 kWh/m^2 and radiation incident on a south-facing surface close to 5 kWh/m^2.

Thermal Simulation and Measurement Procedures

In carrying out the previously mentioned comparison between the thermal behavior of a lightweight housing unit and that of a similar heavyweight structure, the interior temperatures and energy requirements of the two different buildings were first predicted by the thermal simulation program 'Quick 4.0' (CENT, 1991), using climatic data from the two locations and building data from the actual structures monitored. Climate data (hourly ambient temperature, relative humidity, and global and diffuse radiation) were taken from the Sede Dov (Tel Aviv) meteorological station in the coastal region (Bitan & Rubin, 1991) and from the Sede-Boker Campus in the Negev Highlands, the site of subsequent physical climatic monitoring. An infiltration rate of 1.0 air change per hour (ACH) for non-ventilated conditions and 6.0 ACH for ventilated conditions were used in the simulations, reflecting the estimated behavior of the actual buildings in terms of construction detailing and wind exposure.

The physical monitoring included measurement of air and surface temperatures, relative humidity, solar radiation, wind speed, and electricity consumption for heating. Both the lightweight and conventional structures which were monitored in the Negev Highlands are situated at the northern edge of the Sede Boqer Campus. The site edges are exposed to ambient conditions, thus freeing the researchers from the task of analyzing microclimatic differentiations common in densely built environments. In all cases, the buildings evaluated were uninhabited during the period of study.

DESCRIPTION OF BUILDINGS

The lightweight structure monitored and simulated was a factory-built residential building measuring 12.4 by 3.7 m (Fig. 2), with an interior height of approximately 2.35 meters. The building is divided into two equal, independent units, and is placed on short legs at a height of approximately 0.5 m above the ground. The walls are made of sandwich panels including external fiber board, 6 cm fiberglass insulation and internal gypsum board. The roofs are slightly sloped with an exterior finish of white painted sheet metal, an insulating layer of 10 cm mineral wool and internal gypsum board ceiling. The floor is also insulated, with fiberglass mats supported by wire mesh. Windows are single glazed with aluminum frames and are positioned in the long facades. It should be noted that although no air-conditioning devices of any sort have been included, the electrical installation has been designed so as to provide for such devices.

The conventional heavyweight structure monitored for comparison is situated on the same site as the lightweight house. Its structural elements are of reinforced concrete, with floor and roof slabs of 15 cm thickness (Fig. 2), the latter insulated with 10 cm of expanded polystyrene. Walls are made of 20 cm silica block with 5 cm expanded polystyrene external insulation faced with acrylic plaster. Window openings face south and an external roller shutter has been included in the original design.

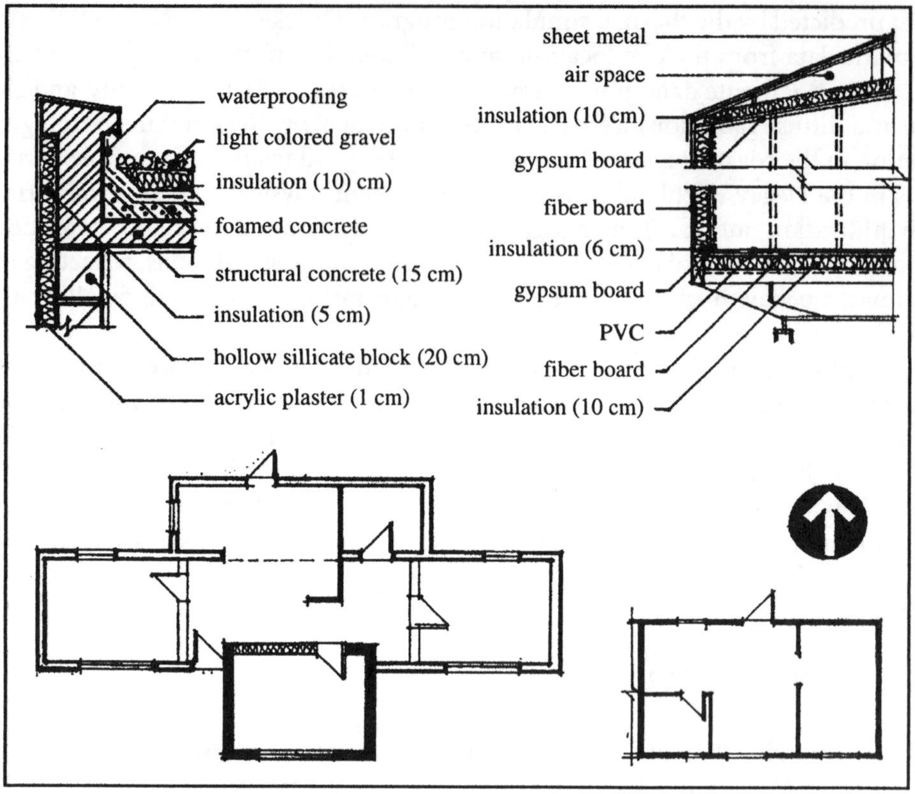

Figure 2. Plan and sectional detail of conventional heavy structure (left) and lightweight structure (right) used in the thermal comparisons.

It may be observed from Figure 2 that the two spaces monitored are similar in most respects relevant to thermal behavior (volume, glazing areas, insulation) other than the obvious difference in the material makeup of the two structures, with the conventional walls and roof providing over ten times as much heat storage capacity as the lightweight ones.

Summer Comparison

Simulation of the two buildings' thermal behavior at the Sede-Boker Campus on a typical summer day (as calculated under ventilation conditions found to be optimal for each) closely predict the observed interior temperature patterns as measured on site (Fig. 3). It was found that while the heavy building maintains conditions defined as comfortable (below 27°C with low ambient humidity) at all hours, interior conditions in the lightweight building were uncomfortable for approximately ten hours per day, reaching a peak temperature of approximately 34°C.

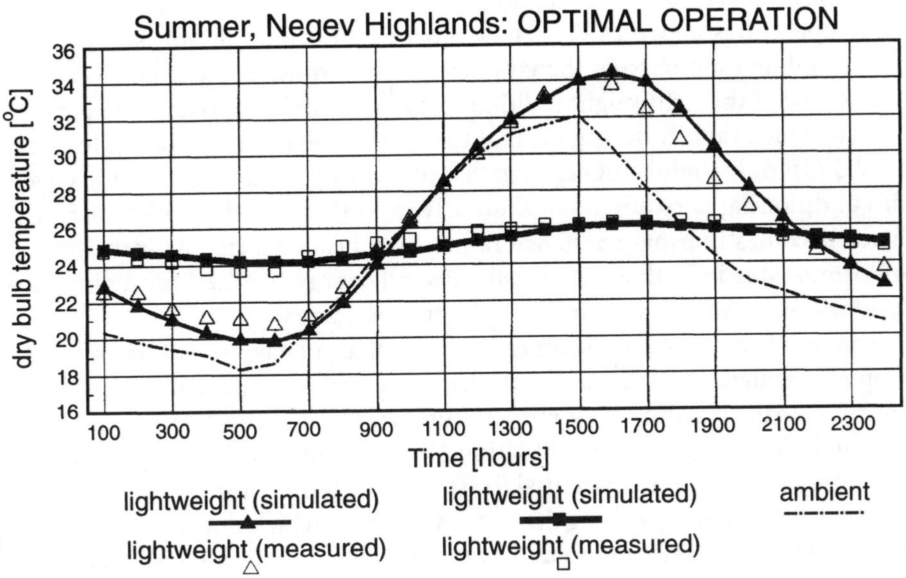

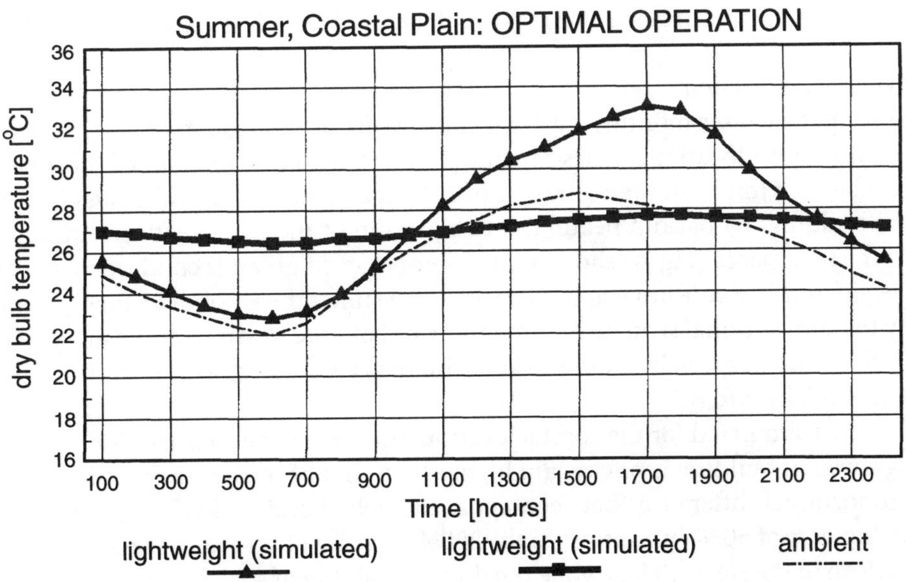

Figure 3. Comparison of internal building temperatures in summer in the Negev Highlands (above) and Coastal Plain (below) regions.

Such a clear distinction cannot be made in the comparison of heavy and lightweight structures in the coastal plain region. Although the relative patterns of temperature fluctuation in the two buildings (Fig. 3) are similar to those in the Negev, the implications for comfort are quite different. Due to the higher relative humidity, temperatures above 24°C such as those generated at all hours in the conventional building are outside the comfort range and necessitate modification. The interior air of the lightweight building, although extremely overheated in the daytime, is cooled rapidly during the night and the resultant daily average temperature of 27.5°C is only slightly higher than that of the heavy building (27.0°C). Given that air conditioning is required continuously in both cases, the critical performance factor becomes thermal resistance rather than heat capacity. As insulation values are comparable in both envelopes, the overall energy required for maintenance of comfortable conditions is similar in the two buildings.

The implication of the summer results is that a heavy structure in the Negev, properly insulated and ventilated at night, can insure comfortable conditions at all hours, since the building mass absorbs sufficient heat in the daytime and releases it during the night when cool ventilation air is available. A lightweight building, on the other hand, has little potential for stabilizing temperature swings and thus requires air conditioning during the day—a necessity which does not exist when conventional techniques common to the area are employed. In the Mediterranean center of the country, night time ventilation is less efficient because ambient air is not as cool, and the high humidity necessitates ventilation during daytime hours as well: thus the importance of a massive structure is much less critical.

Winter Comparison

During the winter months, low temperatures dictate a substantial heating requirement in both locations. For simulation, as well as measurement, of thermal performance, a minimum temperature of 20°C was maintained at all hours in each of the buildings by backup heating. A comparison of the daily heating requirements of the two spaces (Fig. 4) shows that under Negev Highlands conditions, the lightweight building consumes approximately three times the amount of energy required by the conventional structure. Simulation and on-site readings are again correlated fairly closely, especially in the case of the heavy building, whose consumption is under 3.5 kWh/day.

When simulated for the coastal location, the energy requirement for heating the lightweight building is once again higher than that of the heavy one. However, the proportional difference (between approximately 8 and 12 kWh/day) amounts to an increase of 50%, in contrast to the 200% increase observed in the previous comparison in the Negev. Thus while the thermal liability of such lightweight construction may be observed in both locations, the implications are far more pronounced in a desert climate than in a Mediterranean one, in winter as well as in summer.

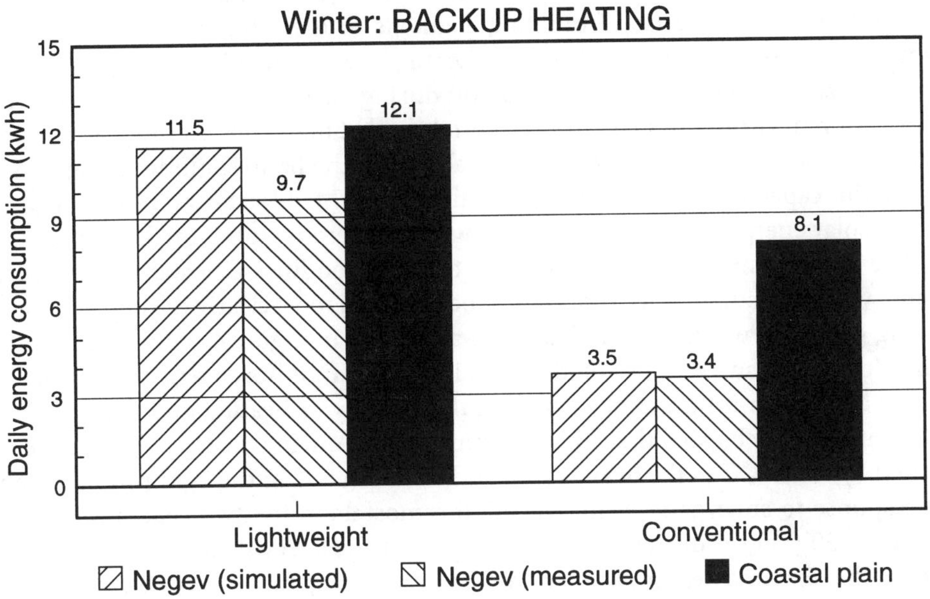

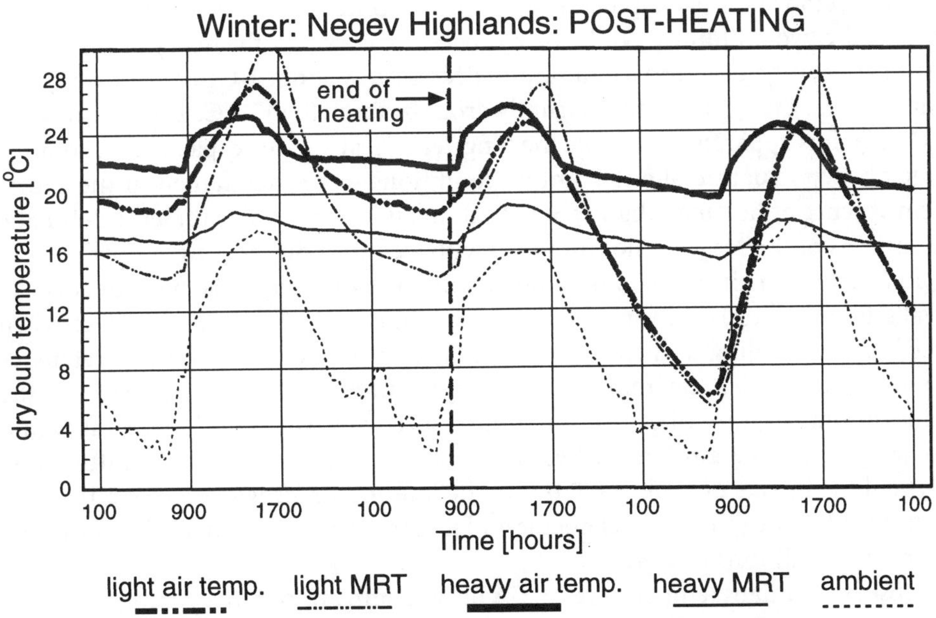

Figure 4. Thermal comparison of conventional and lightweight buildings in winter, as expressed by backup heating requirements (above, for both Negev and Coastal regions) and internal air and radiant temperatures following the cessation of heating (below, for Negev region).

The explanation for this difference lies in a combination of climatic variables, and the selective response of building elements to these thermal stimuli. While temperatures are lower in the Negev, the heating requirement in this region may be largely offset by the sun's intense radiation during daylight hours if such solar energy is stored within the structure and re-released at night. Such a process requires sufficient thermal mass, which is largely absent in the lightweight building. Once again, this capacity to store heat is less critical in the Mediterranean central region, where solar energy is less available and night time temperatures are less extreme.

The mechanism by which a massive structure modifies thermal performance under widely fluctuating conditions may be observed most clearly during the monitoring period in which backup heating was ceased. In Figure 4, interior temperatures (of air and surfaces, as noted below) in the two monitored structures in Sede-Boker are shown for a period including the last day of heating and the two days following. It may be seen that daily air temperatures during the heated period fluctuated by some 4°C in the heavy room and by about 8°C in the light room, mainly in response to solar gain. Although the thermostats in the two spaces were calibrated to 20°C at the outset of the heating period, the heavy structure cumulatively stored this daytime heat input such that its minimum temperature had by the end of several weeks gradually risen by about 2°C, while the light building's nightly minimum actually fell slightly during this time.

Upon removal of backup heating, a radically different response may be seen in each of the two buildings. While the temperature curve for the heavy room retains a similar daytime peak and its minimum decreases by about 1°C per day, the nightly minimum in the light building abruptly drops by more than 12°C, to approximately 6°C. It is thus apparent that while the long-term heat storage capacity of a conventional heavy structure allows comfort conditions to be maintained in the short-term absence of heating, a lightweight building is uninhabitable under the climatic conditions of the Negev if not provided with a continuous heating supply.

In addition to air temperature, the temperature of each of the interior surfaces (walls, floor, ceiling) enclosing the space was monitored. The average of these, when weighted according to their area and distance from a centrally located reference point, approximates the 'mean radiant temperature' (MRT) of the space, or the uniform temperature of a theoretical enclosure with which an inhabitant would exchange the same amount of heat by radiation. This radiant temperature, and particularly its rate of change relative to the interior air, provides a measure of the heat stored in the structure's elements and illustrates their thermal behavior under changing conditions (Fig. 4).

Observed radiant temperatures in the heavy room are extremely stable, fluctuating by under 2°C over a daily cycle when heat is provided. Conversely, surfaces in the light room fluctuate by more than 15°C, an amplitude twice as large as that of the interior air. When backup heat is removed, the heavy MRT remains nearly constant, with a slight downward trend parallel to that of the interior air. However, the lightweight structure's MRT falls at night together with the interior air temperature, such that the two are virtually identical during most hours of the day.

It is thus seen that the light surface reacts to stimuli in an extremely sensitive manner, changing temperature at the slightest input of energy. The heavy surface, with its high capacity, can store large quantities of energy without a significant temperature change, resulting in a stabilization of operative temperatures. The degree to which such a stabilization improves the thermal environment is directly related to *a)* the instability of ambient conditions, and *b)* the comfort which is afforded by a 'balancing' of energy flows in and out of the building over the course of the climatic cycle. Since the arid desert climate rates high on both counts, both the need and the opportunity exist for the exploitation of thermal mass (Pearlmutter *et al.*, 1993).

Implications for Comfort and Energy Use

It has been demonstrated that in summer, tenants of lightweight buildings in the desert will either be faced with overheated conditions, or consume considerable amounts of energy to cool the buildings down to comfort limits. This is accentuated in the case of actual occupied dwellings with inherent heat loads generated by inhabitants, electric appliances and lighting. Thus the initial expense of air conditioning equipment, unnecessary with conventional construction, must be considered with the introduction of the lightweight building type. In winter, such buildings are likely to cause inflated heating expenses, and in the absence of heating are uninhabitable.

Though lightweight construction may be a thermally acceptable compromise in Mediterranean climatic regions such as the coastal plain, the building's lack of thermal inertia thus makes it extremely problematic in a desert environment characterized by extreme temperature swings. The commonly held belief that high levels of *thermal insulation,* typically provided in lightweight buildings, will save on air-conditioning and heating expenses, should therefore be reexamined in the context of arid regions. In particular, such a claim assumes the proper installation and maintenance of mechanical air-conditioning systems—an assumption which may be questionable in light of the poor infrastructure of many sparsely populated arid regions.

Complaints by the residents of such housing, due to both overheating in the summer and lack of sufficient heating in the winter, have already become common. Officials have, since the height of the immigration wave, cancelled orders which were earlier made for both local and imported lightweight housing. One of the reasons which was belatedly given for this was the issue of climatic unsuitability (Chernobroda, 1992).

The roots of this mismatch between housing provision and environmental conditions may be traced to policies which were formulated not on the local level, but at the 'higher' level of central government. The negative implications of such 'top-down' development—in which decisions concerning a sparsely-populated peripheral area are made by central authorities without sufficient regard for local conditions—have been documented repeatedly in the Negev and elsewhere (Stohr, 1981; Rahamimoff, 1981; Etzion, 1985; Gradus & Stern, 1985; Meir, 1992). It is suggested that the type of inappropriate planning presented in this case study may be avoided if the design of the built environment is approached from the outset with an eye toward climatic contextualism, in which the specific environmental constraints and opportunities of each location are regarded as a high priority (Meir, 1989).

Recommendations and Prospects for Sustainable Solutions

Technical design and planning guidelines for housing in arid regions have been developed in various parts of the world, and detailed information which may help avoid inappropriate solutions such as those described above is becoming readily available on the local level (see for example Meir *et al.*, 1990). However such guidelines, and the climate-responsive strategies which they encourage, are rarely integrated and applied to the extent necessary for shaping the built environment. The obstacles which are responsible for this inadequacy no doubt stem from a host of factors: ineffective dessemination of information, conflicting political interests, and perceived short-term economic realities. As illustrated in this case study, the effect of these obstacles may be intensified by the nature of existing political structures and a lack of local decision-making power. The following recommendations are a number of possible courses of action toward appropriate planning and design:
1) *Effective targeting of information* - While planners and architects are often considered responsible for determining the nature of housing and the built environment, the interests of such professionals are dictated to a great extent by others—on the one hand 'from above,' by policy makers and officials in government or private institutions which provide funding and approval for development projects, and on the other hand 'from below' by consumers, who both directly and indirectly determine market preferences. Thus the educational efforts and transfer of information from research institutions and others in the field must be adapted and directed not only toward design professionals, but to these other groups as well.
2) *Incorporation of climatic concerns in master plans and local building regulations* - While some building regulations do provide measures relating to energy-efficiency, they may be limited in scope or minimally effective. The Israel building code, for example, includes minimum standards for thermal insulation. However, both the level of insulation required and the level of enforcement are questionable—and as illustrated in our case study, insulation alone is not necessarily a sufficient measure

of energy-efficiency or climatic appropriateness. In order to better adapt legal requirements to local conditions and needs, the development of community master plans and by-laws may be carried out in conjunction with neighborhood associations and local residents, with the issuing of building permits contingent on the approval of local review committees. Such measures may be effective in achieving environmental goals which are particular to the locale, especially if they are developed into a cohesive 'design language' for buildings and communities.
3) *Economic incentives* - The ultimate effectiveness of legislation is often limited by market forces, and by the inclination of officials to enforce regulations as well as the willingness of housing consumers to abide by them. Various economic tools, including selective taxation, energy credits and rebates, fines for environmental damage and natural resource accounting, have been developed for encouraging energy conservation and ecological preservation in various sectors of economic activity and development.

If a common thread is to be found in all of the above recommendations, it is perhaps expressed in the emerging paradigm of *sustainable community development*—the realization that today's needs, for housing as for other commodities, can and must be met without exhausting the environment's capacity to provide for generations to come. Although still in its infancy in terms of actual impact on the built environment, this outlook is nevertheless entering the public consciousness. This may be surmised, on the one hand, from the increased interest shown in sustainable development by such 'mainstream' bodies as the American Institute of Architects or the European Commission. More important, though, may be the 'grass-roots' circulation of ideas concerning sustainability through the seemingly chaotic channels of worldwide electronic communications.

The 'bottom-up' development effect engendered by such a massive and decentralized exchange of information will most likely find expression in new patterns of physical habitation. It is to be expected that an information age society will become increasingly independent of geographic "centers," with advanced telecommunications allowing the settlement of more and more of the "periphery." If the lessons of sustainability are implemented, than our generation may in fact be the first to alter the trend of marginalization, and accept the challenge of utilization, of the arid frontier.

Acknowledgement

The monitoring of lightweight housing at the Sede Boqer Campus was partially funded by the Department of Research and Development of the Israel Ministry of Energy and Infrastructure, as part of a comprehensive program for monitoring the thermal performance of existing structures in the Negev and Arava regions.

References

Abelson, R. (1992) Advanced building sites. *In* Y. Golani and M. Garon (eds.) *Planning and Housing in Israel in the Wake of Rapid Changes.* Ministries of the Interior and of Construction and Housing, Jerusalem, pp. 155–160.

Becker, R. and G. Lux (1984) Potential application in Israel of foreign lightweight building techniques. Building Research Station, Technion–I.I.T., Haifa.

Becker, R. and Y. Rector (1982) The practical performance of lightweight housing constructed in Israel. Building Research Station, Technion–I.I.T., Haifa.

Bitan, A. and S. Rubin (1991) *Climatic Atlas of Israel for Physical and Environmental Planning and Design.* University of Tel Aviv, Israel Meteorological Service and Ministry of Energy and Infrastructure, Tel Aviv.

CENT (1991) *User's and Reference Manual for QUICK—A Thermal Design Tool and Load Calculation Computer Program, Release 4.0.* Centre for Experimental and Numerical Thermoflow, Pretoria.

Central Bureau of Statistics (1996) *Statistical Abstract of Israel.* Jerusalem.

Chernobroda, D. (1992) Sharon denies importation of prefabs. *Mivnim—Monthly Magazine for the Construction Industry in Israel* 113:8. (in Hebrew)

Collins, I. (1991) Housing ministry incentives for innovation confuse the building industry. *Mivnim—Monthly Magazine for the Construction Industry in Israel* 104:8–12 (in Hebrew).

Eldor, S and M. Evans (1992) Housing policies for immigrant absorption. *In* Y. Golani and M. Garon (eds.) *Planning and Housing in Israel in the Wake of Rapid Changes.* Ministries of the Interior and of Construction and Housing, Jerusalem, pp. 149–154.

Etzion, Y. (1985) Desert architecture—The architecture of the extremes. *In* Y. Gradus (ed.) *Desert Development: Man and Technology in Sparselands.* Reidel, Dordrecht, *etc.*, pp. 81–102.

Faitelson, Y., Y. Margalit and D. Pilzer (1992) The national outline scheme for emergency housing. *In* Y. Golani and M. Garon (eds.) *Planning and Housing in Israel in the Wake of Rapid Changes.* Ministries of the Interior and of Construction and Housing, Jerusalem, pp. 161–168.

Fialkoff, C. (1992) Israel's housing policy during a period of massive immigration. *In* Y. Golani and M. Garon (eds.) *Planning and Housing in Israel in the Wake of Rapid Changes.* Ministries of the Interior and of Construction and Housing, Jerusalem, pp. 169–178.

Gradus, Y. (1984) The emergence of regionalism in a centralized system: The case of Israel. *Environment and Planning* 9:87–100.

Gradus, Y. and E. Stern (1985) From preconceived to responsive planning: Cases of settlement design in arid environments. *In* Y. Gradus (ed.) *Desert Development: Man and Technology in Sparselands.* Reidel, Dordrecht, *etc.*, pp. 41–59.

Kidma, K. (1990) The wonder drug of lightweight building, (in Hebrew). *Mivnim—Monthly Magazine for the Construction Industry in Israel* 98:44–51.

Kolka, G. (1991) Housing ministry calls local authorities' resistance to lightweight construction a hindrance to immigrant housing. *Mivnim—Monthly Magazine for the Construction Industry in Israel* 105:68. (in Hebrew)

Meir, I.A. (1989) Climatic sub-regions and design contextualism. *Building & Environment* 24.3:245–251.

Meir, I.A., Y. Etzion and D. Faiman (1990) *Energy Aspects of Design in Arid Zones.* Blaustein Institute for Desert Research, Ben-Gurion University of the Negev, Sede Boqer & Israel Ministry of Energy and Infrastructure, Jerusalem.

Meir, I.A. (1992) Urban space evolution in the desert—The case of Beer-Sheva. *Building & Environment* 27(1): 1–11.

Milard, R. (1991) Foreign builders in our midst. *Mivnim—Monthly Magazine for the Construction Industry in Israel* 111:17–18. (in Hebrew)

Ministry of Housing (1991) Statistical Report on Development in the Building Industry. Jerusalem.

Pearlmutter, D., E. Erell and Y. Etzion (1993) Monitoring the thermal performance of an insulated earth-sheltered structure: A hot-arid zone case study. *Architectural Science Review* 36.1:3–12.27.

Rahamimoff, A. (1981) Extreme conditions and responsive architecture. *In* L. Berkofsky, D. Faiman, and J. Gale (eds.) *Settling the Desert.* Gordon & Breach Science Publishers, New York, pp.213–240.

Sigora, Y. (1991) The lack of quality standards for lightweight construction. *Mivnim—Monthly Magazine for the Construction Industry in Israel* 105:55–57. (in Hebrew)

Stern, E., Y. Gradus, A. Meir, S. Krakover, and H. Tsoar (1986) *Atlas of the Negev.* Dept. of Geography, Ben-Gurion University of the Negev, Beer-Sheva.

Stohr, W.B. (1981) Development from below: The bottom-up and periphery-inward development paradigm. *In* W.B. Stohr and D.R. Fraser Taylor (eds.) *Development from Above or Below?* John Wiley & Sons Ltd, pp. 39–70.

Tal-Yosef, Z. (1991) Building for the immigration wave—Approaches to planning in a period of turbulence. Ministry of Construction and Housing Staff Report, Beer-Sheva.

ABOUT THE AUTHORS

Architect David Pearlmutter was born in 1962 and raised in west Texas. He received his B.Arch. degree from the University of Texas School of Architecture in 1986, and M.A. in Geography and Environmental Development at Ben-Gurion University of the Negev in 1996.
Since his arrival to Israel in 1987, he has been with the Center for Desert Architecture and Urban Planning, working both on architectural design projects and research concerning climatic aspects of building design in arid regions. He is currently investigating techniques for improving the microclimate of public and private urban spaces.

Architect Isaac A. Meir was born in Thessaloniki, Greece in 1957. He immigrated to Israel in 1975 and studied at the Faculty of Architecture and Town Planning of the Technion–I.I.T., where he received his B.Arch. degree in 1981 and his M.Sc. degree in 1984. He joined the Center for Desert Architecture and Urban Planning in 1986, lecturing and publishing on appropriate design for arid regions. He received the Dori Award for Technology (with Y. Etzion and D. Faiman, in 1992) and was a Visiting Lecturer at the Architectural Association School of Architecture, London. His current research concerns 'low-tech' solutions for arid zones and alternative information dissemination approaches.

Both authors were co-recipients of the Sheba Award for Excellence of the Ben-Gurion Fund (with E. Erell, in 1988), and co-editors of Architecture of the Extremes in 1994, with Y. Etzion and E. Erell.

David Pearlmutter and Isaac A. Meir
The Center for Desert Architecture and Urban Planning
The Jacob Blaustein Institute for Desert Research
Ben-Gurion University of the Negev
Sede-Boker Campus
Israel 84990

The GeoJournal Library

1. B. Currey and G. Hugo (eds.): *Famine as Geographical Phenomenon.* 1984 ISBN 90-277-1762-1
2. S.H.U. Bowie, F.R.S. and I. Thornton (eds.): *Environmental Geochemistry and Health.* Report of the Royal Society's British National Committee for Problems of the Environment. 1985 ISBN 90-277-1879-2
3. L.A. Kosiński and K.M. Elahi (eds.): *Population Redistribution and Development in South Asia.* 1985 ISBN 90-277-1938-1
4. Y. Gradus (ed.): *Desert Development.* Man and Technology in Sparselands. 1985 ISBN 90-277-2043-6
5. F.J. Calzonetti and B.D. Solomon (eds.): *Geographical Dimensions of Energy.* 1985 ISBN 90-277-2061-4
6. J. Lundqvist, U. Lohm and M. Falkenmark (eds.): *Strategies for River Basin Management.* Environmental Integration of Land and Water in River Basin. 1985 ISBN 90-277-2111-4
7. A. Rogers and F.J. Willekens (eds.): *Migration and Settlement.* A Multiregional Comparative Study. 1986 ISBN 90-277-2119-X
8. R. Laulajainen: *Spatial Strategies in Retailing.* 1987 ISBN 90-277-2595-0
9. T.H. Lee, H.R. Linden, D.A. Dreyfus and T. Vasko (eds.): *The Methane Age.* 1988 ISBN 90-277-2745-7
10. H.J. Walker (ed.): *Artificial Structures and Shorelines.* 1988 ISBN 90-277-2746-5
11. A. Kellerman: *Time, Space, and Society.* Geographical Societal Perspectives. 1989 ISBN 0-7923-0123-4
12. P. Fabbri (ed.): *Recreational Uses of Coastal Areas.* A Research Project of the Commission on the Coastal Environment, International Geographical Union. 1990 ISBN 0-7923-0279-6
13. L.M. Brush, M.G. Wolman and Huang Bing-Wei (eds.): *Taming the Yellow River: Silt and Floods.* Proceedings of a Bilateral Seminar on Problems in the Lower Reaches of the Yellow River, China. 1989 ISBN 0-7923-0416-0
14. J. Stillwell and H.J. Scholten (eds.): *Contemporary Research in Population Geography.* A Comparison of the United Kingdom and the Netherlands. 1990 ISBN 0-7923-0431-4
15. M.S. Kenzer (ed.): *Applied Geography.* Issues, Questions, and Concerns. 1989 ISBN 0-7923-0438-1
16. D. Nir: *Region as a Socio-environmental System.* An Introduction to a Systemic Regional Geography. 1990 ISBN 0-7923-0516-7
17. H.J. Scholten and J.C.H. Stillwell (eds.): *Geographical Information Systems for Urban and Regional Planning.* 1990 ISBN 0-7923-0793-3
18. F.M. Brouwer, A.J. Thomas and M.J. Chadwick (eds.): *Land Use Changes in Europe.* Processes of Change, Environmental Transformations and Future Patterns. 1991 ISBN 0-7923-1099-3
19. C.J. Campbell: *The Golden Century of Oil 1950–2050.* The Depletion of a Resource. 1991 ISBN 0-7923-1442-5
20. F.M. Dieleman and S. Musterd (eds.): *The Randstad: A Research and Policy Laboratory.* 1992 ISBN 0-7923-1649-5
21. V.I. Ilyichev and V.V. Anikiev (eds.): *Oceanic and Anthropogenic Controls of Life in the Pacific Ocean.* 1992 ISBN 0-7923-1854-4

22. A.K. Dutt and F.J. Costa (eds.): *Perspectives on Planning and Urban Development in Belgium.* 1992 ISBN 0-7923-1885-4
23. J. Portugali: *Implicate Relations.* Society and Space in the Israeli-Palestinian Conflict. 1993 ISBN 0-7923-1886-2
24. M.J.C. de Lepper, H.J. Scholten and R.M. Stern (eds.): *The Added Value of Geographical Information Systems in Public and Environmental Health.* 1995 ISBN 0-7923-1887-0
25. J.P. Dorian, P.A. Minakir and V.T. Borisovich (eds.): *CIS Energy and Minerals Development.* Prospects, Problems and Opportunities for International Cooperation. 1993 ISBN 0-7923-2323-8
26. P.P. Wong (ed.): *Tourism vs Environment: The Case for Coastal Areas.* 1993 ISBN 0-7923-2404-8
27. G.B. Benko and U. Strohmayer (eds.): *Geography, History and Social Sciences.* 1995 ISBN 0-7923-2543-5
28. A. Faludi and A. der Valk: *Rule and Order. Dutch Planning Doctrine in the Twentieth Century.* 1994 ISBN 0-7923-2619-9
29. B.C. Hewitson and R.G. Crane (eds.): *Neural Nets: Applications in Geography.* 1994 ISBN 0-7923-2746-2
30. A.K. Dutt, F.J. Costa, S. Aggarwal and A.G. Noble (eds.): *The Asian City: Processes of Development, Characteristics and Planning.* 1994 ISBN 0-7923-3135-4
31. R. Laulajainen and H.A. Stafford: *Corporate Geography.* Business Location Principles and Cases. 1995 ISBN 0-7923-3326-8
32. J. Portugali (ed.): *The Construction of Cognitive Maps.* 1996 ISBN 0-7923-3949-5
33. E. Biagini: *Northern Ireland and Beyond.* Social and Geographical Issues. 1996 ISBN 0-7923-4046-9
34. A.K. Dutt (ed.): *Southeast Asia: A Ten Nation Region.* 1996 ISBN 0-7923-4171-6
35. J. Settele, C. Margules, P. Poschlod and K. Henle (eds.): *Species Survival in Fragmented Landscapes.* 1996 ISBN 0-7923-4239-9
36. M. Yoshino, M. Domrös, A. Douguédroit, J. Paszynski and L.D. Nkemdirim (eds.): *Climates and Societies – A Climatological Perspective.* A Contribution on Global Change and Related Problems Prepared by the Commission on Climatology of the International Geographical Union. 1997 ISBN 0-7923-4324-7
37. D. Borri, A. Khakee and C. Lacirignola (eds.): *Evaluating Theory-Practice and Urban-Rural Interplay in Planning.* 1997 ISBN 0-7923-4326-3
38. J.A.A. Jones, C. Liu, M-K. Woo and H-T. Kung (eds.): *Regional Hydrological Response to Climate Change.* 1996 ISBN 0-7923-4329-8
39. R. Lloyd: *Spatial Cognition.* Geographic Environments. 1997 ISBN 0-7923-4375-1
40. I. Lyons Murphy: *The Danube: A River Basin in Transition.* 1997 ISBN 0-7923-4558-4
41. H.J. Bruins and H. Lithwick (eds.): *The Arid Frontier.* Interactive Management of Environment and Development. 1998 ISBN 0-7923-4227-5

The GeoJournal Library

42. G. Lipshitz: *Core versus Periphery: Migration to and within Israel, 1948–1995.* 1998 ISBN 0-7923-4850-8
43. S. Musterd, W. Ostendorf and M. Breebaart: *Multi-Ethnic Metropolis: Patterns and Policies.* 1998 ISBN 0-7923-4854-0
44. B.K. Maloney (ed.): *Human Activities and the Tropical Rainforest.* Past, Present and Possible Future. 1998 ISBN 0-7923-4858-3

KLUWER ACADEMIC PUBLISHERS – DORDRECHT / BOSTON / LONDON